AF333038

Qing-Hua Qin
Qing-Sheng Yang

Macro-Micro Theory on Multifield Coupling Behavior of Heterogeneous Materials

Qing-Hua Qin
Qing-Sheng Yang

Macro-Micro Theory on Multifield Coupling Behavior of Heterogeneous Materials

With 78 figures

高等教育出版社
HIGHER EDUCATION PRESS

Springer

AUTHORS:

Prof. Qing-Hua Qin
Department of Engineering
Australian National University
ACT 2601, Canberra, Austrlia
E-mail: qinghua.qin@anu.edu.au

Prof. Qing-Sheng Yang
Department of Engineering Mechanics
Beijing University of Technology
100022, Beijing, China
E-mail: qsyang@bjut.edu.cn

ISBN 13 978-7-04-022350-7 **Higher Education Press, Beijing**
ISBN 13 978-3-540-78258-2 **Springer Berlin Heidelberg New York**
e ISBN 978-3-540-78259-9 **Springer Berlin Heidelberg New York**

Library of Congress Control Number: 2008920889

Springer is a part of Springer Science+Business Media
springer.com

The use of general descriptive names, registered names, trademarks, etc. in this publication does not imply, even in the absence of a specific statement, that such names are exempt from the relevant protective laws and regulations and therefore free for general use.

Cover design: Frido Steinen-Broo, EStudio Calamar, Spain
Printed on acid-free paper

Preface

Intelligent material with multifield coupling properties is an important aspect of modern science and technology with applications in many industrial fields such as biomedical, electronic and mechanical engineering.

It is well known that most engineering materials, composite materials in particular, are heterogeneous. The heterogeneity is either designed to meet engineering requirements for specific properties and functions or a natural evolution to adapt the historical architecture to changes in long term loadings and environment. Typical examples include functionally gradient materials and biomaterials. Functionally gradient materials are designed according to specific functions required by users. Biomaterials, on the other hand, remodel themselves to adapt to changes in the natural environment. Obviously, there are many heterogeneous materials in engineering including composites, defective materials and natural biomaterials. Heterogeneous materials exhibit complex properties at both microscopic and macroscopic level due to their anisotropy and interaction between components. Generally, there are two approaches used in investigating heterogeneous materials. One is the continuum mechanics approach, where the materials are assumed to be approximately homogeneous and continuous media. The other is the micromechanics approach, used for investigating the deformation and stress of heterogeneous materials by considering the interactions of the components in the microscopic scale.

In recent years, research in macro-micro mechanics of composite materials has resulted in a great many publications including journal papers and monographs. Up to the present, however, no systematic treatment of macro-micro theory of heterogeneous multifield composites has been available. The objective of this book is to fill this gap, so that the reader can obtain a sound basic knowledge of the solution methods of multifield composites. This volume details the development of linear theories of multifield materials and presents up-to-date results on magneto-electro-elastic composites. The book consists of eight chapters. Chapters 1, 2, 5, and 7 were written by Qing-Sheng Yang, and the remaining four chapters were completed by Qing-Hua Qin. Chapter 1 describes basic concepts and solution methods of heterogeneous multifield composites. Chapter 2 introduces the essentials of homogenization approaches for heterogeneous composites. Chapter 3 deals with basic equations and solutions of linear piezoelectricity, and extensions to include magnetic effects are discussed in Chapter 4. Chapter 5 is concerned with basic equations, variational principles, and finite element solution of thermo-electro-chemo-elastic problems. Applications of multifield theories to bone remodelling process are detailed in Chapter 6. Chapter 7 examines general homogenization schemes of heterogeneous multifield composites. In Chapter 8, the final chapter, a detailed discussion of various micromechanics models of defective piezoelectricity is provided.

The main contents of this book were collected from the authors' most recent research outcomes and the research achievements of others in this field. Different parts of the research presented here were partially conducted by the authors at the Department of Engineering, Australian National University; and the Department of Mechanics of Tianjin University, the Department of Engineering Mechanics at Beijing University of Technology. Support from these universities, the National Science Foundation of China, and the Australian Research Council is gratefully acknowledged.

We are indebted to a number of individuals in academic circles and organizations who have contributed in different, but important, ways to the preparation of this book. In particular, we wish to extend our appreciation to our postgraduate students for their assistance in preparing this book. Special thanks go to Ms. Jianbo Liu of Higher Education Press for her commitment to the publication of this book. Finally, we wish to acknowledge the individuals and organizations cited in the book for permission to use their materials.

The authors would be grateful if readers would be so kind as to send us reports of any typographical and other errors, as well as their more general comments.

Qing-Hua Qin, Canberra, Australia
Qing-Sheng Yang, Beijing, China
May 2007

Contents

Chapter 3   Thermo-electro-elastic problems ·················· 59

Chapter 5   Thermo-electro-chemo-mechanical coupling ·············· **183**

Chapter 6   Thermo-electro-elastic bone remodelling ·············· **207**

Chapter 7   Effective coupling properties of heterogeneous materials ································ 255

Chapter 1   Introduction

1.1   Heterogeneous materials

In classical continuum mechanics, materials are viewed as ideal, continuous, homogeneous media. The aim of continuum mechanics is to describe the response of homogeneous materials to external forces using approximate constitutive relations without microstructural considerations. In fact, all materials are inhomogeneous in the microscopic scale. Manufactured composites, natural soils and rocks as well as biological tissues are typical examples. The continuum is a model of materials in the macroscopic scale. Therefore, the homogeneity of materials depends on the scale of measurement. The magnitude of the micro-scale used differs for specific materials. In general, the approximate range of the micro-scale is 10^{-7} m to 10^{-4} m.

Heterogeneous materials exist in both synthetic products and nature. Synthetic examples include aligned and chopped fiber composites, particulate composites, interpenetrating multiphase composites, cellular solids, colloids, gels, foams, microemulsions, block copolymers, fluidized beds, and concrete. Some examples of natural heterogeneous materials are polycrystals, soils, sandstone, granular media, earth's crust, sea ice, wood, bone, lungs, blood, animal and plant tissue, cell aggregates, and tumors [1]. These heterogeneous materials have a legible microstructure. Figs.1.1 to 1.3 show microscopic pictures of some inhomogeneous materials.

It is noted that an important class of heterogeneous media is composites which are manufactured mixtures of two or more constituents, firmly (as a rule, but not always) bonded together [2]. The composites have inhomogeneous properties for different domains or different directions due to the inhomogeneity of their microstructures. This is an important feature and merit of heterogeneous materials. The microstructures of the composite materials can be designed to meet various desired properties and functions. The materials may possess very high properties in one or two directions and very weak properties in other directions, depending on the design for structural performance. Because of their excellent designable characteristics, composite materials are increasingly applied to industrial fields, for example, aeronautics and astronautics, electronics, chemical engineering, biomedical fields and so on.

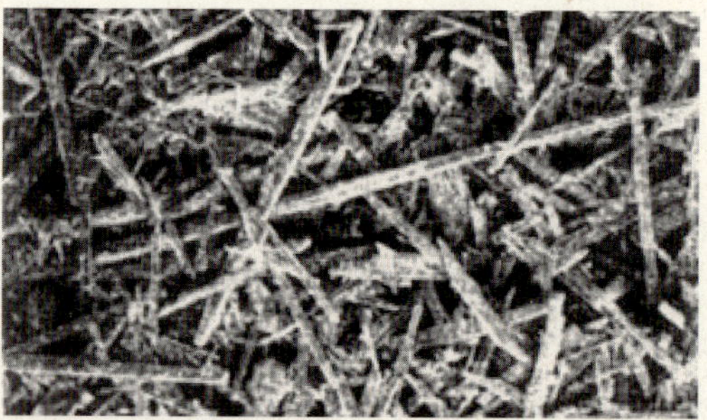

Fig. 1.1   Fiber reinforced composite

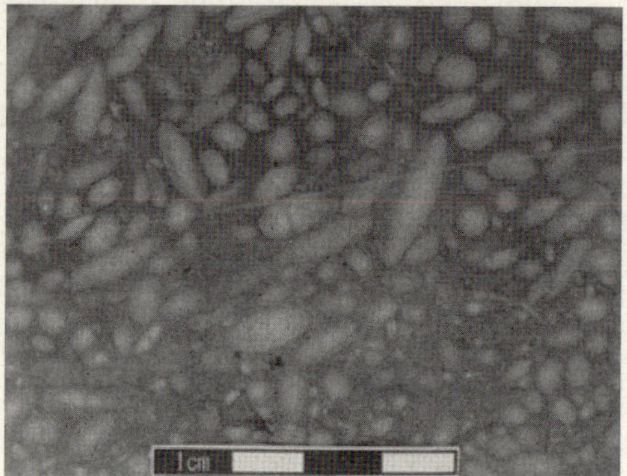

Fig. 1.2   Microstructure of concrete

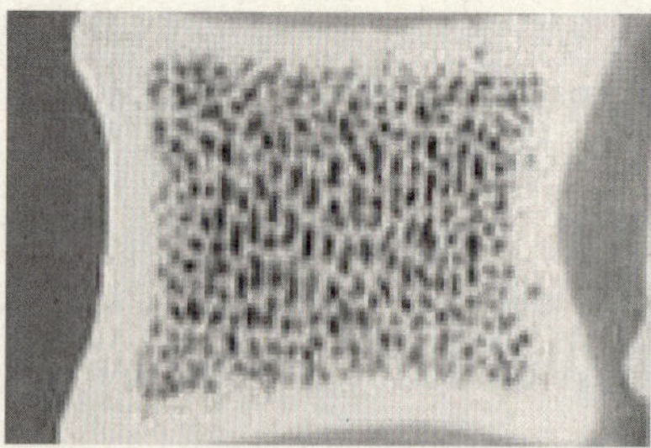

Fig. 1.3   Microstructure of a bone

Heterogeneous materials often exhibit very complex properties, presenting new challenges and opportunities to scientists and engineers. In recent years several new composite materials have been developed which display not only good mechanical properties but also some new functions such as thermal, electric, magnetic, photic, and chemical effects. At the same time, composite materials can create new functions and performance which are absent in their constituents. Such multiple physical properties are usually coupled with each other. Consequently, the coupling properties and deformation behavior of heterogeneous materials are topics of great interest for qualitative and quantitative investigation.

1.2   Multifield coupling properties of heterogeneous materials

A number of heterogeneous materials can fulfill the transfer between mechanical and non-mechanical energy (thermal, electrical, chemical energy, etc). Such materials are usually called *intelligent materials*. These materials can be used in adaptive structures, sensors, and actuators. Intelligent materials are sensitive to variables of the external environment, adjusting their shape or size to adapt to changes in that environment. This multifield coupling behavior is a unique characteristic of intelligent materials. For instance, piezoelectric ceramics, piezoelectric polymers, and some biological tissues (e.g. bone, skin, etc) exhibit thermo-electro-elastic coupling properties [3]. Electric current and heat flow will be excited when the material is subject to a mechanical loading, and vice versa.

As an example, a composite material consisting of a piezoelectric phase and a piezomagnetic phase exhibits considerable multifield coupling properties, i.e. both electro-mechanical and magneto-mechanical coupling. In addition, it displays a remarkably large coupling coefficient between static electric and magnetic fields, which is absent in either constituent. The magnetoelectric coupling in the composite is created through the interaction between the piezoelectric phase and the piezomagnetic phase, which is called a *product property*. The product property of composites offers great engineering opportunities to develop new materials.

In a different example, biological tissues, a form of natural material, can perform energy transfer between chemical and mechanical energy. In this process electric and thermal effects are coupled. This phenomenon can also be found in clay, gel, and so on, and can be described by thermo-electro-chemo-mechanical coupling theory.

Research into heterogeneous media has a long history. Two approaches have been adopted: macro-mechanical and micro-mechanical approaches. Macromechanics deals with material as a homogeneous continuum based on the approximate constitutive model, ignoring heterogeneity of the microstructure. The macroscopic or averaged properties of heterogeneous materials are studied. However, the macroscopic properties of materials depend on micro-structural information, such as the geometric and physical properties of the constituents and the behavior of their interface. Micromechanics has been developed to investigate the relations between the effective properties and micro-structures of heterogeneous materials and the interactions among the constituents[4,5]. As the characteristic length of microstructure is far less than the characteristic length of the whole body, a homogenization is carried out to capture the macroscopic behavior of the materials, as shown in Fig.1.4. Denoting y as the microscopic scale and x as the structural scale, since $y \ll x$, the composite is replaced by the homogenized continuum.

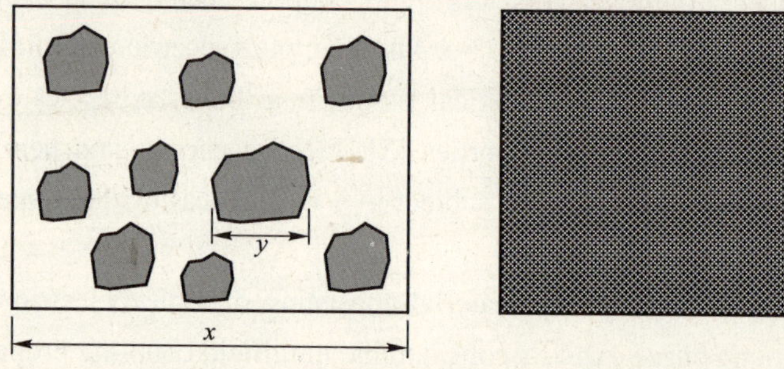

(a) Heterogeneous material with microstructure   (b) Homogenized continuum

Fig.1.4   Homogenization of heterogeneous materials

In the frame of micromechanics, the emphasis is placed on the bridging of effective properties and micro-structure parameters of materials. Effective properties that can be measured experimentally include effective elastic stiffness, conductivity of electricity and heat, permeability coefficient of fluid and coupling coefficients among physical fields. An understanding of the relations of effective properties and microstructure of materials is very vital in the design of new composite materials.

1.3 Overview and structure of the book

The multifield coupling behavior of the heterogeneous material is a multi-disciplinary subject. This book focuses on the multifield coupling properties of several intelligent materials, investigating them by means of macro- and micro-mechanics. The first group of materials involved is artificially intelligent materials, such as piezoelectric solids, piezomagnetic materials, and electric activity polymers which are sensitive to stimuli from the external environment. The second group of materials includes natural materials, such as biological materials (bone, soft tissue, articular cartilage). These materials exhibit thermo-electro-chemo-mechanical coupling effects. Investigation of the behavior of such materials can contribute to understanding of the interaction of the fields and mechanism of deformation, growth, aging and rebuilding of the biological system.

This book is divided into two parts: macromechanics and micromechanics. Macromechanical analysis is covered in Chapters 3 to 6. The phenomenological theory of continuous media is applied in the investigation of multifield coupling behavior of heterogeneous materials. In Chapter 3 the linear theory and general solutions of piezoelectric materials are described. In Chapter 4 electro-elastic coupling theory is extended to magneto-electro-elastic coupling problems. In Chapter 5 we discuss fundamental equations and analytical methods of thermo-electro-chemo-mechanical coupling problem. Chapter 6 involves applications of thermo-electro-elastic coupling in bone remodeling.

Micro-mechanical analysis focuses on the connection between macro-properties and micro-structure parameters, devoting attention to establishing analytical methods for the effective coupling properties of materials. Micro-mechanical analysis is dealt with in Chapters 2, 7, and 8. Chapter 2 discusses the homogenization theory of microstructure and the method of calculation of the effective properties of heterogeneous elastic materials. In Chapter 7, we introduce the homogenization methods in the general sense, including the direct average method, the indirect average method, and the mathematical homogenization method. In Chapter 8, a micro-mechanical model of thermo-piezoelectric solid is described, and the effective properties of thermo-piezoelectric materials with micro-defects are computed.

References

[1] Torquato S. Random heterogeneous materials, microstructure and macroscopic properties. New York: Springer-Verlag, 2002.

[2] Muhammad S. Heterogeneous materials. New York: Springer, 2003.

[3] Qin Qinghua. Fracture mechanics of piezoelectric materials. Southampton: WIT Press, 2001.

[4] Nemat-Nasser S, Hori M. Micromechanics: overall properties of heterogeneous materials. Amsterdam: Elsevier, 1993.

[5] Mura T. Micromechanics of defects in solids. Dordrecht: Martinus Nijhoff, 1987.

Chapter 2　Homogenization theory for heterogeneous materials

In this chapter we discuss the characteristics of heterogeneous media, basic concepts, and methods of homogenization of microstructures of the materials. Because there is much literature on this topic, for example references [1,2], the chapter presents a brief review of the current state and new developments of homogenization theory.

2.1　Microstructure of heterogeneous materials

Heterogeneous materials such as composites, solids with micro-defects, rocks, and natural biomaterials consist of combinations of different media that form regions large enough to be regarded as continua, which are usually firmly bonded together at the interface. Their microstructure can be observed by means of electric scanning microscopes. Generally, for a heterogeneous composite, continuous constituents or phase can be referred to as a *matrix*, and a discrete phase as an *inclusion* which is embedded in the matrix. The inclusion may be a particle, a fiber, a micro-void, or a micro-crack. The overall (effective or macroscopic) properties of composite materials depend on the geometric and physical properties of the phases.

The microstructure of heterogeneous materials may be very disordered and complex in that the distribution, size and shape of inclusions are random. Moreover, there are local fluctuations of the phase volume fraction in a composite. Therefore, mathematical description of the microstructure of a composite is a difficult issue.

From a practical point of view, it is considered that a composite material is an assembly of periodic unit cells. A unit cell is also called a *representative volume element* (RVE), as shown in Figs.2.1 and 2.2. A necessary characteristic of composite materials is *statistical homogeneity* (SH). A strict definition of this concept must be expressed in terms of *n*-point probabilities and ensemble averages. Suffice it to say for practical purposes that in a composite displaying SH all global geometrical characteristics, such as volume fraction, two-point correlations, etc., are the same in any RVE, regardless of its position [3].

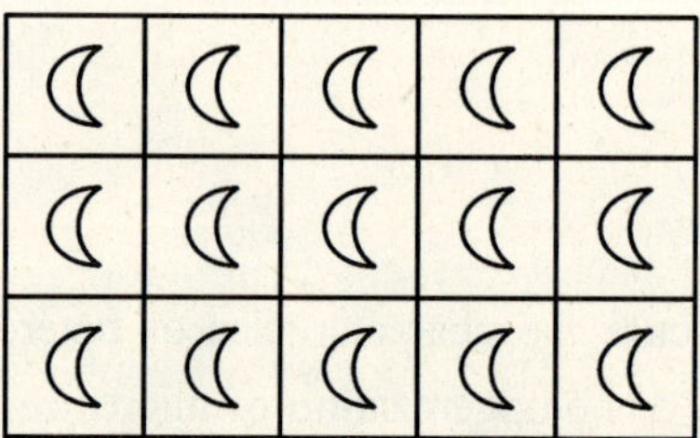

Fig.2.1   Composite with periodic cell

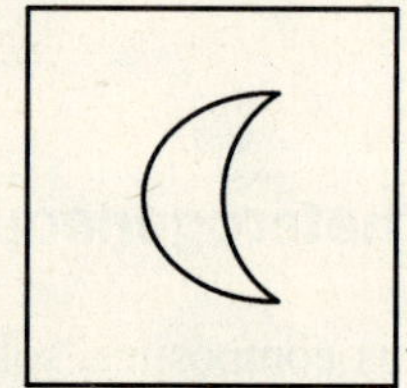

Fig.2.2   Representative volume element (RVE)

Boundary conditions imposed on a deformable body are called homogeneous boundary conditions if either one of

$$u_i(s) = \varepsilon_{ij}^0 x_j \tag{2.1.1}$$

$$T_i(s) = \sigma_{ij}^0 n_j \tag{2.1.2}$$

is satisfied, where ε_{ij}^0 are constant strains and σ_{ij}^0 are constant stresses, x_j are the coordinates and n_j are the components of the outward normal of the boundary. Homogeneous boundary conditions applied to a medium displaying SH produce statistically homogeneous fields within the body. The statistically homogeneous fields are statistically indistinguishable within different RVEs in a heterogeneous body. This implies that their statistical moments such as average, variance, etc. are the same when taken over any RVE within the heterogeneous body. In particular, statistical homogeneity implies that the body average and the RVE average are the same.

A homogeneous material which has the effective properties of composite material is referred to as an *effective medium*. For a SH statistically homogeneous medium, the mechanical behavior of a RVE must be equivalent to the mechanical behavior of the effective medium.

2.2 Periodic boundary conditions

2.2.1 General considerations

The SH composites usually consist of periodic cells, as shown in Fig.2.1. In this case, the microscopic displacement field and stress field are the periodic solutions and a RVE is a periodic cell, as illustrated in Fig.2.2. Therefore the boundary conditions of a RVE should reflect the periodicity of the microstructure. Without loss of generality, the strict periodic conditions of the displacement and stress field can be expressed by [4]

$$u_i(\mathbf{y}) = u_i(\mathbf{y}+\mathbf{Y}), \qquad \forall \mathbf{y} \in \Omega \tag{2.2.1}$$

$$\sigma_{ij}(\mathbf{y}) = \sigma_{ij}(\mathbf{y}+\mathbf{Y}), \qquad \forall \mathbf{y} \in \Omega \tag{2.2.2}$$

where $\mathbf{Y} = (Y_1, Y_2, Y_3)$ is the periodicity, Ω is the domain of the RVE. A typical periodic deformation of a composite is illustrated in Fig.2.3. For $\forall \mathbf{y}^0 \in \Gamma$, the periodic displacement boundary condition of the RVE can be written as

$$u_i(\mathbf{y}^0) = u_i(\mathbf{y}^0 + \mathbf{Y}), \qquad \forall \mathbf{y}^0 \in \Gamma \tag{2.2.3}$$

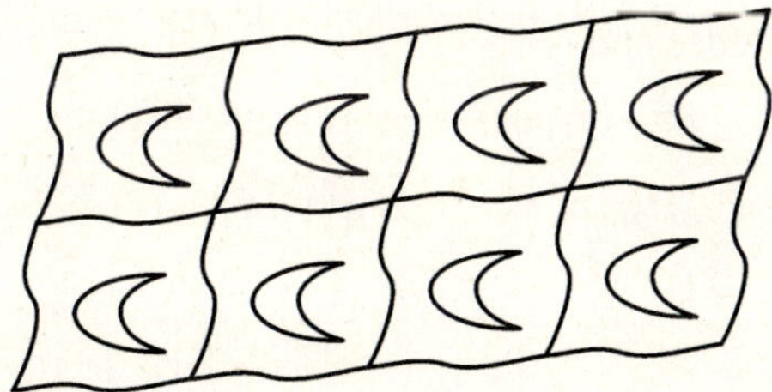

Fig.2.3   Typical deformation of a composite

where Γ is the boundary of the RVE. The stress periodicity of the RVE requires an anti-periodic traction boundary condition

$$T_i(\mathbf{y}^0) = -T_i(\mathbf{y}^0 + \mathbf{Y}), \qquad \forall \mathbf{y}^0 \in \Gamma \tag{2.2.4}$$

where $\mathbf{y}^0 + \mathbf{Y}$ is the boundary of the periodic RVE.

For a 2D square or rectangular RVE, as shown in Fig.2.4, the periodic displacement boundary conditions can be expressed by

$$u_1(y_1^0, y_2) = u_1(y_1^0 + Y_1, y_2) \qquad (2.2.5a)$$

$$u_2(y_1^0, y_2) = u_2(y_1^0 + Y_1, y_2) \qquad (2.2.5b)$$

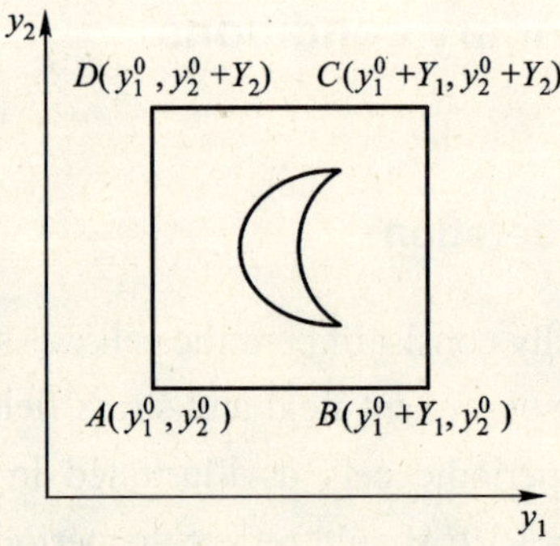

Fig.2.4   The periodic RVE

on the left and right opposite sides, and

$$u_1(y_1, y_2^0) = u_1(y_1, y_2^0 + Y_2) \qquad (2.2.6a)$$

$$u_2(y_1, y_2^0) = u_2(y_1, y_2^0 + Y_2) \qquad (2.2.6b)$$

on the upper and lower opposite sides. The anti-periodicity of the traction boundary conditions requires

$$\sigma_{11}(y_1^0, y_2) = -\sigma_{11}(y_1^0 + Y_1, y_2) \qquad (2.2.7a)$$

$$\sigma_{12}(y_1^0, y_2) = -\sigma_{12}(y_1^0 + Y_1, y_2) \qquad (2.2.7b)$$

on the left and right sides and

$$\sigma_{22}(y_1, y_2^0) = -\sigma_{22}(y_1, y_2^0 + Y_2) \qquad (2.2.8a)$$

$$\sigma_{21}(y_1, y_2^0) = -\sigma_{21}(y_1, y_2^0 + Y_2) \qquad (2.2.8b)$$

on the upper and lower sides.

2.2.2   Symmetric and periodic boundary conditions

The periodic conditions described by Eq.(2.2.5) to Eq.(2.2.8) can be simplified to ordinary boundary conditions as the RVE is symmetric. This case can reflect many model composites in which the inclusion has, in the 2D state, the shape of a circle, ellipse, or rectangle, as shown in Fig.2.5.

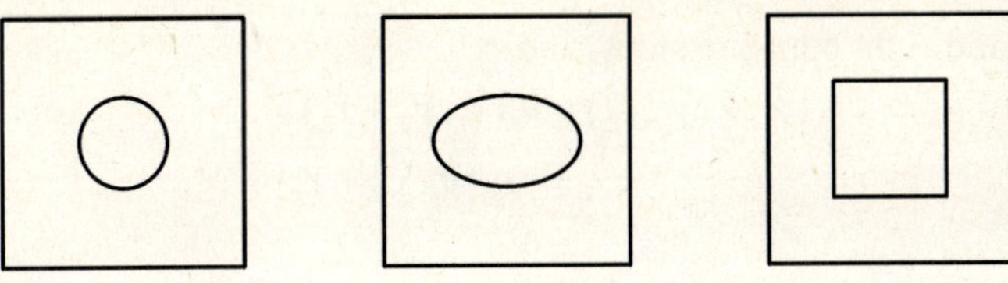

Fig.2.5   Symmetric and periodic RVEs

Firstly, consider the normal (extension and contraction) deformation modes of the RVE. The periodicity and symmetry of the RVE require

$$u_1(y_1^0, y_2) = u_1(y_1^0 + Y_1, y_2)$$
$$= -u_1(y_1^0 + Y_1, y_2) \qquad (2.2.9)$$
$$= 0$$

on the left and right opposite sides, and

$$u_2(y_1, y_2^0) = u_2(y_1, y_2^0 + Y_2)$$
$$= -u_2(y_1, y_2^0 + Y_2) \qquad (2.2.10)$$
$$= 0$$

on the upper and lower opposite sides. Eqs.(2.2.9) and (2.2.10) imply that the normal displacements on all external edges of the RVE are fixed, as shown in Fig.2.6. Clearly, these constraints can satisfy the anti-periodic and symmetric requirements of the traction boundary conditions.

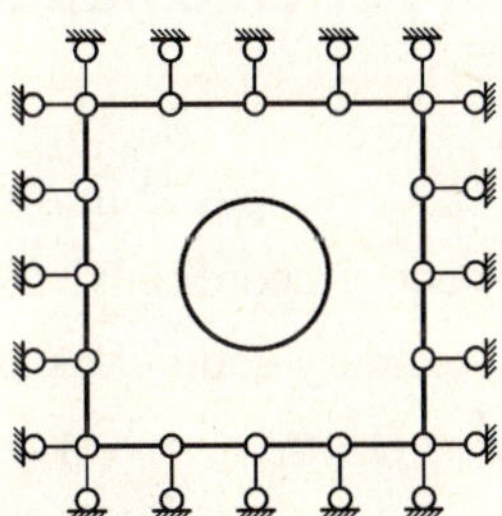

Fig.2.6   Constraints on RVE for normal deformation

Secondly, considering the pure shear deformation of the RVE, an anti-symmetric deformation mode occurs. Then we can obtain

$$u_2(y_1^0, y_2) = u_2(y_1^0 + Y_1, y_2)$$
$$= -u_2(y_1^0 + Y_1, y_2) \qquad (2.2.11)$$
$$= 0$$

on the left and right opposite sides, and

$$u_1(y_1, y_2^0) = u_1(y_1, y_2^0 + Y_2)$$
$$= -u_1(y_1, y_2^0 + Y_2) \qquad (2.2.12)$$
$$= 0$$

on the upper and lower opposite sides. Eqs.(2.2.11) and (2.2.12) mean that the tangent displacements on the boundary of the RVE are fixed, as shown in Fig.2.7.

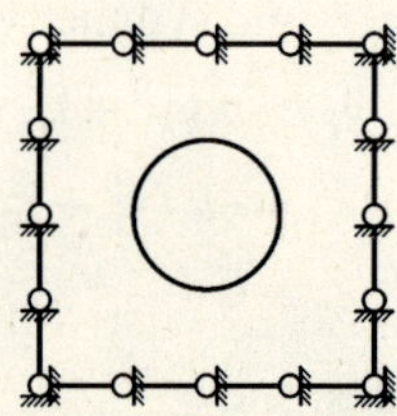

Fig.2.7   Constraints on RVE for pure shear deformation

The periodic and symmetric boundary conditions can be applied to the two-scale expansion method where the initial strains are loaded [5, 6], which is discussed in the following section. For a symmetric RVE, FE analysis of only half a quarter of the RVE is sufficient.

2.3   Implementation of periodic boundary conditions in FE analysis

Generally, displacement can be decomposed into two parts: constant displacement and periodic displacement. Accordingly, alternative to Eqs.(2.2.5) and (2.2.6), the general periodic boundary conditions can be rewritten as

$$u_1(y_1^0, y_2) = u_1(y_1^0 + Y_1, y_2) + c_1 \qquad (2.3.1a)$$
$$u_2(y_1^0, y_2) = u_2(y_1^0 + Y_1, y_2) + c_2 \qquad (2.3.1b)$$

on the left and right opposite sides, and

$$u_1(y_1, y_2^0) = u_1(y_1, y_2^0 + Y_2) + e_1 \qquad (2.3.2a)$$
$$u_2(y_1, y_2^0) = u_2(y_1, y_2^0 + Y_2) + e_2 \qquad (2.3.2b)$$

on the upper and lower opposite sides, where c_1, c_2 and e_1, e_2 are constant displacements. These boundary conditions produce a periodic strain field and therefore a periodic stress field. However, the displacements lose their periodicity.

2.3.1 Multi-point constraints

In finite element analysis, there are different methods for imposing periodic boundary conditions. For example, appropriate multi-point constraints are imposed on the boundary of a RVE to produce periodic boundary conditions [7]. For a square or rectangular RVE, identical displacement functions must be specified for corresponding nodes on the opposite edges. For example, the pairs of nodes on the opposite edges of the RVE can be linked by a constraint equation so that opposite edges have identical deformations. The periodic boundary conditions of a 2-D rectangular RVE, as shown in Fig.2.8, have been defined by Hohe and Becker [8].

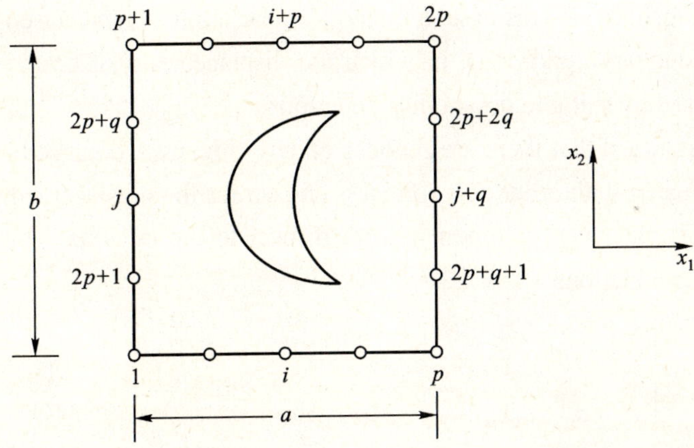

Fig.2.8   Multi-point constraints of a RVE

It is assumed that there are $2p$ nodes on the upper and lower sides, $2q$ nodes on the other two sides. The multi-point constraint for RVE can be expressed by

$$u_{(p)l} - u_{(i)l} = u_{(2p)l} - u_{(i+p)l}, \qquad i = 1,3,\cdots,p \qquad (2.3.3a)$$

on the upper and lower sides, and

$$u_{(2p+q)l} - u_{(j)l} = u_{(2p+2q)l} - u_{(j+q)l}, \qquad j = (2p+1),(2p+2),\cdots,(2p+q)$$

$$(2.3.3b)$$

on the right and left sides, and $u_{(i)l}$ are values of displacement components u_l at node i, $l = 1,2,3$.

For the problem of bending of a plate, the periodicity of the rotations will be described as

$$\varphi_{(i)} = \varphi_{(i+p)}, \qquad i = 1,3,\cdots,p \tag{2.3.4a}$$

$$\varphi_{(j)} = \varphi_{(j+q)}, \qquad j = (2p+1),(2p+2),\cdots,(2p+q) \tag{2.3.4b}$$

where $\varphi_{(i)}$ is the rotation with respect to the x_3-axis at node i.

2.3.2   Polynomial interpolations

The multi-point constraints satisfying periodic conditions can be implemented in standard finite element programs, such as ABAQUS software. In some cases, however, it is difficult to express periodic boundary conditions by multi-point constraints, especially in the case of quite arbitrary FE mesh and/or arbitrary boundaries of a RVE. In this case, the boundary node correspondence cannot be easily established. For this case, a method can be implemented for enacting the periodic boundary conditions, in which the displacements of boundary nodes are expressed by suitable polynomial functions.

It is assumed that there are p nodes on any side, even a curved side of the RVE. Then a $(p{-}1)$th order polynomial is chosen for the displacements. Denote u as the displacement component with respect to the x axis. Therefore we can obtain p equations

$$\begin{cases} u_1 = a_0 + a_1 x_1 + a_2 y_1 \cdots a_{p-1} x_1^m y_1^n \\ \vdots \\ u_p = a_0 + a_1 x_p + a_2 y_p \cdots a_{p-1} x_p^m y_p^n \end{cases} \qquad m+n = p-1 \tag{2.3.5}$$

where $u_1,\cdots,u_p$ are nodal displacement components with respect to the x axis and $x_1,\cdots,x_p$ are the coordinates of the nodes on the boundary. Similar equations of displacement components v and w, with respect to the y and z axes, respectively, can be obtained. These lead to displacement constraints in the FE equations prior to solving. These displacement constraints can be introduced into the FE equations by the Lagrange multiplier method or the penalty method [9, 10].

2.3.3   Specified strain states

Boundary constraints can be directly imposed on the RVE according to the deformation modes of the RVE. For a 2D problem, for instance, there are three kinds of basic deformation modes, two normal modes and one in-plane shear mode, as shown in Fig.2.9. The three modes of deformation are used in the FE calculation of effective stiffness coefficients of a composite by the direct average method which is described in the following section. Fig.2.9 shows the boundary constraints corresponding to the following states of simple strain:

$$\varepsilon = \begin{bmatrix} \varepsilon_{11} \\ \varepsilon_{22} \\ 2\varepsilon_{12} \end{bmatrix} = \begin{bmatrix} 1 \\ 0 \\ 0 \end{bmatrix}, \qquad \varepsilon = \begin{bmatrix} \varepsilon_{11} \\ \varepsilon_{22} \\ 2\varepsilon_{12} \end{bmatrix} = \begin{bmatrix} 0 \\ 1 \\ 0 \end{bmatrix}, \qquad \varepsilon = \begin{bmatrix} \varepsilon_{11} \\ \varepsilon_{22} \\ 2\varepsilon_{12} \end{bmatrix} = \begin{bmatrix} 0 \\ 0 \\ 1 \end{bmatrix} \qquad (2.3.6)$$

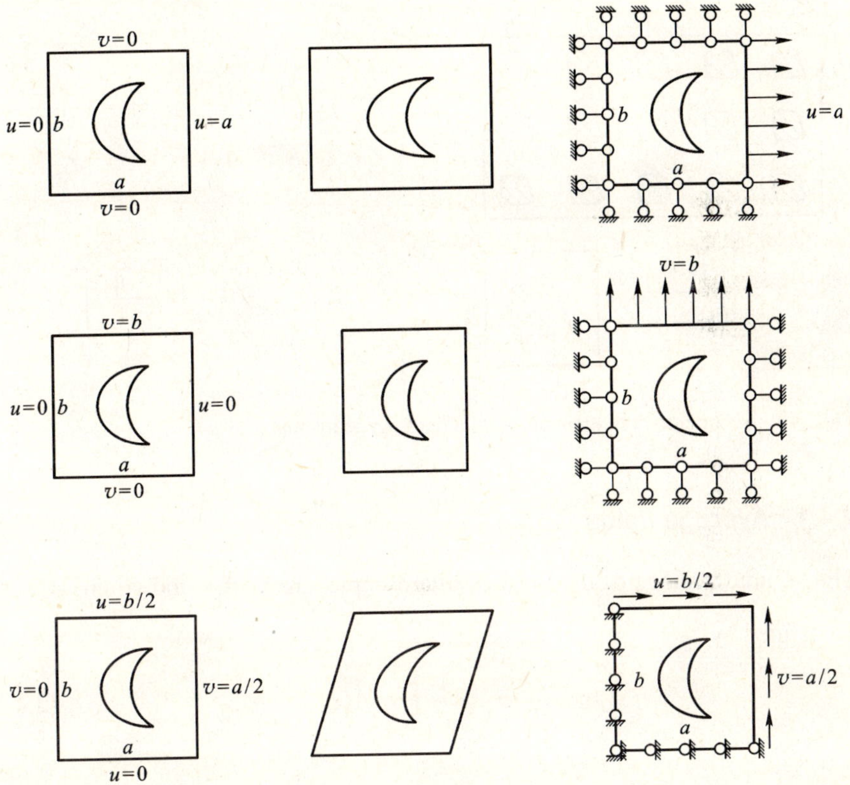

Fig.2.9   Three deformational modes of a 2D RVE

2.4   Effective fields and effective properties

Microstructural materials such as various kinds of composites, are bodies with structural hierarchy, where the characteristic length of the entire body is much greater than the characteristic length of the microstructure. There is a complex interaction between phases. Thus, if only quantities on the macroscopic scale need to be determined, an optional method is that the microstructure is homogenized for reasons of a more efficient analysis. Homogenization is a method for finding the macroscopic fields and properties based on the microstructural parameters and local properties of heterogeneous media. Effective properties represent the overall behavior and depend on the phase properties and microstructure information of heterogeneous materials. A schematic of a homogenization is presented in Fig.2.10.

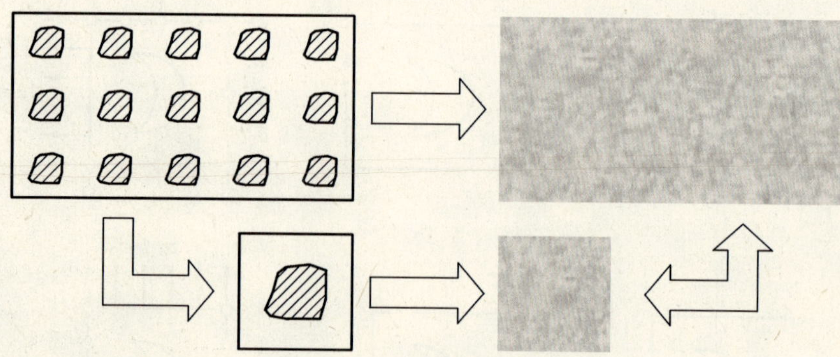

Fig.2.10   Schematic homogenization of a heterogeneous material

2.4.1   Average fields

The volume average of the local or microscopic stress σ_{ij} and strain ε_{ij} can be defined by

$$\bar{\sigma}_{ij} = \frac{1}{V} \int_{\Omega} \sigma_{ij} \mathrm{d}\Omega \tag{2.4.1}$$

and

$$\bar{\varepsilon}_{ij} = \frac{1}{V} \int_{\Omega} \varepsilon_{ij} \mathrm{d}\Omega \tag{2.4.2}$$

where Ω denotes a RVE and V is the volume of the RVE, the superscript bar denotes the volume average of the quantity, e.g. macroscopic or effective quantity. For an elastic body, the volume average of the strain energy can be expressed by

$$\bar{w} = \frac{1}{V}\int_{\Omega} w \, \mathrm{d}\Omega = \frac{1}{V}\int_{\Omega}\frac{1}{2}\sigma_{ij}\varepsilon_{ij}\,\mathrm{d}\Omega$$

$$= \frac{1}{V}\int_{\Omega}\frac{1}{2}c_{ijkl}\varepsilon_{ij}\varepsilon_{kl}\,\mathrm{d}\Omega \tag{2.4.3}$$

$$= \frac{1}{V}\int_{\Omega}\frac{1}{2}f_{ijkl}\sigma_{ij}\sigma_{kl}\,\mathrm{d}\Omega$$

where $\dfrac{1}{2}\sigma_{ij}\varepsilon_{ij} = w$ is the strain energy density, c_{ijkl} are the local stiffness coefficients and $f_{ijkl}(\boldsymbol{f} = \boldsymbol{c}^{-1})$ are local compliance coefficients which are different from phase to phase. Additionally, the macroscopic strain energy should satisfy

$$\bar{w} = \frac{1}{2}\bar{\sigma}_{ij}\bar{\varepsilon}_{ij} \tag{2.4.4}$$

2.4.2 Effective properties

The effective properties which are represented by the effective stiffness $\bar{c}_{ijkl}$ or compliance $\bar{f}_{ijkl}$ of composites, in terms of the average stress and strain, can be defined as

$$\bar{\sigma}_{ij} = \bar{c}_{ijkl}\bar{\varepsilon}_{kl}\,, \qquad \bar{\varepsilon}_{ij} = \bar{f}_{ijkl}\bar{\sigma}_{kl} \tag{2.4.5}$$

or according to the equivalence of the strain energy, defined as

$$\frac{1}{2}\bar{\sigma}_{ij}\bar{\varepsilon}_{ij} = \frac{1}{V}\int_{\Omega}\frac{1}{2}\sigma_{ij}\varepsilon_{ij}\,\mathrm{d}\Omega \tag{2.4.6}$$

that is

$$\frac{1}{2}\bar{c}_{ijkl}\bar{\varepsilon}_{ij}\bar{\varepsilon}_{kl} = \frac{1}{V}\int_{\Omega}\frac{1}{2}c_{ijkl}\varepsilon_{ij}\varepsilon_{kl}\,\mathrm{d}\Omega \tag{2.4.7}$$

This relation was obtained by Hill [11] and is referred to as Hill's principle [12]. The principle has been generalized into nonlinear and inelastic materials [13].

The linearity of the stress-strain relation into an elastic body leads to

$$\bar{c}_{ijkl} = \frac{\partial^2 \bar{w}}{\partial \bar{\varepsilon}_{ij}\partial \bar{\varepsilon}_{kl}} \tag{2.4.8}$$

then an explicit form of the effective stiffness components can be obtained [8]

$$\overline{c}_{ijkl} = \begin{cases} 2\overline{w}(\varepsilon_{ij})\dfrac{1}{\overline{\varepsilon}_{ij}^{2}}, & i=j,k=l,i=k \\[2.5ex] \overline{w}(\varepsilon_{ij})\dfrac{1}{2\overline{\varepsilon}_{ij}^{2}}, & i\neq j,k\neq l,i=k,j=l \\[2.5ex] \left[\overline{w}(\varepsilon_{ij},\varepsilon_{kl})-\overline{w}(\varepsilon_{ij})-\overline{w}(\varepsilon_{kl})\right]\dfrac{1}{\overline{\varepsilon}_{ij}\overline{\varepsilon}_{kl}}, & i=j,k=l,i\neq k \\[2.5ex] \left[\overline{w}(\varepsilon_{ij},\varepsilon_{kl})-\overline{w}(\varepsilon_{ij})-\overline{w}(\varepsilon_{kl})\right]\dfrac{1}{4\overline{\varepsilon}_{ij}\overline{\varepsilon}_{kl}}, & i\neq j,k\neq l(i\neq k\ or\ j\neq l) \\[2.5ex] \left[\overline{w}(\varepsilon_{ij},\varepsilon_{kl})-\overline{w}(\varepsilon_{ij})-\overline{w}(\varepsilon_{kl})\right]\dfrac{1}{2\overline{\varepsilon}_{ij}\overline{\varepsilon}_{kl}}, & i=j,k\neq l \end{cases} \qquad (2.4.9)$$

where $\overline{w}(\varepsilon_{ij},\varepsilon_{kl})$ denotes the strain energy density for a reference strain state, where only ε_{ij} and ε_{kl} have non-zero values. Indices mean that no summation has to be performed.

Homogeneous boundary conditions are usually used to evaluate overall material properties. For homogeneous traction σ_{ij}^{0} on the boundary $\varGamma$ of the RVE, we have

$$\overline{\sigma}_{ij} = \sigma_{ij}^{0} \qquad (2.4.10)$$

and

$$\overline{\varepsilon}_{ij} = \overline{f}_{ijkl}\sigma_{kl}^{0} \qquad (2.4.11)$$

Thus to find the effective compliance $\overline{f}_{ijkl}$ the average strain $\overline{\varepsilon}_{ij}$ must be computed for a composite subjected to a homogeneous traction boundary condition.

For a homogeneous displacement condition on the boundary $\varGamma$ of the RVE, we have

$$\overline{\varepsilon}_{ij} = \varepsilon_{ij}^{0} \qquad (2.4.12)$$

and

$$\overline{\sigma}_{ij} = \overline{c}_{ijkl}\varepsilon_{kl}^{0} \qquad (2.4.13)$$

accordingly, to determine $\overline{c}_{ijkl}$ the average stress $\overline{\sigma}_{ij}$ must be computed for heterogeneous material subjected to a homogeneous displacement boundary condition.

It is worthwhile to note that the volume average of stress, strain and strain energy density can be expressed by phase volume fractions. For a general function F, the volume average can be written as

$$\overline{F} = \frac{1}{V} \int_{\Omega} F \mathrm{d}\Omega = \frac{1}{V} \left[\int_{\Omega_1} F \mathrm{d}\Omega + \int_{\Omega_2} F \mathrm{d}\Omega + \cdots \right]$$

$$= \frac{V_1}{V} \overline{F}^{(1)} + \frac{V_2}{V} \overline{F}^{(2)} + \cdots \qquad (2.4.14)$$

$$= v_1 \overline{F}^{(1)} + v_2 \overline{F}^{(2)} + \cdots$$

where $\Omega_1, \Omega_2, \cdots$ ($\Omega_1 + \Omega_2 + \cdots = \Omega$) are subdomains which represent the domains occupied by phase 1, 2, $\cdots$ of the composite material, and $V_1, V_2, \cdots$ are their volume, while

$$v_1 = \frac{V_1}{V}, \quad v_2 = \frac{V_2}{V}, \cdots \qquad (2.4.15)$$

are referred to as volume fractions of the corresponding phases and $v_1 + v_2 + \cdots = 1$. For n-phase composite, the stress, strain and strain energy can be expressed by

$$\overline{\sigma}_{ij} = \sum_{i=1}^{n} v_i \overline{\sigma}_{ij}^{(i)} \qquad (2.4.16a)$$

$$\overline{\varepsilon}_{ij} = \sum_{i=1}^{n} v_i \overline{\varepsilon}_{ij}^{(i)} \qquad (2.4.16b)$$

$$\overline{w}_{ij} = \sum_{i=1}^{n} v_i \overline{w}_{ij}^{(i)} \qquad (2.4.16c)$$

where the superscript (i) corresponds with phase i. In a word, the average of the stress, strain and the strain energy density can be calculated by a volume average method. The average properties of a composite can be obtained by using any one of the two averaged quantities mentioned above.

As the shape of an inclusion is ellipsoidal, the stress or strain in the inclusion is uniform. In this case, the effective properties can be expressed by the so-called concentration factor of the stress or strain. It is assumed that Hooke's law holds in each elastic phase

$$\sigma_{ij}^{(r)} = c_{ijkl}^{r} \varepsilon_{kl}^{(r)}, \qquad r = 0, 1, \cdots n \qquad (2.4.17a)$$

$$\varepsilon_{ij}^{(r)} = f_{ijkl}^{r} \sigma_{kl}^{(r)}, \qquad r = 0, 1, \cdots n \qquad (2.4.17b)$$

Substituting Eq.(2.4.17a) into Eq.(2.4.16a), and using Eq.(2.4.6), we can obtain

$$\overline{c} = c^0 + \sum_{r=1}^{n} v_r (c^r - c^0) \varepsilon^{(r)} \overline{\varepsilon}^{-1} \qquad (2.4.18a)$$

Substituting Eq.(2.4.17b) into Eq.(2.4.16b), and using Eq.(2.4.5), we have

$$\overline{f} = f^0 + \sum_{r=1}^{n} v_r (f^r - f^0)\sigma^{(r)}\overline{\sigma}^{-1} \tag{2.4.18b}$$

It is assumed that there is a relation between average strain and local strain

$$\varepsilon^{(r)} = A^r \overline{\varepsilon} \tag{2.4.19a}$$

Similarly, the average stress and local stress have the relation

$$\sigma^{(r)} = B^r \overline{\sigma} \tag{2.4.19b}$$

Thus the effective stiffness $\overline{c}$ and compliance $\overline{f}$ of the composite can be written as

$$\overline{c} = c^0 + \sum_{r=1}^{n} v_r (c^r - c^0)A^r \tag{2.4.20a}$$

$$\overline{f} = f^0 + \sum_{r=1}^{n} v_r (f^r - f^0)B^r \tag{2.4.20b}$$

where A^r and B^r are referred to as the concentration factors of stress and strain, respectively. They are functions in term of the properties of the constituents and the shape of inclusions.

For an isotropic composite, such as particle reinforced composites, the stress-strain relation can be expressed by two independent engineering constants as follows

$$\overline{\sigma}_{kk} = 3k\overline{\varepsilon}_{kk}, \quad \overline{s}_{ij} = 2\mu\overline{e}_{ij} \tag{2.4.21}$$

where $\overline{\sigma}_{kk} = \overline{\sigma}_{11} + \overline{\sigma}_{22} + \overline{\sigma}_{33}, \overline{\varepsilon}_{kk} = \overline{\varepsilon}_{11} + \overline{\varepsilon}_{22} + \overline{\varepsilon}_{33}$, $\overline{s}_{ij}$, and $\overline{e}_{ij}$ are deviatoric parts of $\overline{\varepsilon}_{ij}$ and $\overline{\sigma}_{ij}$, respectively. k is effective bulk modulus and μ is shear modulus. Under this situation, Eq. (2.4.20a) can be rewritten as

$$k = k_0 + \sum_{r=1}^{n} v_r (k_r - k_0)\varepsilon_{kk}^{(r)}\overline{\varepsilon}_{jj}^{-1} \tag{2.4.22a}$$

$$\mu = \mu_0 + \sum_{r=1}^{n} v_r (\mu_r - \mu_0)e_{ij}^{(r)}\overline{e}_{ij}^{-1}, \quad ij \text{ no sum} \tag{2.4.22b}$$

2.4.3   Homogenization methods

There are different homogenization approaches. Direct homogenization is based on volume average of field quantities such as stress, strain and energy density. Then the effective properties can be found according to the definition of the effective properties of the composite. The average and the calculation of local field quantities can be performed by a numerical procedure, FEM or

BEM [14,15], for instance, while the geometry and the properties of micro-structures can be arbitrary.

Indirect homogenization follows the idea of the equivalent inclusion method based on Eshelby's eigenstrain solution for a single inclusion embedded into an infinite matrix [16]. This method does not use the average of field quantities, and the effective properties can be derived in terms of the volume fraction and geometry of the inclusion as well as the properties of the constituents. The self-consistent scheme [17-19], the generalized self-consistent scheme [20], the differential method [21, 22] and the Mori-Tanaka method [23-25] have been developed along the lines of this approach and are used widely to find the properties of various composite materials. However, the arbitrary microstructural morphology that is frequently encountered in actual materials cannot be deterministically treated with these models. The constitutive responses of the constituent phases are also some what restricted, and predictions with large property mismatches are not very reliable. Additionally, due to the lack of a proper representation of microscopic stresses and strains, these models cannot capture the effect of local inhomogeneities. A survey of indirect homogenization methods and applications in predictions of the effective properties of composites has been presented by Hashin [3].

Alternatively to direct and indirect homogenizations, the variational method is unique in that it can give the upper and lower bounds of the elastic moduli [26-29]. This method gives improved results over earlier bounds [30,31].

A relatively new approach for homogenization of microstructure consists of mathematical homogenization based on a two-scale expansion of the displacement field. This originated for analyzing physical systems containing two or more length scales [32-35]. It is suitable for multi-phase materials in which the natural scales are the microscopic scale characterized by inter-heterogeneity or local discontinuity spacing and the macroscopic scale characterizing the overall dimensions of the structure. This method can be called mathematical homogenization.

Books covering different homogenization methods have been written by Mura [36], Nemat-Nasser and Hori [37]. A critical review of different homogenization methods and applications in cellular sandwich structures can be found in a recent article by Hohe and Becker [38].

In the following sections we focus on the direct method, indirect method and two-scale expansion method. Special attention is given to application in conjunction with FE analysis and the implementation of the methods.

2.5   Direct homogenization

In direct homogenization, the direct averages of the microscopic fields, such as stress, strain and strain energy density, are calculated by a volume or surface averaging process. The effective properties of the composites are then predicted according to relations of macroscopic stress, strain and strain energy density.

The effective quantities of the stress, strain and strain energy density can be calculated from the corresponding boundary values by a surface averaging procedure. For the strain $\varepsilon_{ij} = \dfrac{1}{2}(u_{i,j} + u_{j,i})$, applying the divergence theorem in Eq.(2.4.2) yields

$$\bar{\varepsilon}_{ij} = \frac{1}{V}\int_{\Omega}\varepsilon_{ij}\mathrm{d}\Omega = \frac{1}{V}\int_{\Gamma}\frac{1}{2}(u_i n_j + u_j n_i)\mathrm{d}\Gamma \tag{2.5.1}$$

where Γ is the boundary of the RVE, n_i is the outward normal vector on Γ.

The surface average of the stresses can be obtained by integration by part of Eq.(2.4.1), that is

$$\bar{\sigma}_{ij} = \frac{1}{V}\int_{\Omega}\sigma_{ij}\mathrm{d}\Omega = \frac{1}{V}\int_{\Gamma}\frac{1}{2}(T_i x_j + T_j x_i)\mathrm{d}\Gamma \tag{2.5.2}$$

where T_i is the traction vector on the surface of the RVE. This implies that $T_i = \sigma_{ij}n_j$ holds. It is shown from Eq.(2.5.2) that average stresses can be calculated by the volume averaging of the stresses or surface averaging of the tractions.

Let us consider, for illustration, a brick shaped RVE as shown in Fig.2.11, which has been used in most research work. The surface averages of tractions can be expressed by, for instance,

$$\bar{\sigma}_{11} = \frac{1}{b}\int_{BC}\sigma_{11}\mathrm{d}\Gamma\,, \qquad \bar{\sigma}_{22} = \frac{1}{a}\int_{DC}\sigma_{22}\mathrm{d}\Gamma \tag{2.5.3a}$$

$$\bar{\sigma}_{12} = \bar{\sigma}_{21} = \frac{1}{b}\int_{BC}\sigma_{12}\mathrm{d}\Gamma = \frac{1}{a}\int_{DC}\sigma_{21}\mathrm{d}\Gamma \tag{2.5.3b}$$

The average strain energy can be expressed by the boundary values, according to the work-energy principle

$$\bar{w} = \frac{1}{V}\int_{\Omega}\frac{1}{2}\sigma_{ij}\varepsilon_{ij}\mathrm{d}\Omega = \frac{1}{V}\int_{\Gamma}T_i u_i \mathrm{d}\Gamma \tag{2.5.4}$$

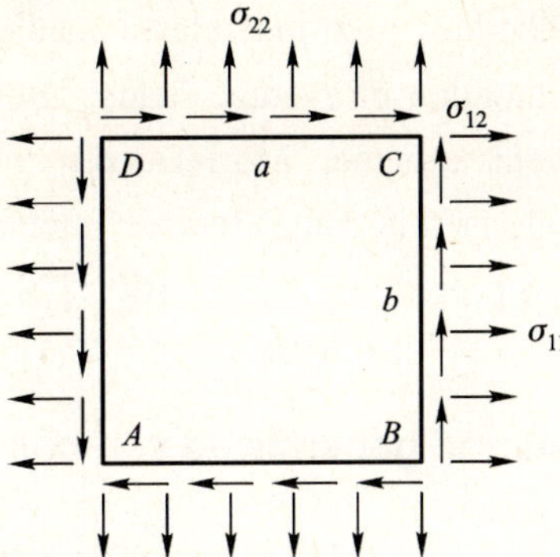

Fig.2.11   Traction condition of a RVE

This relation can be proved and be generalized mathematically. In fact, using Green's theorem, we can obtain

$$
\begin{aligned}
\int_{\Omega}\sigma_{ij}\varepsilon_{ij}\mathrm{d}\Omega &= \frac{1}{2}\int_{\Omega}\sigma_{ij}(u_{i,j}+u_{j,i})\mathrm{d}\Omega \\
&= \int_{\Omega}\sigma_{ij}u_{i,j}\mathrm{d}\Omega \\
&= \int_{\Omega}\left[(\sigma_{ij}u_{i})_{,j}-\sigma_{ij,j}u_{i}\right]\mathrm{d}\Omega \\
&= -\int_{\Omega}\sigma_{ij,j}u_{i}\mathrm{d}\Omega + \int_{\Gamma}\sigma_{ij}u_{i}n_{j}\mathrm{d}\Gamma
\end{aligned}
\tag{2.5.5}
$$

Noting that $\sigma_{ij,j}=-f_{i}$ (in Ω) and $\sigma_{ij}n_{j}=T_{i}$ (on Γ_{t}), then Eq.(2.5.4) becomes

$$
\int_{\Omega}\sigma_{ij}\varepsilon_{ij}\mathrm{d}\Omega = \int_{\Omega}f_{i}u_{i}\mathrm{d}\Omega + \int_{\Gamma_{t}}T_{i}u_{i}\mathrm{d}\Gamma
\tag{2.5.6}
$$

This is the work-energy principle: the strain energy stored in the RVE is equal to the work of external forces. In the case of a free body force, i.e. $f_{i}=0$, the total strain energy can be represented by the work of the traction of the boundary surface.

Therefore, it is concluded that the average of stress, strain and strain energy density can be calculated by either a volume or a surface averaging process. Once two of the three quantities are found, the effective properties of the composite can be predicted.

2.6   Indirect method

Indirect homogenization in this book refers to various homogenization methods derived from Eshelby's inclusion theory. An elastic solution has been obtained for a single inclusion embedded in infinite elastic medium [16]. Such a method does not involve the calculation of average fields. The self-consistent scheme, generalized self-consistent scheme, Mori-Tanaka method and differential method are developed along this route. Indirect methods are widely used to prediction of the effective properties of composites [39,40].

2.6.1   Self-consistent and generalized self-consistent scheme

The self-consistent and generalized self-consistent schemes provide methods to calculate stress or strain concentration factors. They are briefly reviewed here for a binary composite with matrix ($r = 0$) and inclusion ($r=1$).

In self-consistent scheme, it is assumed that a typical inclusion (fiber, particle or micro-void) is embedded in an infinite effective medium subjected to a uniform strain $\bar{\varepsilon}$ at an infinite boundary. Denote $\bar{c}$ as the effective stiffness of the composite to be found. According to Eq. (2.4.12), $\bar{\varepsilon}$ is the effective strain of the composite. The corresponding effective stress is

$$\bar{\sigma} = \bar{c}\bar{\varepsilon} \tag{2.6.1}$$

The strain in the inclusion consists of two parts, uniform strain $\bar{\varepsilon}$ and a perturbing strain ε^{pt}, and the stress in the inclusion is $\bar{S} + S^{pt}$, that is

$$\varepsilon^{(1)} = \bar{\varepsilon} + \varepsilon^{pt} \tag{2.6.2a}$$

$$\sigma^{(1)} = \bar{\sigma} + \sigma^{pt} \tag{2.6.2b}$$

Using the equivalent inclusion principle yields

$$\bar{\sigma} + \sigma^{pt} = c^{1}(\bar{\varepsilon} + \varepsilon^{pt}) = c(\bar{\varepsilon} + \varepsilon^{pt} - \varepsilon^{*}) \tag{2.6.3a}$$

and

$$\varepsilon^{pt} = S\varepsilon^{*} \tag{2.6.3b}$$

where S is the Eshelby tensor and ε^{*} the equivalent eigenstrain.

Solving Eqs. (2.6.2) and (2.6.3), we find

$$\varepsilon^{(1)} = \left[I + Sc^{-1}(c^{1} - \bar{c}) \right]^{-1} \bar{\varepsilon} \tag{2.6.4a}$$

where I is the unit tensor. A comparison of Eq.(2.6.4a) with Eq.(2.3.20a) gives the strain concentration factor

$$A = \left[I + S\bar{c}^{-1}(c^{1} - \bar{c}) \right]^{-1} \tag{2.6.4b}$$

The effective properties can be found by substitution of Eq.(2.6.4b) into Eq.(2.4.20a). As the homogeneous traction boundary condition is applied on the infinite boundary of the effective medium, we can obtain the stress concentration factor and effective compliance of the composite.

In this model, the strain subjected on the effective medium is the effective strain of the composite to be found. The scheme is self-consistent.

It is noted that the strain concentration factor is a function of unknown effective stiffness $\bar{c}$. An iteration procedure should be used to solve the effective properties. In addition, the use of Eshelby's solution means that the shape of the inclusion is assumed to be an ellipsoidal. The self-consistent scheme can be applied to calculating the effective properties of the composite. But the complex interaction of inclusions cannot be considered in this model, which can therefore lead to inaccurate prediction of effective properties. In particular, a wrong result will be obtained when the inclusion volume fraction is greater than 0.5.

The generalized self-consistent scheme is a modification of the self-consistent model. It is assumed that a RVE is embedded in an infinite effective medium subjected to homogeneous boundary conditions. This is a novel model and gives a reasonable result although the operation of the scheme is more complex.

2.6.2   Mori-Tanaka method

Mori and Tanaka [23] have given a solution of back stress in matrix of composites. This result can be applied in extension of Eshelby's solution for a single inclusion to a composite with a finite inclusion volume fraction.

For a finite-fraction inclusion problem with eigenstrain ε^*, although there are complex interactions of phases, the average stress can be expressed by

$$<\sigma>_m = c<\varepsilon>_m = -v_1 c(S\varepsilon^* - \varepsilon^*) \tag{2.6.5}$$

where $<\varepsilon>_m$ is the average strain in the matrix. v_1 is the volume fraction of inclusion.

There are different variations to applying Mori-Tanaka's average stress conception to composites with inclusion of a finite fraction. Weng [24] gave the following reexamination of the Mori-Tanaka(M-T) method.

For a binary composite subjected to a homogeneous boundary condition [Eq. (2.1.2)], denoting $r=0$ matrix and $r=1$ inclusion, the effective stress is σ^0. For the sameshaped pure matrix applied to the same boundary condition, the corresponding strain ε^0 can be expressed by

$$\sigma^0 = c^0 \varepsilon^0 \tag{2.6.6}$$

where c^0 is the stiffness of the matrix. Due to the existence of an inclusion, the strain in the real matrix of the composite differs from that in a pure matrix. Let $\tilde{\varepsilon}$ be the perturbing strain and $\tilde{\sigma}$ corresponding perturbing stress. Thus $\varepsilon^0 + \tilde{\varepsilon}$ and $\sigma^0 + \tilde{\sigma}$ are strain and stress in the real matrix with

$$\sigma^0 + \tilde{\sigma} = c^0(\varepsilon^0 + \tilde{\varepsilon}) \tag{2.6.7}$$

The strain and stress in the inclusion are different from those in the matrix. The differences are ε' and σ', respectively. Thus $\varepsilon^0 + \tilde{\varepsilon} + \varepsilon'$ and $\sigma^0 + \tilde{\sigma} + \sigma'$ are the strain and stress in the inclusion. The equivalent inclusion principle yields

$$\sigma^{(1)} = \sigma^0 + \tilde{\sigma} + \sigma' = c^1(\varepsilon^0 + \tilde{\varepsilon} + \varepsilon') = c^0(\varepsilon^0 + \varepsilon + \varepsilon' - \varepsilon^*) \tag{2.6.8a}$$

$$\varepsilon' = S\varepsilon^* \tag{2.6.8b}$$

$$\tilde{\varepsilon} = -v_1(S - I)\varepsilon^* \tag{2.6.8c}$$

where v_1 is the volume fraction of the inclusion. Eq.(2.6.8c) arises from Mori-Tanaka's concept of average stress [Eq.(2.6.5)]. Solving Eq.(2.6.8) yields

$$\varepsilon^* = H\varepsilon^0 \tag{2.6.9}$$

where $H = \left[c^0 + \Delta c(v_1 I - v_0 S) \right]^{-1} \Delta c$, $\Delta c = c^1 - c^0$, $v_0 = 1 - v_1$ is the volume fraction of matrix.

Accordingly, the effective strain $\bar{\varepsilon}$ is

$$\begin{aligned}
\bar{\varepsilon} &= (1 - v_1)\varepsilon^{(0)} + v_1 \varepsilon^{(1)} \\
&= (1 - v_1)(\varepsilon^0 + \tilde{\varepsilon}) + v_1(\varepsilon^0 + \tilde{\varepsilon} + \varepsilon') \\
&= \varepsilon^0 + v_1 \varepsilon^* \\
&= (I + v_1 H)\varepsilon^0
\end{aligned} \tag{2.6.10a}$$

and the effective stiffness of the composite is

$$\bar{c} = c^0 (I + v_1 H)^{-1} \tag{2.6.10b}$$

Benveniste [25] has presented another explanation of the Mori-Tanaka method by using the concept of the concentration factor. Denote $\tilde{A}$ as the strain concentration factor for the composite with a dilute inclusion. Here $\tilde{A}$ is independent of the volume fraction of the inclusion. Denoting A as the strain concentration factor for the circumstance of a finite volume fraction of inclusion, the relation exists

$$\varepsilon^{(1)} = A\bar{\varepsilon} \tag{2.6.11}$$

Therefore, the effective stiffness of the composite can be determined by

Eq. (2.4.20a) as

$$\bar{c} = c^0 + v_1(c^1 - c^0)A \tag{2.6.12}$$

To calculate the concentration factor A, introduce a new tensor G satisfying

$$\varepsilon^{(1)} = G\varepsilon^{(0)} \tag{2.6.13}$$

Using relation $\bar{\varepsilon} = v_0\varepsilon^{(0)} + v_1\varepsilon^{(1)}$ yields

$$A = G[v_0 I - v_1 G]^{-1} \tag{2.6.14}$$

At the limit state, the concentration factor should satisfy the following condition

$$A\big|_{v_1 \to 0} = \tilde{A}, \qquad A\big|_{v_1 \to 1} = I \tag{2.6.15}$$

Obviously, by setting only $G = \tilde{A}$, the above mentioned limit condition will be satisfied. Thus the Mori-Tanaka method can be summarized as: an inclusion is embedded in an infinite matrix subjected to uniform strain $\bar{\varepsilon}$. The strain concentration factor can be calculated by [see Eq.(2.6.4b), but $\bar{c}$ is replaced by c^0]

$$\tilde{A} = \left[I + S(c^0)^{-1}(c^1 - c^0) \right]^{-1} \tag{2.6.16}$$

Then the effective stiffness is

$$\bar{c} = c^0 + v_1(c^1 - c^0)\tilde{A}\left[v_0 I + v_1\tilde{A} \right]^{-1} \tag{2.6.17}$$

2.6.3 Self-consistent FEM and M-T FEM

The self-consistent scheme and M-T method are close mathematically. However, they are applicable only to ellipsoidal shaped inclusions. This is a critical limit for their practical utility. The self-consistent finite element and M-T finite element method (FEM) are numerical procedures used to solve the effective properties of composites. The self-consistent model or M-T model in conjunction with the FEM can be applied to dealing with composites with arbitrary shaped inclusions.

It is assumed that a typical inclusion is embedded in an infinite effective medium subjected to a uniform strain $\bar{\varepsilon}$. This boundary value problem can be solved by FEM and then the average strain in the inclusion can be obtained. Consequently, the strain concentration factor can be found and effective stiffness can be calculated by Eq.(2.4.20a). In this process, the unknown effective properties must be used in FE calculation, requiring an iteration procedure [41].

The self-consistent FEM can deal with arbitrary shapes of inclusions. The influence of shape of the inclusion can be considered. Also, it can deal with nonlinear material and the effect of interfacial properties.

Similarly, the M-T FEM can be used to numerically deal with arbitrary shapes of inclusion. In M-T FEM, a typical inclusion is embedded in an infinite matrix medium subjected to a homogeneous strain boundary condition. A FE procedure is applied to calculate the strain concentration factor $\tilde{A}$, then the effective properties of the composite can be found by using Eq. (2.6.17).

2.6.4 Differential method

The differential method has a long history in physics. In considering the interaction of phases, the differential method has been applied to composite and cracked solids.

Denote $\bar{c}$ as the effective stiffness of a composite with volume V_0 and inclusion volume fraction v_1. Add the volume δV of the inclusion to the composite so that the inclusion volume fraction is $v_1 + \delta v_1$ and the effective stiffness is $\bar{c} + \delta \bar{c}$. To keep a constant volume V_0 of the composite, the volume δV is subtracted from the composite before adding the inclusion. Thus the concentration of the inclusion is

$$v_1 V_0 + \delta V - v_1 \delta V = (v_1 + \delta v_1) V_0 \tag{2.6.18a}$$

that is

$$\frac{\delta V}{V} = \frac{\delta v_1}{1 - v_1} \tag{2.6.18b}$$

The average stress is

$$\bar{\sigma} = (\bar{c} + \delta \bar{c}) \bar{\varepsilon} \tag{2.6.19a}$$

then we have

$$\bar{\varepsilon} = \frac{V_0 - \delta V}{V_0} \varepsilon + \frac{\delta V}{V_0} \varepsilon^{(1)} \tag{2.6.19b}$$

$$\bar{\sigma} = \frac{V_0 - \delta V}{V_0} \sigma + \frac{\delta V}{V_0} \sigma^{(1)} \tag{2.6.19c}$$

where ε and σ denote the average stress and strain in the instantaneous composite, respectively. $\varepsilon^{(1)}$ and $\sigma^{(1)}$ represent the average stress and strain of the added inclusion, respectively. If there are very few inclusions, the strain concentration factor can be calculated by Eshelby's solution for a dilute inclusion problem

$$\varepsilon^{(1)} = A\bar{\varepsilon} \qquad (2.6.19d)$$

where $A = \left[I + S\bar{c}^{-1}(c^1 - \bar{c}) \right]^{-1}$, S is the Eshelby tensor.

Substituting Eqs. (2.6.19b), (2.6.19c) and (2.6.19d) into Eq.(2.6.19a) yields

$$\delta\bar{c} = (c^1 - \bar{c})A\frac{\delta V}{V_0} \qquad (2.6.20)$$

Using Eq. (2.6.18b), and setting $\delta v_1 \to 0$, we can obtain

$$\frac{dc}{dv_1} = \frac{1}{1 - v_1}(c^1 - c)A \qquad (2.6.21a)$$

This is a differential equation for effective stiffness. Its initial condition is

$$\bar{c}\big|_{v_1 = 0} = c^0 \qquad (2.6.21b)$$

Eq.(2.6.21a) is a nonlinear equation which can be solved by a numerical procedure.

For a spherical particulate reinforced composite, the form of Eq. (2.6.21a) is

$$\frac{d\bar{k}}{dv_1} = \frac{k_1 - \bar{k}}{1 - v_1}\frac{\bar{k} + k^*}{k_1 + k^*} \qquad (2.6.22a)$$

$$\frac{d\bar{\mu}}{dv_1} = \frac{\mu_1 - \bar{\mu}}{1 - v_1}\frac{\bar{\mu} + \mu^*}{\mu_1 + \mu^*} \qquad (2.6.22b)$$

where $\bar{k}$ and $\bar{\mu}$ are the bulk modulus and shear modulus of the isotropic composite. k_1 and μ_1 are the bulk and shear moduli of the inclusion, respectively, and

$$k^* = \frac{4}{3}\bar{\mu}, \qquad \mu^* = \frac{\bar{\mu}}{6}\frac{9\bar{k} + 8\bar{\mu}}{\bar{k} + 2\bar{\mu}} \qquad (2.6.22c)$$

The initial condition Eq.(2.6.21b) becomes

$$v_1 = 0, \qquad \bar{k} = k_0, \qquad \bar{\mu} = \mu_0 \qquad (2.6.22d)$$

where k_0 and μ_0 are the bulk and shear moduli of the matrix.

If we take approximately values

$$k^* = \frac{4}{3}\mu_0, \qquad \mu^* = \frac{\mu_0}{6}\frac{9k_0 + 8\mu_0}{k_0 + 2\mu_0} \qquad (2.6.22e)$$

an approximate solution of Eqs.(2.6.22a,b) can be obtained

$$\bar{k} = k_0 + \frac{v_1(k_1 - \bar{k})}{1 + (1 - v_1)\dfrac{k_1 - k_0}{k_1 + k^*}} \tag{2.6.22f}$$

$$\bar{\mu} = \mu_0 + \frac{v_1(\mu_1 - \bar{\mu})}{1 + (1 - v_1)\dfrac{\mu_1 - \mu_0}{\mu_1 + \mu^*}} \tag{2.6.22g}$$

2.7   Variational method

The variational method is used to determine the bound of effective properties of a composite. It gives the upper and lower bounds of the effective properties of the composite according to the stationary principle of the energy.

Consider a composite, volume V, subjected to a homogeneous strain boundary condition [Eq.(2.1.1)]. Denoting ε_{ij} as the virtual strain which satisfies the displacement boundary condition and the geometric equations, the strain energy can be written as

$$\tilde{U} = \frac{1}{2}\int_{V_0} c_{ijkl}^0 \varepsilon_{ij}\varepsilon_{kl}\mathrm{d}V + \frac{1}{2}\int_{V_1} c_{ijkl}^1 \varepsilon_{ij}\varepsilon_{kl}\mathrm{d}V \tag{2.7.1}$$

where V_0 is the domain occupied by a matrix, V_1 is the region of the inclusion and $V_0 + V_1 = V$.

The effective strain energy corresponding to the average strain and stress is

$$U = \frac{1}{2}\bar{c}_{ijkl}\varepsilon_{ij}^0 \varepsilon_{kl}^0 V \tag{2.7.2}$$

According to minimum potential principle, the real strain should satisfy

$$U \leqslant \tilde{U} \tag{2.7.3}$$

This equation will lead to the upper bound of the effective stiffness.

If the composite is subjected to a homogeneous stress boundary condition [Eq.(2.1.2)], denoting σ_{ij} as the stress which satisfies the equilibrium equation and traction boundary condition, the complementary energy is

$$\tilde{\Gamma} = \frac{1}{2}\int_{V_0} f_{ijkl}^0 \sigma_{ij}\sigma_{kl}\mathrm{d}V + \frac{1}{2}\int_{V_1} f_{ijkl}^1 \sigma_{ij}\sigma_{kl}\mathrm{d}V \tag{2.7.4}$$

The real complementary energy can be expressed by means of effective stress and strain

$$\Gamma = \frac{1}{2}\bar{f}_{ijkl}\sigma^0_{ij}\sigma^0_{kl}V \tag{2.7.5}$$

The minimum complementary energy principle leads to

$$\Gamma \leqslant \tilde{\Gamma} \tag{2.7.6}$$

This equation can be used to find the lower bound of the effective stiffness.

Let us consider two special cases.

(1) Constant strain: A uniform uniaxial strain is applied to a composite in one direction, and the strains of matrix and fiber are assumed to be the same. This state stands for a uniaxial deformation of a unidirectional fiber composite. The minimum potential energy principle leads to

$$E_{11} \leqslant v_0 E^m + v_1 E^f \tag{2.7.7}$$

where E_{11} is effective modulus along the axial or single direction. $E^i\,(i=m,f)$ are the elastic moduli of the constituents. The equation above indicates that the upper bound of effective stiffness can be expressed by a mixture law. This result is referred to as the Voigt approximation [30].

(2) Constant stress: It is assumed that a homogeneous traction boundary condition is applied to a composite , and the stresses in the matrix and fiber of the composite are the same. The minimum complementary energy principle yields

$$\frac{1}{E_{22}} \leqslant \frac{v_0}{E^m} + \frac{v_1}{E^f} \tag{2.7.8}$$

This is referred to as the Reuss approximation [31]. It is usually used to predict the transverse effective modulus of composites.

2.8 Two-scale expansion method

It is assumed that an elastic body is an assembly of periodic microscopic unit cells. There are two coordinate systems: global coordinate $x=(x_1,x_2,x_3)$ and local coordinate $y=(y_1,y_2,y_3)$. The global coordinate x is related to the local coordinate y as

$$y = x/\varepsilon \tag{2.8.1}$$

where ε is a very small positive number denoting the ratio between the dimension of a unit cell and a structural body. When subjected to structural level loads and displacements, the resulting evolutionary variables, e.g. deformation and stresses, vary from point to point at the macroscopic scale x. Furthermore, a high level of heterogeneity in the microstructure causes a rapid variation of these variables in a small neighborhood ε of the macroscopic point x. In the present homogenization theory, periodic repetition of the microstructure about a macroscopic point x has been assumed, therefore the field functions depend periodically on $y = x/\varepsilon$. This characteristic is often termed Y-periodicity, where Y corresponds to a RVE.

2.8.1   Expansion of the displacement field

The displacement field can be asymptotically expanded as

$$u_i = u_i^\varepsilon(x) = u_i^0(x,y) + \varepsilon u_i^1(x,y) + \varepsilon^2 u_i^2(x,y) + \cdots \qquad (2.8.2)$$

The superscript ε denotes association of the function with the two length scales. Note that

$$\frac{\partial F^\varepsilon(x,y)}{\partial x_i} = \frac{\partial F(x,y)}{\partial x_i} + \frac{1}{\varepsilon}\frac{\partial F(x,y)}{\partial y_i} \qquad (2.8.3)$$

where F is a general function, for the strain tensor ε_{ij}, we have

$$\varepsilon_{ij} = \frac{1}{2}\left(\frac{\partial u_i}{\partial x_j} + \frac{\partial u_j}{\partial x_i}\right)$$

$$= \frac{1}{\varepsilon}\varepsilon_{ij}^{-1}(x,y) + \varepsilon_{ij}^0(x,y) + \varepsilon\varepsilon_{ij}^1(x,y) + \cdots \qquad (2.8.4)$$

where

$$\varepsilon_{ij}^{-1}(x,y) = \frac{1}{2}\left(\frac{\partial u_i^0}{\partial y_j} + \frac{\partial u_j^0}{\partial y_i}\right) \qquad (2.8.5a)$$

$$\varepsilon_{ij}^0(x,y) = \frac{1}{2}\left(\frac{\partial u_i^0}{\partial x_j} + \frac{\partial u_j^0}{\partial x_i}\right) + \frac{1}{2}\left(\frac{\partial u_i^1}{\partial y_j} + \frac{\partial u_j^1}{\partial y_i}\right) \qquad (2.8.5b)$$

$$\varepsilon_{ij}^1(x,y) = \frac{1}{2}\left(\frac{\partial u_i^1}{\partial x_j} + \frac{\partial u_j^1}{\partial x_i}\right) + \frac{1}{2}\left(\frac{\partial u_i^2}{\partial y_j} + \frac{\partial u_j^2}{\partial y_i}\right) \qquad (2.8.5c)$$

The elastic coefficients c_{ijkl} are periodic functions of x and depend on ε, that is

$$c_{ijkl}^\varepsilon = c_{ijkl}(x/\varepsilon) = c_{ijkl}(y) \qquad (2.8.6)$$

Thus the stress can be expressed as

$$\sigma_{ij}^{\varepsilon} = c_{ijkl}^{\varepsilon}\varepsilon_{kl}$$

$$= \frac{1}{\varepsilon}c_{ijkl}^{\varepsilon}\varepsilon_{kl}^{-1}(\boldsymbol{x},\boldsymbol{y}) + c_{ijkl}^{\varepsilon}\varepsilon_{kl}^{0}(\boldsymbol{x},\boldsymbol{y}) + \varepsilon c_{ijkl}^{\varepsilon}\varepsilon_{kl}^{1}(\boldsymbol{x},\boldsymbol{y}) + \cdots$$

$$= \frac{1}{\varepsilon}\sigma_{ij}^{-1}(\boldsymbol{x},\boldsymbol{y}) + \sigma_{ij}^{0}(\boldsymbol{x},\boldsymbol{y}) + \varepsilon\sigma_{ij}^{1}(\boldsymbol{x},\boldsymbol{y}) + \cdots \tag{2.8.7}$$

The stress-strain relations can be expressed by

$$\sigma_{ij}^{n}(\boldsymbol{x},\boldsymbol{y}) = c_{ijkl}^{\varepsilon}\varepsilon_{kl}^{n}(\boldsymbol{x},\boldsymbol{y}), \quad n = -1,0,1 \tag{2.8.8}$$

From Eqs.(2.8.5) and (2.8.8), the stresses have the following forms

$$\sigma_{ij}^{-1} = c_{ijkl}^{\varepsilon}\frac{\partial u_{k}^{0}}{\partial y_{l}} \tag{2.8.9a}$$

$$\sigma_{ij}^{0} = c_{ijkl}^{\varepsilon}\left(\frac{\partial u_{k}^{0}}{\partial x_{l}} + \frac{\partial u_{k}^{1}}{\partial y_{l}}\right) \tag{2.8.9b}$$

$$\sigma_{ij}^{1} = c_{ijkl}^{\varepsilon}\left(\frac{\partial u_{k}^{1}}{\partial x_{l}} + \frac{\partial u_{k}^{2}}{\partial y_{l}}\right) \tag{2.8.9c}$$

2.8.2   Establishment of basic equations of elastic microstructure

The elastic problem with a periodic microstructure is described as:

$$\sigma_{ij,j}^{\varepsilon} + f_{i} = 0 \qquad (\text{in } \Omega) \tag{2.8.10a}$$

$$\sigma_{ij}^{\varepsilon}n_{j} = T_{i} \qquad (\text{on } \Gamma_{t}) \tag{2.8.10b}$$

$$u_{i}^{\varepsilon} = \bar{u}_{i} \qquad (\text{on } \Gamma_{u}) \tag{2.8.10c}$$

Substituting Eq.(2.8.7) into Eq.(2.8.10), and equating the powers of ε, we obtain

$$\frac{\partial \sigma_{ij}^{-1}}{\partial y_{j}} = 0 \tag{2.8.11a}$$

$$\frac{\partial \sigma_{ij}^{-1}}{\partial x_{j}} + \frac{\partial \sigma_{ij}^{0}}{\partial y_{j}} = 0 \tag{2.8.11b}$$

$$\frac{\partial \sigma_{ij}^{0}}{\partial x_{j}} + \frac{\partial \sigma_{ij}^{1}}{\partial y_{j}} + f_{i} = 0 \tag{2.8.11c}$$

For solving the system of Eq.(2.8.11), an important result is introduced here. For a Y periodic function $\Phi = \Phi(\boldsymbol{x},\boldsymbol{y})$, the equation

$$-\frac{\partial}{\partial y_{i}}\left[a_{ij}(\boldsymbol{y})\frac{\partial \Phi}{\partial y_{j}}\right] = F \tag{2.8.12}$$

has a unique solution if the mean value of F is equal to zero, i.e.

$$\bar{F} = \frac{1}{V} \int_Y F \mathbf{dy} = 0 \tag{2.8.13}$$

where V is the volume of the unit cell. Application of this condition to Eq.(2.8.11a) leads to

$$\sigma_{ij}^{-1} = 0 \tag{2.8.14}$$

and then from Eq.(2.8.8) and Eq.(2.8.5a), we have

$$u_i^0(\mathbf{x}, \mathbf{y}) = u_i^0(\mathbf{x}) \tag{2.8.15}$$

This shows that u_i^0 is a function of the global coordinate $\mathbf{x}$ only.

The expansion of the displacement field can be rewritten as

$$u_i = u_i^\varepsilon(\mathbf{x}) = u_i^0(\mathbf{x}) + \varepsilon u_i^1(\mathbf{x}, \mathbf{y}) + \varepsilon^2 u_i^2(\mathbf{x}, \mathbf{y}) + \cdots \tag{2.8.16}$$

We can regard u_i^0 as the macroscopic displacement, while $u_i^1, u_i^2, \cdots$ are the microscopic displacements. The physical interpretation of Eq.(2.8.16) thus is that the real displacement u_i oscillates rapidly around the mean displacement u_i^0 due to the inhomogeneity from the microscopic point of view. $u_i^1, u_i^2, \cdots$ are the perturbing displacements on the level of the microstructure.

Substituting Eq.(2.8.14) into Eq.(2.8.11b), we can obtain the microscopic equilibrium equation

$$\frac{\partial \sigma_{ij}^0}{\partial y_j} = 0 \quad (\text{in } \Omega) \tag{2.8.17}$$

Taking the mean of Eq. (2.8.11c) over Ω and applying Eq.(2.8.13) to the second term $\dfrac{\partial \sigma_{ij}^1}{\partial y_j}$, leads to the macroscopic equilibrium equation

$$\frac{\partial \bar{\sigma}_{ij}^0}{\partial x_j} + f_i = 0 \quad (\text{in } \Omega) \tag{2.8.18}$$

where $\bar{\sigma}_{ij}^0$ are the macroscopic stresses.

2.8.3   Determination of effective properties of material with microstructure

It is assumed that the displacement fields u_i^0 and u_i^1 are related by

$$u_i^1 = -\psi_i^{kl}(\mathbf{x}, \mathbf{y}) \frac{\partial u_k^0}{\partial x_l} \tag{2.8.19}$$

Substituting Eq.(2.8.19) into (2.8.9b) yields

$$\sigma_{ij}^{0} = \left(c_{ijkl} - c_{ijmn}\frac{\partial \psi_{m}^{kl}}{\partial y_{n}} \right)\frac{\partial u_{k}^{0}}{\partial x_{l}} \tag{2.8.20}$$

Then integrating on the RVE leads to the effective stress-strain relations for an elastic medium

$$\bar{\sigma}_{ij}^{0} = \bar{c}_{ijkl}\frac{\partial u_{k}^{0}}{\partial x_{l}} \tag{2.8.21}$$

where

$$\bar{\sigma}_{ij}^{0} = \frac{1}{V}\int_{Y}\sigma_{ij}^{0}(\boldsymbol{x},\boldsymbol{y})\mathrm{d}Y \tag{2.8.22}$$

$$\bar{c}_{ijkl} = \frac{1}{V}\int_{Y}\left[c_{ijkl} - c_{ijmn}\frac{\partial \psi_{m}^{kl}}{\partial y_{n}} \right]\mathrm{d}Y \tag{2.8.23}$$

$\bar{c}_{ijkl}$ are the homogenized elastic coefficients. It can been seen from Eq.(2.8.23) that the function $\psi(\boldsymbol{x},\boldsymbol{y})$ must be calculated before determination of the homogenized properties. Generally, the evaluation of $\psi(\boldsymbol{x},\boldsymbol{y})$ can be performed by the FEM.

2.8.4 Variational forms

The variational forms of the abovementioned equations can be established to calculate the effective properties of a composite in conjunction with the FEM. The variational form of Eq.(2.8.11a) is

$$\int_{\Omega^{\varepsilon}}\frac{\partial \sigma_{ij}^{-1}}{\partial y_{j}}\delta u_{i}^{0}\mathrm{d}\Omega = \int_{\Omega^{\varepsilon}}\left(c_{ijkl}^{\varepsilon}\frac{\partial u_{k}^{0}}{\partial y_{l}} \right)_{,j}\delta u_{i}^{0}\mathrm{d}\Omega = 0 \tag{2.8.24}$$

where δu_{i}^{1} can be viewed as arbitrary virtual displacements. For a Y-periodic function $\phi(\boldsymbol{y})$, we define a mean operator as follows:

$$\lim_{\varepsilon \to 0}\int_{\Omega^{\varepsilon}}\phi\left(\frac{\boldsymbol{x}}{\varepsilon}\right)\mathrm{d}\Omega = \frac{1}{V}\int_{\Omega}\int_{Y}\phi(\boldsymbol{y})\mathrm{d}Y\mathrm{d}\Omega \tag{2.8.25}$$

Since the homogenization method consists of finding the limit of the solutions to Eqs.(2.8.11a)~(2.8.11c) as ε tends to zero, we have the form of Eq.(2.8.24)

$$\lim_{\varepsilon \to 0}\int_{\Omega^{\varepsilon}}\left(c_{ijkl}^{\varepsilon}\frac{\partial u_{k}^{0}}{\partial y_{l}} \right)_{,j}\delta u_{i}^{0}\mathrm{d}\Omega = \frac{1}{V}\int_{\Omega}\int_{Y}\left(c_{ijkl}\frac{\partial u_{k}^{0}}{\partial y_{l}} \right)_{,j}\delta u_{i}^{0}\mathrm{d}Y\mathrm{d}\Omega = 0 \tag{2.8.26}$$

Using the divergence theorem in Eq.(2.8.26) yields

$$\frac{1}{V}\int_{\Omega}\int_{Y}\left(c_{ijkl}\frac{\partial u_k^0}{\partial y_l}\right)_{,j}\delta u_i^0\,\mathrm{d}Y\mathrm{d}\Omega$$

$$=\frac{1}{V}\int_{\Omega}\oint_{s}c_{ijkl}\frac{\partial u_k^0}{\partial y_l}n_j\delta u_i^0\,\mathrm{d}s\mathrm{d}\Omega=0 \tag{2.8.27}$$

Thus

$$\frac{\partial u_k^0}{\partial y_j}=0 \tag{2.8.28}$$

It is shown again that u_i^0 is a function of x only.

Substituting Eq. (2.8.9b) into the variational form of Eq. (2.8.11b) yields

$$\int_{\Omega^{\varepsilon}}\frac{\partial\sigma_{ij}^0}{\partial y_j}\delta u_i^1\,\mathrm{d}\Omega=\int_{\Omega^{\varepsilon}}c_{ijkl}^{\varepsilon}\left(\frac{\partial u_k^0}{\partial x_l}+\frac{\partial u_k^1}{\partial y_l}\right)_{,j}\delta u_i^1\,\mathrm{d}\Omega=0 \tag{2.8.29}$$

Then

$$\lim_{\varepsilon\to0}\int_{\Omega^{\varepsilon}}c_{ijkl}^{\varepsilon}\left(\frac{\partial u_k^0}{\partial x_l}+\frac{\partial u_k^1}{\partial y_l}\right)_{,j}\delta u_i^1\,\mathrm{d}\Omega$$

$$=\frac{1}{V}\int_{\Omega}\int_{Y}c_{ijkl}\left(\frac{\partial u_k^0}{\partial x_l}+\frac{\partial u_k^1}{\partial y_l}\right)_{,j}\delta u_i^1\,\mathrm{d}Y\mathrm{d}\Omega=0 \tag{2.8.30}$$

Integrating by parts, and noting that the virtual displacements $\delta u_i^1=0$ at the boundary of the RVE, and u_i^0 is a function of x only, we have

$$\int_{\Omega}\frac{\partial u_k^0}{\partial x_l}\left(\int_{Y}c_{ijkl}^{\varepsilon}\frac{\partial\delta u_i^1}{\partial y_j}\mathrm{d}Y\right)\mathrm{d}\Omega+\int_{\Omega}\int_{Y}c_{ijkl}\frac{\partial u_k^1}{\partial y_l}\frac{\partial\delta u_i^1}{\partial y_j}\mathrm{d}Y\mathrm{d}\Omega=0 \tag{2.8.31}$$

Introducing the function $\psi(x,y)$ which satisfies

$$\int_{Y}c_{ijpq}\frac{\partial\psi_p^{kl}}{\partial y_q}\frac{\partial\delta u_i^1}{\partial y_j}\mathrm{d}Y=\int_{Y}c_{ijkl}\frac{\partial\delta u_i^1}{\partial y_j}\mathrm{d}Y \tag{2.8.32}$$

and substituting Eq.(2.8.32) into Eq.(2.8.31), we have

$$\int_{\Omega}\frac{\partial u_k^0}{\partial x_l}\int_{Y}c_{ijpq}\frac{\partial\psi_p^{kl}}{\partial y_q}\frac{\partial\delta u_i^1}{\partial y_j}\mathrm{d}Y\mathrm{d}\Omega+\int_{\Omega}\int_{Y}c_{ijkl}\frac{\partial u_k^1}{\partial y_l}\frac{\partial\delta u_i^1}{\partial y_j}\mathrm{d}Y\mathrm{d}\Omega=0 \tag{2.8.33}$$

Applying the divergence theorem to Eq.(2.8.33) leads to

$$\int_{\Omega}\oint_{s}c_{ijpq}\psi_p^{kl}n_q\frac{\partial u_k^0}{\partial x_l}\frac{\partial\delta u_i^1}{\partial y_j}\mathrm{d}s\mathrm{d}\Omega+\int_{\Omega}\oint_{s}c_{ijkl}u_p^1n_q\frac{\partial\delta u_i^1}{\partial y_j}\mathrm{d}s\mathrm{d}\Omega=0 \tag{2.8.34}$$

From Eq.(2.8.34), we can again obtain Eq.(2.8.19). This explains the reason for the assumption made in the previous section.

2.8.5   Finite element formulation

The interpolation of the FE form for the function $\psi(x, y)$ can be expressed by

$$\psi_i^{kl} = (N_\alpha \psi_\alpha)_i^{kl} = (N\psi)_i^{kl}, \quad \alpha = 1, 2, \cdots, M \tag{2.8.35}$$

where N is a shape function, ψ stands for the nodal generalized coordinates, and M denotes the total number of degrees of freedom in a FE system. Then the derivatives in Eq.(2.8.19) can be expressed as

$$\frac{\partial \psi_p^{kl}}{\partial y_q} = (B_q \psi)_p^{kl} \tag{2.8.36}$$

$$\frac{\partial \delta u_i^1}{\partial y_j} = (B_j \psi)_i^{kl} \frac{\partial u_k^0}{\partial x_l} \tag{2.8.37}$$

where B_i are the derivatives of the shape function N with respect to y_i. Note that the function u_i^0 is independent of y.

We can rewrite Eq.(2.8.32) in the standard form of FE

$$\left(\int_Y B^{\mathrm{T}} cB \mathrm{d}Y \right) \psi^{kl} = \int_Y B^{\mathrm{T}} c^{kl} \mathrm{d}Y \tag{2.8.38}$$

where c is the stress-strain matrix, B is the discrete displacement-strain matrix depending on the element shape functions, c^{kl} is a vector of a column of kl (kl=11,22,33,23,31,12) of the stress-strain matrix c, and ψ^{kl} is the characteristic displacement vector associated with the kl mode. There are six equations to be solved for different strain states. A conventional FE can be used to calculate Eq.(2.8.38).

Therefore, the homogenized elastic constants defined by Eq.(2.8.23), can be expressed as

$$\bar{c} = \frac{1}{V} \int_Y c(I - B\psi) \mathrm{d}Y \tag{2.8.39}$$

where

$$\psi = \begin{bmatrix} \psi^{11} & \psi^{22} & \psi^{33} & \psi^{23} & \psi^{31} & \psi^{12} \end{bmatrix} \tag{2.8.40}$$

In summary, ψ^{kl} in Eq.(2.8.38) is solved by the FEM and then the effective properties can be calculated from Eq.(2.8.39).

2.9   An approximate estimation of effective properties

Here let us approximately evaluate the effective properties based on the two-scale expansion method. Consider two specific cases: the constant strain model and the constant stress model.

Analyzing the basic assumption made in the two-scale expansion method, Eq.(2.8.19), and the effective stiffness, Eq. (2.8.23), we can see that the first term in Eq. (2.8.23) is the well-known rule of mixture, while the second term is a correction term due to the heterogeneity of the microstructure.

In the constant strain model, it is assumed that the strains undergone in each phase have the same values. Thus no perturbing displacements exist, that is

$$u_i^1 = -\psi_i^{kl}(\boldsymbol{x}, \boldsymbol{y})\frac{\partial u_k^0}{\partial x_l} = 0 \tag{2.9.1}$$

Then the effective stiffness Eq.(2.8.23) can be reduced as

$$\bar{\bar{c}}_{ijkl} = \frac{1}{V}\int_Y \left[c_{ijkl} - c_{ijmn}\frac{\partial \psi_m^{kl}}{\partial y_n} \right]\mathrm{d}Y$$

$$= \frac{1}{V}\int_Y c_{ijkl}\mathrm{d}Y \tag{2.9.2}$$

This is the known rule of mixture. A simple expression under uniaxial state is

$$\bar{E}_{11} = \frac{1}{V}\int_Y E_{11}\mathrm{d}V = v_1 E_{11}^{(1)} + v_2 E_{11}^{(2)} + \cdots \tag{2.9.3}$$

where E_{11} is the Young's modulus, v_i and $E_{11}^{(i)}$ are the volume fraction and the Young's modulus of phase i, respectively. Eq.(2.9.3) is referred to as the Voigt approximation [30] and usually is used to predict the effective axial modulus of unidirectional fiber composite material. Eq.(2.9.3) gives the upper bound of the elastic modulus.

For estimating the approximation of effective properties in the constant stress state, we can represent the relation by

$$\frac{1}{2}\left(\frac{\partial u_i^0}{\partial x_j} + \frac{\partial u_j^0}{\partial x_i} \right) = f_{ijkl}\sigma_{kl} \tag{2.9.4}$$

and

$$u_i^1 = -\psi_i^{kl}(\boldsymbol{x}, \boldsymbol{y})\frac{\partial u_k^0}{\partial x_l} = -\psi_i^{kl}(\boldsymbol{x}, \boldsymbol{y})f_{klpq}\sigma_{pq} \tag{2.9.5}$$

The strain is

$$\varepsilon_{ij}^0 = \left(f_{ijpq} - f_{ijmn} \frac{\partial \psi_m^{kl}}{\partial y_n} f_{klpq} \right) \sigma_{pq} \tag{2.9.6}$$

Taking integration over the RVE, we can obtain

$$\overline{\varepsilon}_{ij} = \overline{f}_{ijpq} \sigma_{pq} \tag{2.9.7}$$

with the homogenized compliance coefficients

$$\overline{f}_{ijpq} = \frac{1}{V} \int_Y \left(f_{ijpq} - f_{ijmn} \frac{\partial \psi_m^{kl}}{\partial y_n} f_{klpq} \right) dY \tag{2.9.8}$$

Eq.(2.9.8) can be interpreted as the correction operating on the rule of mixture for the compliance. In the constant stress model, it is assumed that the stresses of each phase are uniform and equal. Applying the equal stress condition to Eq.(2.9.8), we can obtain

$$\overline{f}_{ijkl} = \frac{1}{V} \int_Y f_{ijkl} dY \tag{2.9.9}$$

For the uniaxial state, the Young's modulus can be expressed by

$$\frac{1}{\overline{E}_{22}} = \frac{1}{V} \int_Y \frac{1}{E_{22}} dV = \frac{v_1}{E_{22}^{(1)}} + \frac{v_2}{E_{22}^{(2)}} + \cdots \tag{2.9.10}$$

This equation is called Reuss's approximation [31] and is usually used to predict the transverse modulus of a unidirectional composite material. It is verified that this equation gives the simple lower bound of the effective elastic modulus of a composite.

It should be noted that Voigt's and Reuss's approximations provide rigorous upper and lower bounds. They are the most simply cases of Hashin and Strikman's variational solutions [26].

2.10   Formulations and implementation for 2D problem

In this section, the detailed FE formulation of a 2D problem is given for the two-scale expansion method. There are three deformation modes of Ψ^{kl} ($kl = 11, 22, 12$),e.g. $\Psi = (\Psi^{11}, \Psi^{22}, \Psi^{12})$,to be calculated for this case.

2.10.1   Formulations

Consider a plane stress problem of an orthotropic elastic body. The stress-strain relation is

$$\begin{bmatrix} \varepsilon_{11} \\ \varepsilon_{22} \\ 2\varepsilon_{12} \end{bmatrix} = \begin{bmatrix} f_{1111} & f_{1122} & 0 \\ f_{2211} & f_{2222} & 0 \\ 0 & 0 & f_{1212} \end{bmatrix} \begin{bmatrix} \sigma_{11} \\ \sigma_{22} \\ \sigma_{12} \end{bmatrix} \tag{2.10.1}$$

where the compliance matrix can be expressed by the engineering constants

$$f_{1111} = \frac{1}{E_{11}}, \quad f_{1122} = f_{2211} = \frac{-\mu_{12}}{E_{11}} \tag{2.10.2a}$$

$$f_{2222} = \frac{1}{E_{22}}, \quad f_{1212} = \frac{1}{G_{12}} \tag{2.10.2b}$$

Similarly, the stiffness matrix can be written by the equation

$$\begin{bmatrix} \sigma_{11} \\ \sigma_{22} \\ \sigma_{12} \end{bmatrix} = \begin{bmatrix} c_{1111} & c_{1122} & 0 \\ c_{2211} & c_{2222} & 0 \\ 0 & 0 & c_{1212} \end{bmatrix} \begin{bmatrix} \varepsilon_{11} \\ \varepsilon_{22} \\ 2\varepsilon_{12} \end{bmatrix} \tag{2.10.3}$$

where

$$c_{1111} = \frac{E_{11}}{1 - \mu_{12}^2 E_{22} / E_{11}}, \quad c_{1122} = c_{2211} = \frac{\mu_{12} E_{22}}{1 - \mu_{12}^2 E_{22} / E_{11}} \tag{2.10.4a}$$

$$c_{2222} = \frac{E_{22}}{1 - \mu_{12}^2 E_{22} / E_{11}}, \quad c_{1212} = G_{12} \tag{2.10.4b}$$

where E_{11} and E_{22} are the Young's moduli, μ_{12} is the Poisson's ratio, and G_{12} is the shear modulus. For an isotropic elastic body, $E_{11} = E_{22} = E$, $\mu_{12} = \mu$, $G_{12} = G = \dfrac{E}{2(1+\mu)}$ and the compliance and stiffness matrix can be written as

$$f = \begin{bmatrix} f_{1111} & f_{1122} & 0 \\ f_{2211} & f_{2222} & 0 \\ 0 & 0 & f_{1212} \end{bmatrix} = \frac{1}{E} \begin{bmatrix} 1 & -\mu & 0 \\ -\mu & 1 & 0 \\ 0 & 0 & 2(1+\mu) \end{bmatrix} \tag{2.10.5}$$

and

$$c = \begin{bmatrix} c_{1111} & c_{1122} & 0 \\ c_{2211} & c_{2222} & 0 \\ 0 & 0 & c_{1212} \end{bmatrix} = \frac{E}{1 - \mu^2} \begin{bmatrix} 1 & \mu & 0 \\ \mu & 1 & 0 \\ 0 & 0 & \dfrac{1-\mu}{2} \end{bmatrix} \tag{2.10.6}$$

The governing equations (2.8.32) and (2.8.23) of the homogenization method are rewritten as follows:

$$\int_Y c_{ijpq} \frac{\partial \psi_p^{kl}}{\partial y_q} \frac{\partial v_i}{\partial y_j} \, dY = \int_Y c_{ijkl} \frac{\partial v_i}{\partial y_j} \, dY \tag{2.10.7}$$

$$\overline{c}_{ijkl} = \frac{1}{V} \int_Y \left[c_{ijkl} - c_{ijmn} \frac{\partial \psi_m^{kl}}{\partial y_n} \right] dY \qquad (2.10.8)$$

where $v_i = \delta u_i^1$ are arbitrary virtual displacements. The equations will be solved for three cases: $kl = 11$, $kl = 22$ and $kl = 12$, respectively. A detailed solving procedure incorporating FEM is now provided for the three cases.

Case kl=11   Expansion of Eq.(2.10.6) with elements of matrix leads to

$$\int_Y \left[\left(c_{1111} \frac{\partial \psi_1^{11}}{\partial y_1} + c_{1122} \frac{\partial \psi_2^{11}}{\partial y_2} \right) \frac{\partial v_1}{\partial y_1} + \right.$$

$$\left(c_{1122} \frac{\partial \psi_1^{11}}{\partial y_1} + c_{2222} \frac{\partial \psi_2^{11}}{\partial y_2} \right) \frac{\partial v_2}{\partial y_2} +$$

$$\left. c_{1212} \left(\frac{\partial \psi_1^{11}}{\partial y_2} + \frac{\partial \psi_2^{11}}{\partial y_1} \right) \left(\frac{\partial v_1}{\partial y_2} + \frac{\partial v_2}{\partial y_1} \right) \right] dY$$

$$= \int_Y \left(c_{1111} \frac{\partial v_1}{\partial y_1} + c_{1122} \frac{\partial v_2}{\partial y_2} \right) dY \qquad (2.10.9)$$

The effective properties in Eq.(2.10.8) become

$$\overline{c}_{1111} = \frac{1}{V} \int_Y \left(c_{1111} - c_{1111} \frac{\partial \psi_1^{11}}{\partial y_1} - c_{1122} \frac{\partial \psi_2^{11}}{\partial y_2} \right) dY , \qquad ij = 11 \qquad (2.10.10)$$

$$\overline{c}_{2211} = \frac{1}{V} \int_Y \left(c_{2211} - c_{2211} \frac{\partial \psi_1^{11}}{\partial y_1} - c_{2222} \frac{\partial \psi_2^{11}}{\partial y_2} \right) dY , \qquad ij = 22 \qquad (2.10.11)$$

Rewriting Eq.(2.10.9) in matrix form, we obtain

$$\int_Y \left[\frac{\partial v_1}{\partial y_1} \quad \frac{\partial v_2}{\partial y_2} \quad \frac{\partial v_1}{\partial y_2} + \frac{\partial v_2}{\partial y_1} \right] \begin{bmatrix} c_{1111} & c_{1122} & 0 \\ c_{2211} & c_{2222} & 0 \\ 0 & 0 & D_{1212} \end{bmatrix} \begin{bmatrix} \dfrac{\partial \psi_1^{11}}{\partial y_1} \\[2mm] \dfrac{\partial \psi_2^{11}}{\partial y_2} \\[2mm] \dfrac{\partial \psi_1^{11}}{\partial y_2} + \dfrac{\partial \psi_2^{11}}{\partial y_1} \end{bmatrix} dY$$

$$= \int_Y \left[\frac{\partial v_1}{\partial y_1} \quad \frac{\partial v_2}{\partial y_2} \quad \frac{\partial v_1}{\partial y_2} + \frac{\partial v_2}{\partial y_1} \right] \begin{bmatrix} c_{1111} \\ c_{1122} \\ 0 \end{bmatrix} dY \qquad (2.10.12)$$

Now we introduce notations for the strains

$$\boldsymbol{\varepsilon}(\boldsymbol{\psi}) = \begin{bmatrix} \dfrac{\partial \psi_1^{11}}{\partial y_1} \\[2ex] \dfrac{\partial \psi_2^{11}}{\partial y_2} \\[2ex] \dfrac{\partial \psi_1^{11}}{\partial y_2} + \dfrac{\partial \psi_2^{11}}{\partial y_1} \end{bmatrix}, \quad \boldsymbol{\varepsilon}(\boldsymbol{v}) = \begin{bmatrix} \dfrac{\partial v_1}{\partial y_1} \\[2ex] \dfrac{\partial v_2}{\partial y_2} \\[2ex] \dfrac{\partial v_1}{\partial y_2} + \dfrac{\partial v_2}{\partial y_1} \end{bmatrix} \qquad (2.10.13)$$

where $\quad \boldsymbol{\psi} = \begin{bmatrix} \psi_1^{11} \\ \psi_2^{11} \end{bmatrix}, \quad \boldsymbol{v} = \begin{bmatrix} v_1 \\ v_2 \end{bmatrix}.$

The stiffness matrix can be written in a compact form

$$\boldsymbol{c} = \begin{bmatrix} \boldsymbol{c}_1 & \boldsymbol{c}_2 & \boldsymbol{c}_3 \end{bmatrix} \qquad (2.10.14)$$

where

$$\boldsymbol{c}_1 = \begin{bmatrix} c_{1111} \\ c_{2211} \\ 0 \end{bmatrix}, \quad \boldsymbol{c}_2 = \begin{bmatrix} c_{1122} \\ c_{2222} \\ 0 \end{bmatrix}, \quad \boldsymbol{c}_3 = \begin{bmatrix} 0 \\ 0 \\ c_{1212} \end{bmatrix} \qquad (2.10.15)$$

Then Eq.(2.10.12) can be written in matrix form as

$$\int_Y \boldsymbol{\varepsilon}^{\mathrm{T}}(\boldsymbol{v}) \boldsymbol{c}\, \boldsymbol{\varepsilon}(\boldsymbol{\psi}) \mathrm{d}Y = \int_Y \boldsymbol{\varepsilon}^{\mathrm{T}}(\boldsymbol{v}) \boldsymbol{c}_1 \mathrm{d}Y \qquad (2.10.16)$$

FE discretization is introduced by interpolation of the function $\boldsymbol{\psi}$ with the form

$$\boldsymbol{\psi} = \sum_{i=1}^{n} \boldsymbol{N}_i \hat{\boldsymbol{\psi}}_i^e = \boldsymbol{N}\hat{\boldsymbol{\psi}}^e \qquad (2.10.17)$$

where n is the number of nodes in an element, $\hat{\boldsymbol{\psi}}^e$ is the degrees of nodal freedom of the element

$$\hat{\boldsymbol{\psi}}^e = \begin{bmatrix} \hat{\boldsymbol{\psi}}_1 & \hat{\boldsymbol{\psi}}_2 & \cdots & \hat{\boldsymbol{\psi}}_n \end{bmatrix}^{\mathrm{T}} \qquad (2.10.18)$$

The shape function matrix $\boldsymbol{N}$ can be expressed by

$$\boldsymbol{N} = \begin{bmatrix} \boldsymbol{N}_1 & \boldsymbol{N}_2 & \cdots & \boldsymbol{N}_n \end{bmatrix} \qquad (2.10.19)$$

Substituting Eq.(2.10.17) into the first of Eq.(2.10.13), the strain can be obtained

$$\boldsymbol{\varepsilon}^e(\boldsymbol{\psi}) = \boldsymbol{L}\boldsymbol{\psi} = \boldsymbol{L}\boldsymbol{N}\hat{\boldsymbol{\psi}}^e = \boldsymbol{B}\hat{\boldsymbol{\psi}}^e \qquad (2.10.20)$$

where $\boldsymbol{B} = \boldsymbol{L}\boldsymbol{N}$ is the element strain matrix, and

$$L = \begin{bmatrix} \dfrac{\partial}{\partial y_1} & 0 \\[3mm] 0 & \dfrac{\partial}{\partial y_2} \\[3mm] \dfrac{\partial}{\partial y_2} & \dfrac{\partial}{\partial y_1} \end{bmatrix} \qquad (2.10.21)$$

is the matrix of linear differential operator which links the relation between displacements and strains for a plane problem.

Similarly, the FE formulations for the function v, referred to as arbitrary virtual displacements, can also be obtained with exactly the same form. Thus we can obtain the FE equation from Eq.(2.10.16)

$$K\hat{\psi} = F \qquad (2.10.22)$$

where

$$K = \sum_{e=1}^{m} K^e , \qquad F = \sum_{e=1}^{m} F^e \qquad (2.10.23)$$

$$K^e = \int_{\Omega^e} B^\mathrm{T} cB \mathrm{d}\Omega , \qquad F^e = \int_{\Omega^e} B^\mathrm{T} c_1 \mathrm{d}\Omega \qquad (2.10.24)$$

The "force" vector F has a physical meaning. c_1 is the stress induced by a specific initial strain ε^0

$$c_1 = c\varepsilon^0 = \begin{bmatrix} C_{1111} & C_{1122} & 0 \\ C_{1122} & C_{2222} & 0 \\ 0 & 0 & C_{1212} \end{bmatrix} \begin{bmatrix} \varepsilon_{11}^0 \\ \varepsilon_{22}^0 \\ 2\varepsilon_{12}^0 \end{bmatrix} = \begin{bmatrix} C_{1111} \\ C_{1122} \\ 0 \end{bmatrix} \qquad (2.10.25)$$

which implies that a uniform initial strain is applied to the RVE at any point

$$\varepsilon^0 = \begin{bmatrix} \varepsilon_{11}^0 \\ \varepsilon_{22}^0 \\ 2\varepsilon_{12}^0 \end{bmatrix} = \begin{bmatrix} 1 \\ 0 \\ 0 \end{bmatrix} \qquad (2.10.26)$$

Thus, Eq.(2.10.22) is solved in order to give the displacement $\hat{\psi}$ and strain $\varepsilon(\hat{\psi})$, and we can calculate the effective properties by Eqs.(2.10.10) and (2.10.11), of which the matrix forms are

$$\overline{C}_{1111} = \frac{1}{V} \int_Y [C_{1111} - c_1^\mathrm{T} \varepsilon(\psi)] \mathrm{d}Y \qquad (2.10.27a)$$

$$\overline{C}_{2211} = \frac{1}{V} \int_Y [C_{2211} - c_2^\mathrm{T} \varepsilon(\psi)] \mathrm{d}Y \qquad (2.10.27b)$$

For m finite elements, the integration can be replaced by summations element by element

$$\overline{c}_{1111} = \frac{1}{V}\sum_{e=1}^{m}\int_{\Omega^e}[c_{1111} - \mathbf{c}_1^{\mathrm{T}}\mathbf{B}\hat{\boldsymbol{\psi}}^e]\mathrm{d}\Omega \qquad (2.10.28a)$$

$$\overline{c}_{2211} = \frac{1}{V}\sum_{e=1}^{m}\int_{\Omega^e}[c_{2211} - \mathbf{c}_2^{\mathrm{T}}\mathbf{B}\hat{\boldsymbol{\psi}}^e]\mathrm{d}\Omega \qquad (2.10.28b)$$

while the integration in an element can be calculated by a numerical integration procedure, such as the Gauss-Legendre rule. It is easy to add the formulations into a standard FE program.

Case $kl = 22$ An identical approach can be used to derive the FE formulations for the case $kl = 22$. The governing equation becomes

$$\int_Y\left[\left(c_{1111}\frac{\partial\psi_1^{22}}{\partial y_1} + c_{1122}\frac{\partial\psi_2^{22}}{\partial y_2}\right)\frac{\partial v_1}{\partial y_1} + \right.$$

$$\left(c_{1122}\frac{\partial\psi_1^{22}}{\partial y_1} + c_{2222}\frac{\partial\psi_2^{22}}{\partial y_2}\right)\frac{\partial v_2}{\partial y_2} +$$

$$\left.c_{1212}\left(\frac{\partial\psi_1^{22}}{\partial y_2} + \frac{\partial\psi_2^{22}}{\partial y_1}\right)\left(\frac{\partial v_1}{\partial y_2} + \frac{\partial v_2}{\partial y_1}\right)\right]\mathrm{d}Y$$

$$= \int_Y\left(c_{2222}\frac{\partial v_2}{\partial y_2} + c_{1122}\frac{\partial v_1}{\partial y_1}\right)\mathrm{d}Y \qquad (2.10.29)$$

The effective properties of the composites are

$$\overline{c}_{2222} = \frac{1}{V}\int_Y\left(c_{2222} - c_{2211}\frac{\partial\psi_1^{221}}{\partial y_1} - c_{2222}\frac{\partial\psi_2^{22}}{\partial y_2}\right)\mathrm{d}Y, \quad ij = 22 \qquad (2.10.30a)$$

$$\overline{c}_{1122} = \frac{1}{V}\int_Y\left(c_{1122} - c_{1111}\frac{\partial\psi_1^{22}}{\partial y_1} - c_{1122}\frac{\partial\psi_2^{22}}{\partial y_2}\right)\mathrm{d}Y, \quad ij = 11 \qquad (2.10.30b)$$

The matrix form of the Eq.(2.10.29) is

$$\int_Y\boldsymbol{\varepsilon}^{\mathrm{T}}(v)\mathbf{c}\boldsymbol{\varepsilon}(\psi)\mathrm{d}Y = \int_Y\boldsymbol{\varepsilon}^{\mathrm{T}}(v)\mathbf{c}_2\mathrm{d}Y \qquad (2.10.31a)$$

The matrix forms of equations for the effective properties are

$$\overline{c}_{2222} = \frac{1}{V}\int_Y[c_{2222} - \mathbf{c}_2^{\mathrm{T}}\boldsymbol{\varepsilon}(\psi)]\mathrm{d}Y \qquad (2.10.31b)$$

$$\overline{c}_{1122} = \frac{1}{V}\int_Y[c_{1122} - \mathbf{c}_1^{\mathrm{T}}\boldsymbol{\varepsilon}(\psi)]\mathrm{d}Y \qquad (2.10.31c)$$

where $\quad \psi = \begin{bmatrix} \psi_1^{22} \\ \psi_2^{22} \end{bmatrix}$.

The finite element equation is

$$K\hat{\psi} = F \tag{2.10.32}$$

where

$$K = \sum_{e=1}^{m} K^e , \quad F = \sum_{e=1}^{m} F^e \tag{2.10.33}$$

and

$$K^e = \int_{\Omega^e} B^{\mathrm{T}} cB\mathrm{d}\Omega , \quad F^e = \int_{\Omega^e} B^{\mathrm{T}} c_2\mathrm{d}\Omega \tag{2.10.34}$$

In this case, the physical meaning of the "force" vector F is nodal forces induced by the uniform initial strain

$$\varepsilon^0 = \begin{bmatrix} \varepsilon_{11}^0 \\ \varepsilon_{22}^0 \\ 2\varepsilon_{12}^0 \end{bmatrix} = \begin{bmatrix} 0 \\ 1 \\ 0 \end{bmatrix} \tag{2.10.35}$$

The formulations for calculation of the effective properties are

$$\overline{c}_{2222} = \frac{1}{V} \sum_{e=1}^{m} \int_{\Omega^e} \left[c_{2222} - c_2^{\mathrm{T}} B\hat{\psi}^e \right] \mathrm{d}\Omega \tag{2.10.36a}$$

$$\overline{c}_{1122} = \frac{1}{V} \sum_{e=1}^{m} \int_{\Omega^e} \left[c_{1122} - c_1^{\mathrm{T}} B\hat{\psi}^e \right] \mathrm{d}\Omega \tag{2.10.36b}$$

Case $kl = 12$ In this case, we have a series of corresponding equations

$$\int_Y \left[\left(c_{1111} \frac{\partial \psi_1^{12}}{\partial y_1} + c_{1122} \frac{\partial \psi_2^{12}}{\partial y_2} \right) \frac{\partial v_1}{\partial y_1} + \right.$$

$$\left(c_{1122} \frac{\partial \psi_1^{12}}{\partial y_1} + c_{2222} \frac{\partial \psi_2^{12}}{\partial y_2} \right) \frac{\partial v_2}{\partial y_2} +$$

$$\left. c_{1212} \left(\frac{\partial \psi_1^{12}}{\partial y_2} + \frac{\partial \psi_2^{12}}{\partial y_1} \right) \left(\frac{\partial v_1}{\partial y_2} + \frac{\partial v_2}{\partial y_1} \right) \right] \mathrm{d}Y$$

$$= \int_Y c_{1212} \left(\frac{\partial v_1}{\partial y_2} + \frac{\partial v_2}{\partial y_1} \right) \mathrm{d}Y \tag{2.10.37}$$

$$\overline{c}_{1212} = \frac{1}{V} \int_Y c_{1212} \left(1 - \frac{\partial \psi_1^{12}}{\partial y_2} - c_{1122} \frac{\partial \psi_2^{12}}{\partial y_1} \right) \mathrm{d}Y , \quad ij = 12 \tag{2.10.38}$$

$$\int_Y \boldsymbol{\varepsilon}^{\mathrm{T}}(v)\boldsymbol{c}\boldsymbol{\varepsilon}(\boldsymbol{\psi})\mathrm{d}Y = \int_Y \boldsymbol{\varepsilon}^{\mathrm{T}}(v)\boldsymbol{c}_3\mathrm{d}Y \qquad (2.10.39)$$

$$\overline{c}_{1212} = \frac{1}{V}\int_Y \left[c_{1212} - \boldsymbol{c}_3^{\mathrm{T}}\boldsymbol{\varepsilon}(\boldsymbol{\psi}) \right]\mathrm{d}Y \qquad (2.10.40)$$

where　$\boldsymbol{\psi} = \begin{bmatrix} \psi_1^{12} \\ \psi_2^{12} \end{bmatrix}$.

The finite element equation is

$$\boldsymbol{K}\hat{\boldsymbol{\psi}} = \boldsymbol{F} \qquad (2.10.41)$$

where

$$\boldsymbol{K} = \sum_{e=1}^{m} \boldsymbol{K}^e, \quad \boldsymbol{F} = \sum_{e=1}^{m} \boldsymbol{F}^e \qquad (2.10.42)$$

$$\boldsymbol{K}^e = \int_{\Omega^e} \boldsymbol{B}^{\mathrm{T}}\boldsymbol{c}\boldsymbol{B}\mathrm{d}\Omega, \quad \boldsymbol{F}^e = \int_{\Omega^e} \boldsymbol{B}^{\mathrm{T}}\boldsymbol{c}_3\mathrm{d}\Omega \qquad (2.10.43)$$

The uniform initial strain in the RVE inducing the nodal "force" vector $\boldsymbol{F}$ is identified as

$$\boldsymbol{\varepsilon}^0 = \begin{bmatrix} \varepsilon_{11}^0 \\ \varepsilon_{22}^0 \\ 2\varepsilon_{12}^0 \end{bmatrix} = \begin{bmatrix} 0 \\ 0 \\ 1 \end{bmatrix} \qquad (2.10.44)$$

The effective properties can be calculated from

$$\overline{c}_{1212} = \frac{1}{V}\sum_{e=1}^{m} \int_{\Omega^e} \left[c_{1212} - \boldsymbol{c}_3^{\mathrm{T}}\boldsymbol{B}\hat{\boldsymbol{\psi}}^e \right]\mathrm{d}\Omega \qquad (2.10.45)$$

It is noted that shear coupling coefficients exist for anisotropic materials. They can be calculated by similar formulations.

2.10.2　FE implementation of homogenization methods

The standard FE program is available for prediction of the effective properties of heterogeneous materials. But specific additional subroutines must be incorporated into a standard FE program to treat the nodal "force" vector, periodic boundary conditions and the calculation of effective properties.

　　A homogenization program named HOMP is developed here, including a direct method and a two-scale expanding method. The program organization is shown in Fig.2.12. The parts in grey denote the subroutines to be added into a standard FE program.

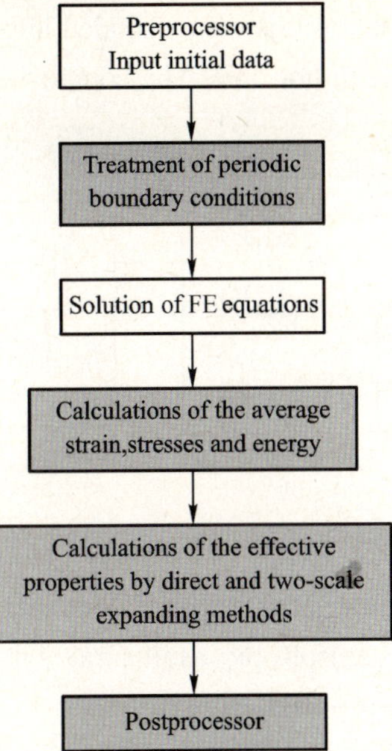

Fig.2.12   A profile of HOMP

2.11   Numerical results

To investigate the effective properties of the composites, using direct methods (including stress, strain and energy) and the two-scale expanding method, we consider the following problems.

Case 1   Circular inclusions are embedded into the isotropic matrix, and the resulting composite is almost transversely isotropic. The microstructure is shown in Fig.2.13.

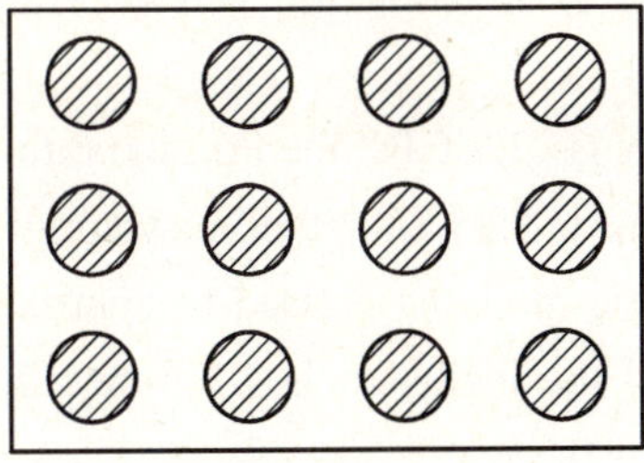
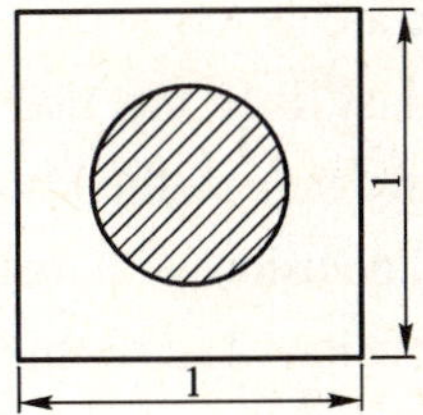

Fig.2.13   Case 1: almost transversely isotropic composite

Case 2   L-shaped inclusions are embedded into the isotropic matrix, and the resulting composite is orthotropic. The composite and RVE are illustrated in Fig.2.14.

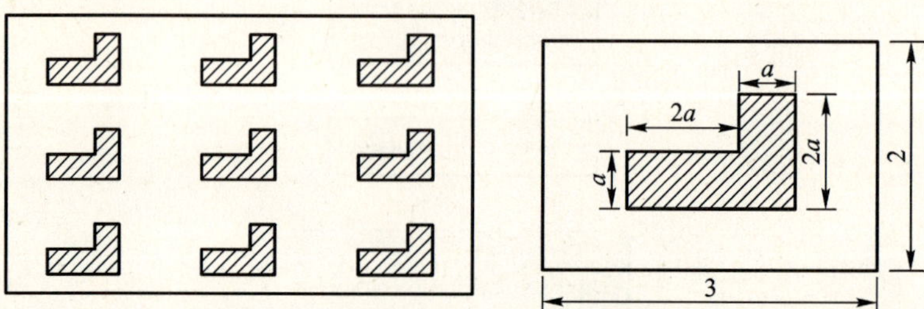

Fig.2.14   Case 2: orthotropic composite

Case 3   Y-shaped inclusions are embedded into the isotropic matrix, and the resulting composite is anisotropic, as illustrated in Fig.2.15.

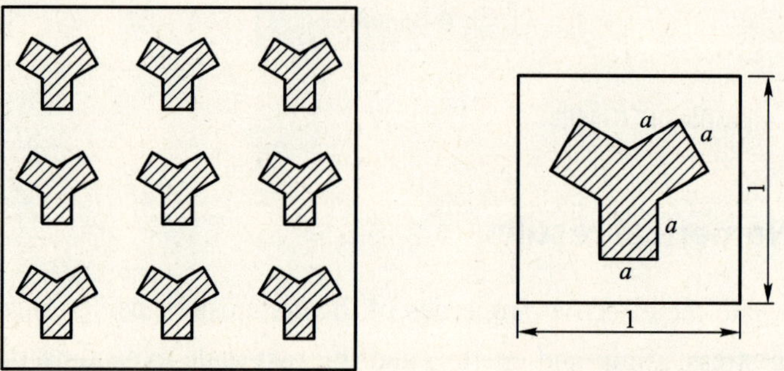

Fig.2.15   Case 3: anisotropic composite

Each case includes a fiber composite, a rigid inclusion medium and a void solid. The properties of the inclusions vary from a very large value (modeling rigid inclusions) to a very small value (simulating the voids). The material constants are as follows:

E-glass fiber: the Young's modulus is 73.1 GPa, the Poisson ratio is 0.22.

Epoxy matrix: the Young's modulus is 3.45 GPa, the Poisson ratio is 0.35.

An inclusion with very large elastic modulus is used to approximate the rigid inclusion. The elastic modulus of the inclusion is 10^4 times that of the matrix. The Poisson ratio of the matrix is 0.35.

An inclusion with a very small elastic modulus is used to model the void. The elastic modulus of the void inclusion is 10^{-6} times that of the matrix. The

Poisson ratio of the matrix is 0.35.

A plane strain model is taken into consideration here. Thus the in-plane or transverse properties of the composites are calculated. For the direct average methods, boundary conditions with specific displacements are imposed and then the FE method is applied in the calculation of the average stress, strain and strain energy density on the RVE with a uniaxial strain state. The resulting effective stiffness matrix of the plane strain problem is

$$\begin{bmatrix} c_{1111} & c_{1122} & 0 \\ c_{1122} & c_{2222} & 0 \\ 0 & 0 & c_{1212} \end{bmatrix} \tag{2.11.1}$$

The engineering constants for an isotropic body ($c_{1111} = c_{2222}$) can be calculated by

$$\mu = \frac{c_{1122}}{c_{1111} + c_{1122}} \tag{2.11.2a}$$

$$E = \frac{c_{1111}(1+\mu)(1-2\mu)}{1-\mu} \tag{2.11.2b}$$

$$G = c_{1212} \tag{2.11.2c}$$

2.11.1 Effective stiffness of isotropic composite

The effective transverse stiffness coefficients of the transversely isotropic composite are listed in Table 2.1. Here ASS denotes the direct average method based on strain and stress fields, ASE the direct average method based on strain energy density and TEM the two-scale expansion method. It is shown that the three methods yield identical stiffness coefficients. This is not surprising because of the same homogenization principle is used in all three methods. The

Table 2.1 Transverse stiffness coefficients for fiber composite

	c_{1111}/GPa			c_{1122}/GPa		c_{1212}/GPa		
	ASS	ASE	TEM	ASS	TEM	ASS	ASE	TEM
0.1	6.3147	6.3147	6.3148	3.2920	3.2920	1.4726	1.4726	1.4727
0.2	7.3218	7.3218	7.3218	3.6171	3.6171	1.6824	1.6824	1.6824
0.3	8.6606	8.6606	8.6606	3.9511	3.9511	1.9255	1.9255	1.9255
0.4	10.4873	10.4873	10.4874	4.2879	4.2879	2.2313	2.2313	2.2314
0.5	13.0754	13.0755	13.0758	4.6347	4.6346	2.6549	2.6550	2.6551
0.6	17.0605	17.0606	17.0608	5.0817	5.0817	3.3356	3.3356	3.3357

engineering constants can be found by Eq.(2.4.2) for comparison with the approximate bounds and experimental data. Fig.2.16 illustrates the transverse Young's modulus E_{22} as a function of the fiber volume fraction. The lower bound was calculated by Ruess's approximation. It is shown that ASS, ASE and TEM provide good consistent results with the experimental data [42]. The experimental data and bounds are listed in Table 2.2.

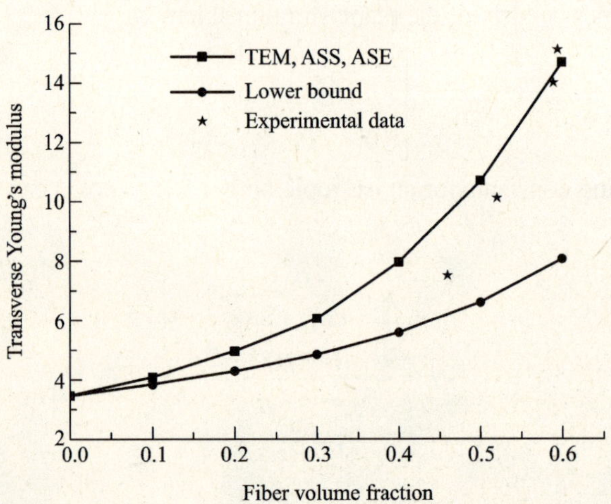

Fig.2.16   Transverse Young's modulus vs fiber volume fraction

Table 2.2   The transverse constants

	E_{22}/GPa			G_{23}/GPa		μ
	UB	LB	Exp. Data	UB	LB	
0.0	3.45	3.45			1.278	1.278	0.35
0.1	4.0582	3.8133			1.4727	1.4133	0.3427
0.2	4.9291	4.2622			1.6824	1.5806	0.3307
0.3	6.0361	4.8309	0.46	7.5	1.9255	1.7929	0.3133
0.4	7.9412	5.5746	0.52	10.1	2.2314	2.0711	0.2902
0.5	10.6499	6.5890	0.59	14	2.6551	2.4514	0.2617
0.6	14.6712	8.0548	0.595	15.1	3.3357	2.9852	0.2295

Fig.2.17 shows the transverse shear modulus G of the composite with different fiber volume fraction. No experimental data for the transverse shear modulus is available for comparison. An approximate estimation for the transverse shear modulus by

$$\frac{1}{G} = \frac{v_m}{G_m} + \frac{v_f}{G_f}$$

(2.11.3)

is plotted in Fig.2.17. It is easy to prove that Eq.(2.11.3) provides a lower bound for the shear modulus.

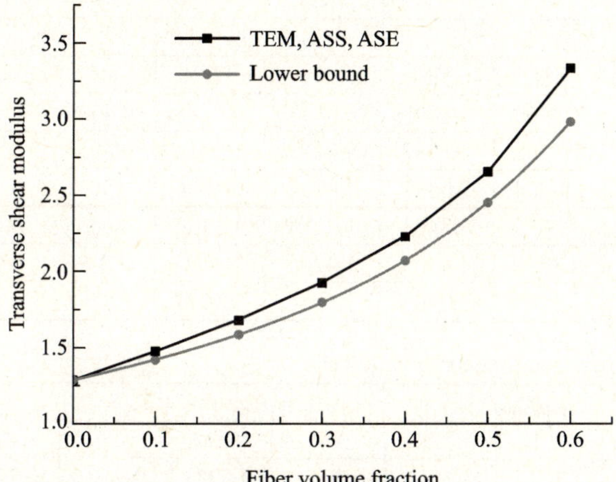

Fig.2.17   Transverse shear modulus vs fiber volume fraction

The transverse Poisson's ratio is shown in Fig.2.18. The nonlinear relation between the effective transverse Poisson's ratio and the fiber volume fraction is demonstrated. No appropriate bounds and experimental data are available for comparison.

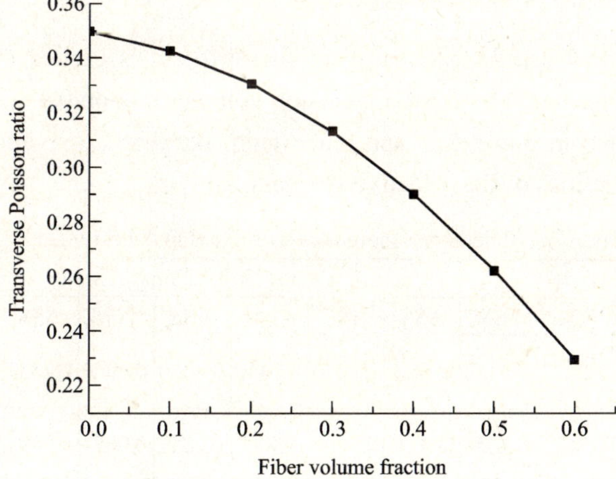

Fig.2.18   Transverse Poisson's ratio vs fiber volume fraction

For rigid and void inclusions, the effective stiffness coefficients are listed in Tables 2.3 and 2.4. It is shown that the three methods yield the same results. In summary, the direct methods and the two-scale expansion method predict the same effective stiffness for a large range of elastic mismatches [43].

Table 2.3   Stiffness coefficients for rigid inclusion medium

v_1	c_{1111}/E_0			c_{1122}/E_0		c_{1212}/E_0		
	ASS	ASE	TEM	ASS	TEM	ASS	ASE	TEM
0.1	1.8560	1.8560	1.8673	0.9678	0.9678	0.4324	0.4311	0.4325
0.2	2.1900	2.1900	2.1967	1.0791	1.0784	0.4952	0.4968	0.5007
0.3	2.6504	2.6504	2.6577	1.1957	1.1986	0.5721	0.5738	0.5821
0.4	3.3123	3.3123	3.3464	1.3128	1.3150	0.6638	0.6724	0.6797
0.5	4.3297	4.3297	4.4308	1.4206	1.4195	0.8096	0.8140	0.8273
0.6	6.1427	6.1427	6.2106	1.5056	1.5047	1.0548	1.0548	1.0766

Table 2.4   Stiffness coefficients for void solid

v_1	c_{1111}/E_0			c_{1122}/E_0		c_{1212}/E_0		
	ASS	ASE	TEM	ASS	TEM	ASS	ASE	TEM
0.1	1.1314	1.1314	1.1314	0.5381	0.5381	0.2763	0.2763	0.2763
0.2	0.8405	0.8405	0.8405	0.3459	0.3459	0.1919	0.1919	0.1919
0.3	0.6388	0.6388	0.6388	0.2221	0.2221	0.1235	0.1235	0.1235
0.4	0.4863	0.4863	0.4863	0.1385	0.1385	0.0731	0.0731	0.0731
0.5	0.3625	0.3625	0.3625	0.0811	0.0811	0.0389	0.0389	0.0389
0.6	0.2537	0.2537	0.2537	0.0413	0.0413	0.0171	0.0171	0.0171

2.11.2   Effective stiffness of anisotropic composite

For Case 2 and Case 3, calculation of the stiffness is carried out. Here the analysis is performed only for the inclusion volume fraction 0.4. The numerical results are listed in Tables 2.5 and 2.6. Again, the ASS, ASE and TEM yield the same predictions of the effective stiffness.

Table 2.5   Stiffness coefficients for composite with fraction 0.4 for Case 2

	c_{1111}			c_{2222}			c_{1122}		c_{1212}		
	ASS	ASE	TEM	ASS	ASE	TEM	ASS	TEM	ASS	ASE	TEM
fiber ($\times 10^4$)	1.3222	1.3222	1.3222	1.1440	1.1440	1.1440	0.4000	0.4000	0.2437	0.2437	0.2437
void ($/E_0$)	0.3621	0.3621	0.3621	0.3007	0.3007	0.3007	0.0651	0.0651	0.0394	0.0394	0.0394
rigid ($/E_0$)	5.0192	5.0192	5.0192	3.9738	3.9738	3.9737	1.1360	1.1360	0.9005	0.9005	0.9005

Table 2.6   Stiffness coefficients for composite with fraction 0.4 for Case 3

	c_{1111}			c_{2222}			c_{1122}		c_{1212}		
	ASS	ASE	TEM	ASS	ASE	TEM	ASS	TEM	ASS	ASE	TEM
fiber ($\times 10^4$)	1.8515	1.8516	1.8516	1.5770	1.5766	1.5770	0.4856	0.4856	0.3425	0.3425	0.3425
void ($/E_0$)	0.2310	0.2310	0.2310	0.1408	0.1408	0.1408	0.0120	0.0120	0.0049	0.0049	0.0049
rigid ($/E_0$)	494.51	494.51	494.51	6.2615	6.2615	6.2614	1.5905	1.5905	2.2646	2.2646	2.2647

2.11.3   Microstructural deformation

This section focuses on the calculation of microstructural deformation of anisotropic composites [44].

Microstructural deformations of an orthotropic composite with L-shaped inclusions are shown in Fig.2.19 where the deformations have been scaled.

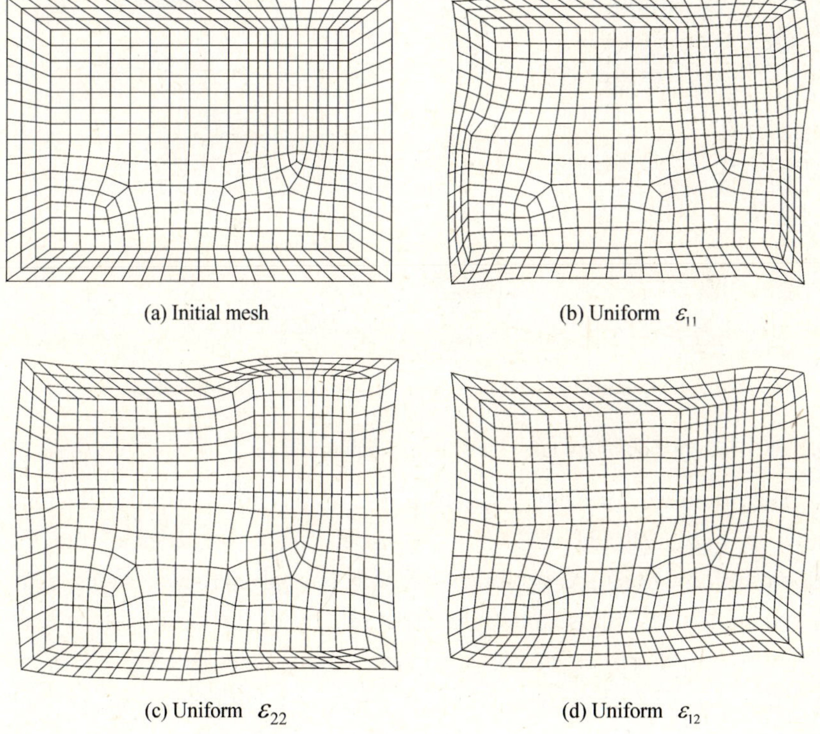

(a) Initial mesh   (b) Uniform ε_{11}

(c) Uniform ε_{22} (d) Uniform ε_{12}

Fig.2.19   Deformation of composite with non-symmetric inclusion

Here three uniform uniaxial strain states are considered. Although the applied strain is uniform, complex deformations of the RVE are found due to the heterogeneity of the microstructure and the periodicity of the boundary condition. It is shown that periodicity of the deformations is exhibited for the applied normal strain and shear strain states. Fig.2.20 shows the deformation of the void solid. Here the voids have the same shape as the fibers shown in Fig.2.19. The periodicity of deformations of the RVE ensures the compatibility of the deformations among the unit cells of the composite.

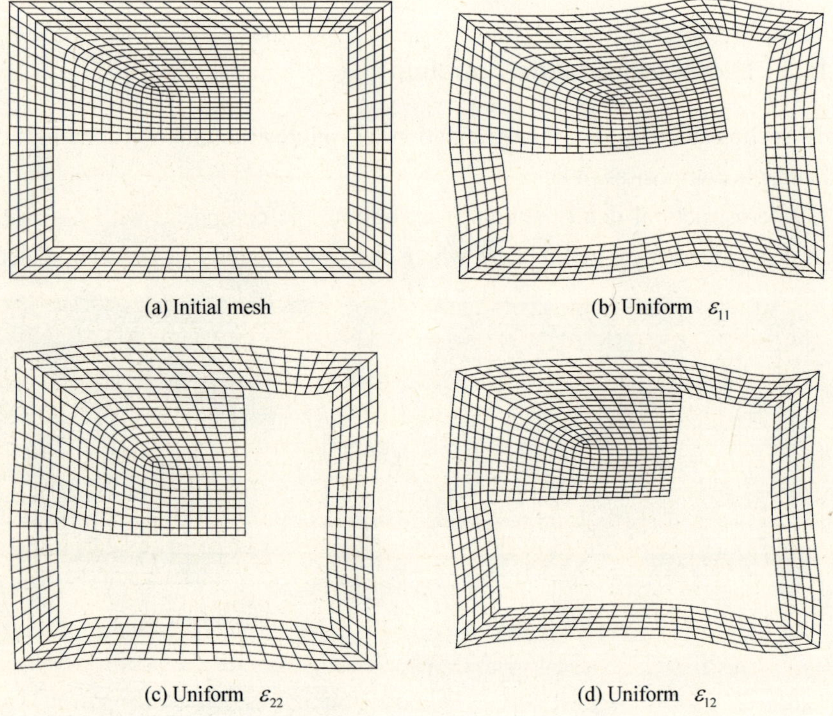

(a) Initial mesh   (b) Uniform ε_{11}

(c) Uniform ε_{22} (d) Uniform ε_{12}

Fig.2.20   Deformation of void solid with non-symmetric hole

The microstructural deformations of the RVE with Y-shaped inclusion are shown in Figs.2.21 and 2.22. For the upper and lower sides of the RVE, periodic deformations are exhibited for both normal and shear states. On the left and right sides, the symmetry and periodicity of the deformations lead to null orthogonal displacements for the normal strain states (see Fig.2.21b, c and Fig.2.22b, c). However, the anti-symmetry and periodicity of the shear deformations result in null tangent displacements on the left and right sides (see Fig.2.21d and Fig.2.22d).

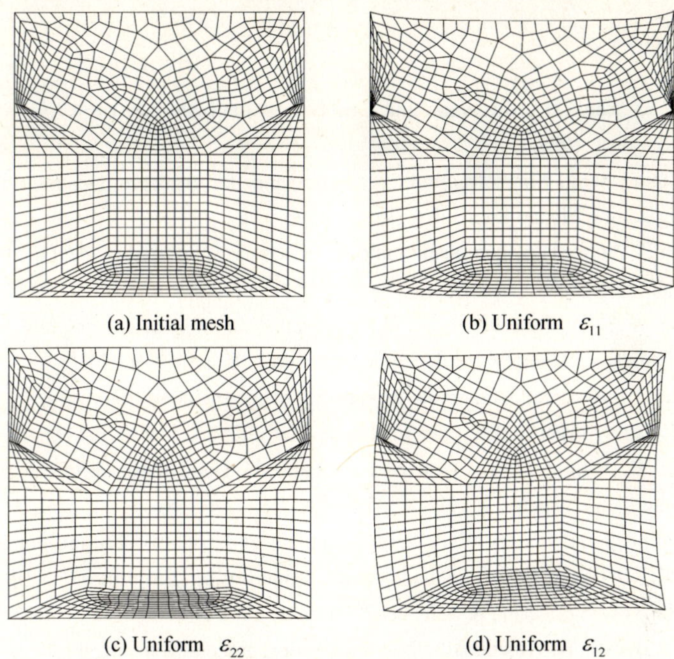

(a) Initial mesh (b) Uniform ε_{11}

(c) Uniform ε_{22} (d) Uniform ε_{12}

Fig.2.21   Deformation of fiber composite with one-symmetric-plane inclusion

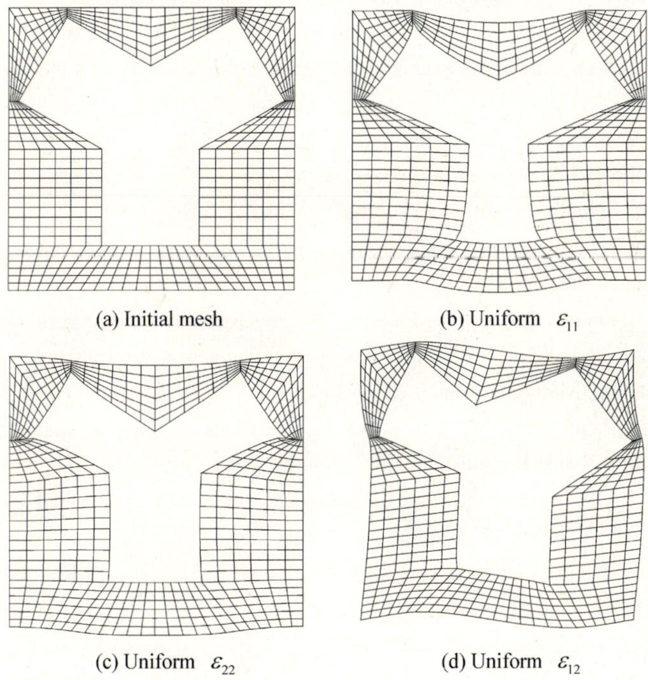

(a) Initial mesh (b) Uniform ε_{11}

(c) Uniform ε_{22} (d) Uniform ε_{12}

Fig.2.22   Deformation of composite with non-symmetric inclusion

References

[1] Du S Y, Wang B. Micromechanics of composite materials (in Chinese). Beijing: Science Press, 1998.

[2] Yang Qingsheng. Microstructural mechanics and design of composites (in Chinese). Beijing: China Tiedao Publishing House, 2000.

[3] Hashin Z. Analysis of composite materials—A survey. Journal of Applied Mechanics, 1983, 50: 481-505.

[4] Havner K S. A discrete model for the prediction of subsequent yield surfaces in polycrystalline plasticity. Int. J. Solids Structures, 1971, 7: 719-730.

[5] Hassani B. A direct method to derive the boundary conditions of homogenization equation or symmetric cell. Comm. Numer. Meth. Engng., 1996, 12: 185-196.

[6] Hassani B, Hinton E. A review of homogenization and topology optimization I, II. Computers and Structures, 1998, 69: 707-738.

[7] Yang Qingsheng, Becker W. A comparative investigation of different homogenization methods for prediction of the macroscopic properties of composites. CMES-Computer Modeling in Engineering & Science, 2004, 6: 319-332.

[8] Hohe J, Becker W. An energetic homogenization procedure for the elastic properties of general cellular sandwich cores. Composites: Part B, 2001, 32: 185-197.

[9] Zienkiewicz O C, Tayloy R L. The finite element method. 4th ed. London: McGraw-Hill.

[10] Cook R D. Finite element modeling for stress analysis. New York: John Wiley & Sons, 1995.

[11] Hill R. Elastic properties of reinforced solids: some theoretical principles. J. Mech. Phys. Solids,1963, 11: 357-372.

[12] Kroner E. Statistical Continuum Mechanics. Berlin: Springer, 1972.

[13] Hazanov S. Hill condition and overall properties of composites. Archive of Applied Mechanics, 1998, 68: 385-394.

[14] Yang Qingsheng, Qin Q H. Modelling the macroscopic elasto-plastic properties of unidirectional composites reinforced by fibre bundles under transverse tension and shear loading. Materials Science & Engineering A, 2003, 344:140-145.

[15] Yang Qingsheng, Qin Q H. Micro-mechanical analysis of composite materials by BEM. Engineering Analysis with Boundary Elements, 2004, 28: 919-926.

[16] Eshelby J D. The determination of the elastic field of an ellipsoidal inclusion and related problems. Proc. R. Soc. London, Ser. A, 1957, 241: 376-396.

[17] Hill R. A self-consistent mechanics of composite materials. J. Mech. Phys. Solids, 1965, 13: 213-222.

[18] Budiansky B. On the elastic moduli of some heterogeneous material. J. Mech. Phys. Solids, 1965, 13: 223-227.

[19] Taya M, Chou T W. On two kinds of ellipsoidal inhomogeneities in an infinite elastic

body: an application to a hybrid composite. Int. J. Solids Strct., 1981, 17: 553-563.

[20] Christensen R M, Lo K H. Solutions for effective shear properties in three phase sphere and cylinder models. J. Mech. Phys. Solids, 1979, 27: 315-330.

[21] Norris A N. A differential scheme for the effective moduli of composites. Mech. of Materials, 1985, 4: 1-16.

[22] Hashin Z. The differential scheme and its application to cracked materials. J Mech Phys Solids, 1988, 36: 719-736.

[23] Mori T, Tanaka K. Average stress in matrix and average elastic energy of materials with misfitting inclusions. Acta Met., 1973, 21: 571-574.

[24] Weng G J. Some elastic properties of reinforced solids with special reference to isotropic matrix containing spherical inclusions. Int. J. Engng. Sci., 1984, 22: 845-856.

[25] Benveniste Y. A new approach to the application of Mori-Tanaka's theory in composite materials. Mech. Materials, 1987, 6: 147-157.

[26] Hashin Z, Shtrikman S. On some variational principles in anisotropic and nonhomogeneous elasticity. J. Mech. Phys. Solids, 1962, 10: 335-342.

[27] Hashin Z, Shtrikman S. A variational approach to the theory of flastic behavior of multiphase materials. J Mech Phys Solids, 1963, 11: 127-140.

[28] Kroner E. Bounds for Effective Elastic Moduli of Disordered Materials. J Mech Phys Solides, 1977, 25: 137-156.

[29] Walpole L J. On bounds for the overall elastic moduli of inhomogeneous systems. J Mech Phys Solids, 1966, 14: 151-162.

[30] Voigt W. On the relation between the elasticity constants of isotropic bodies. Ann. Phys. Chem., 1889, 274: 573-587.

[31] Reuss A. Determination of the yield point of polycrystals based on the yield condition of single crystals (German). Z. Angew. Math. Mech. 1929, 9: 49-58.

[32] Bensoussan A, Lion J L, Papanicolaou G. Asymptotic analysis for periodic structures. Amsterdam: North Holland, 1978.

[33] Sanchez-Palencia E. Non-hongeneous media and vibration theory. Lecture notes in Physics, 1980, 127.

[34] Peng X, Cao J. A dual homogenization and finite element approach for material characterization of textile composites. Composites: Part B, 2002, 33: 45-56.

[35] Ghosh S, Lee K, Moorthy S. Two scale analysis of heterogeneous elastic-plastic materials with asymptotic homogenization and Voronoi cell finite element model. Comput. Methods Appl. Mech. Engng., 1996, 132: 63-116.

[36] Mura T. Micromechanics of defects in solids. Dordrecht: Martin Nijhoff Publishers, 1987.

[37] Nemat-Nasser S, Hori M. Micromechanics: overall properties of heterogeneous materials. Amsterdam:North Holland, 1993.

[38] Hohe J, Becker W. Effective stress-strain relations for two dimensional cellular sandwich cores: homogenization, material models, and properties. Appl Mech Rev,

2002, 55(1): 61-87.

[39] Christensen R M. A critical evaluation for a class of micromechanics models. J Mech Phys Solids,1990, 38:379-406.

[40] Chou T W, Nomura S, Taya M. A self-consistent approach to the elastic stiffness of short-fiber composites. J Composite Materials, 1980, 14: 178-188.

[41] Yang Qingsheng, Tang L M, Chen H R. Self-consistent finite element method: a new method for predicting effective properties of inclusion media. Finite Elements in Analysis and Design, 1994, 17:247-257.

[42] Tsai S W. Structural behavior of composite materials. NASA CR-71, 1964.

[43] Yang Qingsheng, Becker W. Effective stiffness and microscopic deformation of an orthotropic plate containing arbitrary holes. Computers & Structures, 2004, 82: 2301-2307.

[44] Yang Qingsheng, Becker W. Numerical investigation for stress, strain and energy homogenization of orthotropic composite with periodic microstructure and non-symmetric inclusions. Computational Materials Science, 2004, 31:169-180.

Chapter 3 Thermo-electro-elastic problems

3.1 Introduction

When piezoelectric material is subjected to a mechanical load, it generates an electric charge. This effect is usually called the "piezoelectric effect". Conversely, when piezoelectric material is stressed electrically by a voltage, its dimensions change. This phenomenon is known as the "inverse piezoelectric effect". Thermo-piezoelectric materials, on the other hand, can produce electric and mechanical fields when they are heated. The coupling properties among thermal, electric, and mechanical fields make piezoelectric materials suitable for widespread use in industrial applications in various fields including the electronics industry, nuclear industry, smart structures, microelectromechanical systems, biomedical devices, and superconducting devices. These applications have generated renewed interest in the coupling behaviour of multi-field materials including thermo-piezoelectric materials. In particular, information regarding thermal stress concentrations around material or geometrical defects in piezoelectric solids will have wide application in analyzing and designing composite structures. Early in 1974, Mindlin [1] was the first to develop the governing equations of a three-dimensional linear thermo-piezoelectric medium. Nowacki [2] subsequently developed some general theorems and mathematical models of thermo-piezoelectricity which can be viewed as the basis of various numerical methods. Dunn [3] studied micromechanics models for effective thermal expansion and pyroelectric coefficients of piezoelectric composites. Benveniste [4] obtained some exact results in the micromechanics of piezoelectric fibrous composites of two, three and four phases. By using seven potential functions, Ashida et al. [5] introduced a technique for three-dimensional asymmetric problems of piezothermoelasticity of the crystal class 6 mm. Altay and Dökmeci [6] introduced a set of Euler-Lagrange equations of discontinuous thermo-piezoelectric fields. Starting with the principle of virtual work and modifying it through Friedrichs's transformation, they presented the fundamental equations of discontinuous thermo-piezoelectric fields in variational form. Noda and Kimura [7] studied the response of a thin piezothermoelastic composite plate subjected to stationary thermal and electric fields. They showed that coupled direct piezoelectric and pyroelectric effects have a significant influence on the response of the deflection. Ashida and Tauchert [8] presented a finite difference formulation for determining the time-varying, axisymmetric, ambient temperature on the face of a piezoelectric circular disk, based on knowledge of the distribution of the induced electric potential difference across the disk thickness. For the fracture analysis of thermo-piezoelectricity, Shang et al. [9] proposed a method for three-dimensional axisymmetric problems of transversely isotropic thermo-piezoelectric materials by means of potential functions and Fourier-Hankel transformations. Fracture and damage behaviours of a cracked piezoelectric solid under coupled thermal, mechanical and electrical loads were studied by Yu and Qin [10,11]. Using techniques of Fourier transformation and extended Stroh formalism, they reduced the temperature field for a single crack problem to a pair of dual integral equations with the aid of an auxiliary function. The electroelastic field was governed by another pair of dual integral equations. With these equations, closed form solutions were obtained for strain energy release rate under thermal, mechanical and electric fields. Based on the above results, several micromechanics models were developed for crack or void-weakened piezoelectric materials, including the dilute, self-consistent, Mori-Tanaka, generalized self-consistent and differential methods [12-14]. More recently, Qin and Mai [15-24] presented a series of Green's functions for thermo-piezoelectric materials with various defects such as crack, hole and inclusion, with application to practical problems.

In this chapter, we begin with discussion of a linear theory of piezoelectricity, followed by an introduction of the two classical solution approaches for electroelastic problems. Then, solutions are presented for analyzing logarithmic singularity of crack-tip fields in homogeneous piezoelectricity. In Section 3.5, a finite element model is developed for electroelastic problems. Extensions of linear electroelastic theory to include thermal effect are discussed in Section 3.6. Fourier transform approach and its application to fracture analysis are presented in Section 3.7. Finally, formulations expressed in terms of cylindrical coordinate systems and their application to penny-shaped crack and piezoelectric fibre push-out problems are discussed.

3.2   Linear theory of piezoelectricity

3.2.1   Basic equations of linear piezoelectricity

In this section, we recall briefly the three-dimensional formulation of linear piezoelectricity that appeared in references [25,26]. Here, a three-dimensional Cartesian coordinate system is adopted where the position vector is denoted by x (or x_i). In this book, both conventional indicial notation x_i and traditional Cartesian notation (x, y, z) are utilized. In the case of indicial notation we invoke the summation convention over repeated Latin indices, which can be of two types with different ranges: $i, j, k=1,2,3$ for lower-case letters and M, $N=1,2,3,4$ for upper-case letters. Moreover, vectors, tensors and their matrix representations are denoted by boldface letters. The corresponding energy principle can be established in a way similar to the case of elastic problems if we take (ε_{ij}, E_m) as the generalized strain tensor and (σ_{ij}, D_m) as the generalized stress tensor. Using the Cartesian coordinate system, the three-dimensional constitutive equations for linear piezoelectricity can be derived by considering the internal energy density U defined by [26]

$$\mathrm{d}U = \sigma_{ij}\mathrm{d}\varepsilon_{ij} + E_m \mathrm{d}D_m \tag{3.2.1}$$

Obviously, Eq.(3.2.1) is a straightforward extension from the elastic energy density $\mathrm{d}U = \sigma_{ij}\mathrm{d}\varepsilon_{ij}$. Thus, the electric entropy per unit volume g can be defined as

$$g = U - E_m D_m \tag{3.2.2}$$

where U, D_m and E_m are the internal energy density, electric displacement and electric field, respectively, and E_m is defined by

$$E_m = -\phi_{,m} \tag{3.2.3}$$

in which ϕ is electric potential, a comma followed by arguments denotes partial differentiation with respect to the arguments. The constitutive relation of piezoelectricity can then be obtained by considering the following Legendre transformation

$$dg = \sigma_{ij} d\varepsilon_{ij} - D_m dE_m \qquad (3.2.4)$$

in which the strain ε_{ij} is defined by

$$\varepsilon_{ij} = \frac{1}{2}(u_{i,j} + u_{j,i}) \qquad (3.2.5)$$

with u_i being elastic displacement. It can be seen from Eq.(3.2.4) that

$$\sigma_{ij} = \left[\frac{\partial g}{\partial \varepsilon_{ij}}\right], \qquad D_m = -\left[\frac{\partial g}{\partial E_m}\right] \qquad (3.2.6)$$

When the function g is expanded with respect to ε_{ij} and E_m within the scope of linear interactions, we have

$$g = \frac{1}{2}\left(\varepsilon_{ij}\frac{\partial}{\partial \varepsilon_{ij}} + E_m\frac{\partial}{\partial E_m}\right)\left(\varepsilon_{kl}\frac{\partial}{\partial \varepsilon_{kl}} + E_n\frac{\partial}{\partial E_n}\right)g \qquad (3.2.7)$$

The following constants can then be defined:

$$c_{ijkl}^{(E)} = \left[\frac{\partial^2 g}{\partial \varepsilon_{ij}\partial \varepsilon_{kl}}\right], \qquad \kappa_{nm}^{(\varepsilon)} = -\left[\frac{\partial^2 g}{\partial E_n\partial E_m}\right], \qquad e_{mij} = -\left[\frac{\partial^2 g}{\partial \varepsilon_{ij}\partial E_m}\right] \qquad (3.2.8)$$

where $c_{ijkl}^{(E)}$ are the elastic moduli measured at a constant electric field, $\kappa_{nm}^{(\varepsilon)}$ the dielectric constants measured at a constant strain, e_{mij} the piezoelectric coefficients, the superscript "E" (or "ε") represents the value of the related variable measured at a given electric field (or strain). When the function g is differentiated according to Eq.(3.2.4) and the above constants are used, we find

$$\sigma_{ij} = c_{ijkl}\varepsilon_{kl} - e_{mij}E_m, \qquad D_n = e_{nij}\varepsilon_{ij} + \kappa_{mn}E_m \qquad (3.2.9)$$

A set of these two equations is the constitutive relation in the coupled system. It should be noted that the superscripts "ε" and "E" appearing in Eq.(3.2.8) have been dropped here. To simplify subsequent writing we omit them in the remainder of this book. Using the notation defined above, the electric entropy function per unit volume can now be expressed as [27]

$$g = \frac{1}{2}c_{ijkl}\varepsilon_{ij}\varepsilon_{kl} - \frac{1}{2}\kappa_{ij}E_iE_j - e_{ijk}E_i\varepsilon_{jk} \qquad (3.2.10)$$

while the related divergence equations and boundary conditions can be derived by considering the modified Biot's variational principle [27]

$$\delta \int_\Omega U d\Omega - \int_\Omega (b_i\delta u_i - b_e\delta\phi)\,d\Omega - \int_\Gamma (\bar{t}_i\delta u_i - \bar{q}_s\delta\phi)\,d\Gamma = 0 \qquad (3.2.11)$$

where δ is the variational symbol, Ω and Γ are the domain and boundary of the material, b_i and b_e are the body force per unit volume and electric charge density, and $\bar{t}_i$ and $\bar{q}_s$ are the applied surface traction and the applied surface charge, respectively. The variational equation (3.2.11) provides the following results

$$\sigma_{ij,j} + b_i = 0, \quad D_{i,i} + b_e = 0 \tag{3.2.12}$$

$$\sigma_{ij}n_j = \bar{t}_i, \quad D_i n_i = -\bar{q}_s \tag{3.2.13}$$

together with the constitutive equation (3.2.9), where n_i is the outer unit normal vector to Γ. Eq.(3.2.12) includes the elastic equilibrium equation and Gauss' law of electrostatics, respectively, Eq.(3.2.13) is boundary condition.

It should be mentioned that four equivalent constitutive representations are commonly used in the stationary theory of linear piezoelectricity to describe the coupled interaction between the elastic and electric variables. Each type has its own different set of independent variables and corresponds to a different thermodynamic function, as listed in Table 3.1. While all equations in Table 3.1 are expressed in terms of tensor, the indices have been omitted for brevity. It should be pointed out that an alternative derivation of formulae is merely a transformation from one type of relation to another. Some relationships between various constants occurring in the four types are given as follows:

$$\beta_{np}\kappa_{pm} = \delta_{nm}, \quad \beta_{nm}^{\varepsilon} - \beta_{nm}^{\sigma} = g_{nkl}h_{nkl}, \quad \kappa_{nm}^{\sigma} - \kappa_{nm}^{\varepsilon} = d_{nkl}e_{mkl}$$

$$c_{ijkl}^{D} - c_{ijkl}^{E} = e_{mij}h_{mkl}, \quad f_{ijkl}^{E} - f_{ijkl}^{D} = d_{mij}g_{mkl}, \quad d_{nij} = \kappa_{nm}^{\sigma}g_{mij} = e_{nkl}f_{klij}^{E} \tag{3.2.14}$$

$$e_{nij} = \kappa_{nm}^{\varepsilon}h_{mij} = d_{nkl}c_{klij}^{E}, \quad g_{nij} = \beta_{nm}^{\sigma}d_{mij} = h_{nkl}f_{klij}^{D}, \quad h_{nij} = \beta_{nm}^{\varepsilon}e_{mij} = g_{nkl}c_{klij}^{D}$$

The material constants can be reduced by the following consideration. According to definition [Eq.(3.2.5)] we may write $\varepsilon_{ij} = \varepsilon_{ji}$. It follows that

Table 3.1 Four types of fundamental electroelastic relation

Independent variable	Constitutive relation	Thermodynamic potentials
ε, E	$\begin{cases} \sigma = c^E\varepsilon - e^{\mathrm{T}}E \\ D = e\varepsilon + \kappa^{\varepsilon}E \end{cases}$	$g_0 = \dfrac{1}{2}c^E\varepsilon^2 - \dfrac{1}{2}\kappa^{\varepsilon}E^2 - e\varepsilon E$
ε, D	$\begin{cases} \sigma = c^D\varepsilon - h^{\mathrm{T}}D \\ E = -h\varepsilon + \beta^{\varepsilon}D \end{cases}$	$g_1 = g_0 + ED$
σ, E	$\begin{cases} \varepsilon = f^{\varepsilon}\sigma + d^{\mathrm{T}}E \\ D = d\sigma + \kappa^{\sigma}E \end{cases}$	$g_2 = g_0 - \sigma\varepsilon$
σ, D	$\begin{cases} \varepsilon = f^D\sigma + g^{\mathrm{T}}D \\ E = -g\sigma + \beta^{\sigma}D \end{cases}$	$g_3 = g_0 + ED - \sigma\varepsilon$

$$c_{ijkm} = c_{ijmk}, \quad c_{ijkm} = c_{jikm}, \quad e_{kij} = e_{kji} \tag{3.2.15}$$

in which the relation $\sigma_{ij} = \sigma_{ji}$ has been used.

In view of these properties, it is useful to introduce the so-called two-index notation or compressed matrix notation [29]. Two-index notation consists of replacing ij or km by p or q, i.e. $c_{ijkm}=c_{pq}$, $e_{ikm}=e_{iq}$, $\sigma_{ij}=\sigma_p$, where i, j, k, m take the values $1\sim3$, and p, q assume the values $1\sim6$ according to the replacements $11 \rightarrow 1$, $22 \rightarrow 2$, $33 \rightarrow 3$, 23 or $32 \rightarrow 4$, 13 or $31 \rightarrow 5$, 12 or $21 \rightarrow 6$. The constitutive equation (3.2.9) then becomes

$$\sigma_p = c_{pq}\varepsilon_q - \varepsilon_{kp}E_k, \quad D_i = e_{iq}\varepsilon_q + \kappa_{ik}E_k \tag{3.2.16}$$

in which

$$\varepsilon_q = \begin{cases} \varepsilon_{ij}, & \text{when } i = j \\ 2\varepsilon_{ij}, & \text{when } i \neq j \end{cases} \tag{3.2.17}$$

In addition, the elastic, piezoelectric and dielectric constants can now be written in matrix form since they all are described by two indices. The arrays for an arbitrarily anisotropic material are

$$c = \begin{bmatrix} c_{11} & c_{12} & c_{13} & c_{14} & c_{15} & c_{16} \\ c_{12} & c_{22} & c_{23} & c_{24} & c_{25} & c_{26} \\ c_{13} & c_{23} & c_{33} & c_{34} & c_{35} & c_{36} \\ c_{14} & c_{24} & c_{34} & c_{44} & c_{45} & c_{46} \\ c_{15} & c_{25} & c_{35} & c_{45} & c_{55} & c_{56} \\ c_{16} & c_{26} & c_{36} & c_{46} & c_{56} & c_{66} \end{bmatrix} \tag{3.2.18}$$

$$e = \begin{bmatrix} e_{11} & e_{12} & e_{13} & e_{14} & e_{15} & e_{16} \\ e_{21} & e_{22} & e_{23} & e_{24} & e_{25} & e_{26} \\ e_{31} & e_{32} & e_{33} & e_{34} & e_{35} & e_{36} \end{bmatrix} \tag{3.2.19}$$

$$\kappa = \begin{bmatrix} \kappa_{11} & \kappa_{12} & \kappa_{13} \\ \kappa_{12} & \kappa_{22} & \kappa_{23} \\ \kappa_{13} & \kappa_{23} & \kappa_{33} \end{bmatrix} \tag{3.2.20}$$

It can be seen that there are $21+18+6=45$ independent constants for this material type. For a transversely isotropic material with x_3 in the poling direction (Class $C_{6v}=6\,\text{mm}$), the related material matrices are simplified as

$$\boldsymbol{c} = \begin{bmatrix} c_{11} & c_{12} & c_{13} & 0 & 0 & 0 \\ c_{12} & c_{11} & c_{13} & 0 & 0 & 0 \\ c_{13} & c_{13} & c_{33} & 0 & 0 & 0 \\ 0 & 0 & 0 & c_{44} & 0 & 0 \\ 0 & 0 & 0 & 0 & c_{44} & 0 \\ 0 & 0 & 0 & 0 & 0 & \frac{1}{2}(c_{11}-c_{12}) \end{bmatrix} \tag{3.2.21}$$

$$\boldsymbol{e} = \begin{bmatrix} 0 & 0 & 0 & 0 & e_{15} & 0 \\ 0 & 0 & 0 & e_{15} & 0 & 0 \\ e_{31} & e_{31} & e_{33} & 0 & 0 & 0 \end{bmatrix} \tag{3.2.22}$$

$$\boldsymbol{\kappa} = \begin{bmatrix} \kappa_{11} & 0 & 0 \\ 0 & \kappa_{11} & 0 \\ 0 & 0 & \kappa_{33} \end{bmatrix} \tag{3.2.23}$$

Thus it is clear that a material with this type of symmetry is described by 10 independent material constants. This category of material is important because polarized ceramics have high piezoelectric coupling. Finally, an isotropic dielectric material has arrays which are similar to the arrays for transversely isotropic materials, except that there are some additional relations among the material constants. They are

$$e_{ip} = 0, \quad \text{for all values of } i \text{ and } p \tag{3.2.24}$$

$$c_{12} = c_{13} = \lambda, \quad c_{11} = c_{33} = \lambda + 2G, \quad c_{44} = c_{66} = G, \quad \kappa_{11} = \kappa_{33} \tag{3.2.25}$$

where $G=E/[2(1+\mu)]$ is the shear modulus of elasticity, $\lambda=2G\mu/(1-2\mu)$ is the Lamé constant and E, μ are the Young's modulus and Poisson's ratio, respectively. In the MKS system the material constants and variables mentioned above are measured in the following units: $[c_{ij}] = \text{Nm}^{-2}$, $[e_{ij}] = \text{Cm}^{-2}$, $[\kappa_{ij}] = \text{C}^2\text{N}^{-1}\text{m}^{-2} = \text{NV}^{-2}$, $[\sigma_{ij}] = \text{Nm}^{-2}$, $[\varepsilon_{ij}] = \text{mm}^{-1}$, $[D_i] = \text{Cm}^{-2} = \text{N(Vm)}^{-1}$, $[E_i] = \text{NC}^{-1} = \text{Vm}^{-1}$, $[\phi] = \text{V}$. For poled barium-titanate (BaTiO$_3$) and lead-zirconate-titanate, these physical constants are of the orders: $c_{ij}=O(10^{11}\,\text{Nm}^{-2})$, $e_{ij}=O(10\,\text{Nm}^{-2})$, $\kappa_{ij}=O(10^{-8}\text{NV}^{-2})$.

Substitution of Eq.(3.2.3) and Eq.(3.2.5) into Eq.(3.2.16), and later into Eq.(3.2.12), results in

$$c_{11}u_{1,11} + \frac{1}{2}(c_{11}+c_{12})u_{2,12} + (c_{13}+c_{44})u_{3,13} + \frac{1}{2}(c_{11}-c_{12})u_{1,22} +$$

$$c_{44}u_{1,33} + (e_{31}+e_{15})\phi_{,13} + b_1 = 0 \tag{3.2.26}$$

$$c_{11}u_{2,22} + \frac{1}{2}(c_{11}+c_{12})u_{1,12} + (c_{13}+c_{44})u_{3,23} + \frac{1}{2}(c_{11}-c_{12})u_{2,11} +$$

$$c_{44}u_{2,33} + (e_{31}+e_{15})\phi_{,23} + b_2 = 0 \tag{3.2.27}$$

$$c_{44}u_{3,11} + (c_{44}+c_{13})(u_{1,31}+u_{2,32}) + c_{44}u_{3,22} + c_{33}u_{3,33} +$$

$$e_{15}(\phi_{,11}+\phi_{,22}) + e_{33}\phi_{,33} + b_3 = 0 \tag{3.2.28}$$

$$e_{15}(u_{3,11}+u_{3,22}) + (e_{15}+e_{31})(u_{1,31}+u_{2,32}) +$$

$$e_{33}u_{3,33} - \kappa_{11}(\phi_{,11}+\phi_{,22}) - \kappa_{33}\phi_{,33} + b_e = 0 \tag{3.2.29}$$

for transversely isotropic materials (class C_{6v}=6 mm) with x_3 as the poling direction and the x_1–x_2 plane as the isotropic plane. This type of material is adopted in the remaining chapters.

3.2.2 Two-dimensional simplification

For most practical problems piezoelectric materials are treated as a two-dimensional problem to simplify the solution process. Here we discuss two special cases which are of some interest:

(1) Plane strain. Without loss of generality we focus on transversely isotropic piezoelectricity. Assuming that the x–y plane is the isotropic plane, one can employ either the x–z or y–z plane for the study of plane electromechanical phenomena. Choosing the former, plain strain conditions require that

$$\varepsilon_{yy} = \varepsilon_{zy} = \varepsilon_{xy} = E_y = 0 \tag{3.2.30}$$

By substitution of Eq.(3.2.30) into Eq.(3.2.16), we have

$$\begin{bmatrix} \sigma_1 \\ \sigma_3 \\ \sigma_5 \\ D_1 \\ D_3 \end{bmatrix} = \begin{bmatrix} c_{11} & c_{13} & 0 & 0 & e_{31} \\ c_{13} & c_{33} & 0 & 0 & e_{33} \\ 0 & 0 & c_{55} & e_{15} & 0 \\ 0 & 0 & e_{15} & -\kappa_{11} & 0 \\ e_{31} & e_{33} & 0 & 0 & -\kappa_{33} \end{bmatrix} \begin{bmatrix} \varepsilon_1 \\ \varepsilon_3 \\ \varepsilon_5 \\ -E_1 \\ -E_3 \end{bmatrix} \tag{3.2.31}$$

or inversely

$$\begin{bmatrix} \varepsilon_1 \\ \varepsilon_3 \\ \varepsilon_5 \\ -E_1 \\ -E_3 \end{bmatrix} = \begin{bmatrix} f_{11} & f_{13} & 0 & 0 & p_{31} \\ f_{13} & f_{33} & 0 & 0 & p_{33} \\ 0 & 0 & f_{55} & p_{15} & 0 \\ 0 & 0 & p_{15} & -\beta_{11} & 0 \\ p_{31} & p_{33} & 0 & 0 & -\beta_{33} \end{bmatrix} \begin{bmatrix} \sigma_1 \\ \sigma_3 \\ \sigma_5 \\ D_1 \\ D_3 \end{bmatrix} \tag{3.2.32}$$

in which f_{ij} is constant of elastic compliance of the material, p_{ij} is the piezoelectric coefficient and β_{ij} is the dielectric impermeability constant. In the constitutive equations (3.2.31) and (3.2.32), $-E_i$ is used instead of E_i because it will allow the construction of a symmetric generalized linear response matrix which will prove to be advantageous. When the constitutive equation (3.2.31) is substituted into Eq.(3.2.12) we obtain

$$c_{11}u_{1,11} + (c_{13} + c_{55})u_{3,13} + c_{55}u_{1,33} + (e_{31} + e_{15})\phi_{,13} + b_1 = 0 \qquad (3.2.33)$$

$$c_{55}u_{3,11} + (c_{55} + c_{13})u_{1,31} + c_{33}u_{3,33} + e_{15}\phi_{,11} + e_{33}\phi_{,33} + b_3 = 0 \qquad (3.2.34)$$

$$e_{15}u_{3,11} + (e_{15} + e_{31})u_{1,31} + e_{33}u_{3,33} - \kappa_{11}\phi_{,11} - \kappa_{33}\phi_{,33} + b_e = 0 \qquad (3.2.35)$$

(2) Anti-plane deformation. In this case only the out-of-plane elastic displacement u_3 and the in-plane electric fields are non-zero, i.e.,

$$\begin{aligned}
u_1 &= u_2 = 0, & u_3 &= u_3(x_1, x_2) \\
E_1 &= E_1(x_1, x_2), & E_2 &= E_2(x_1, x_2), & E_3 &= 0
\end{aligned} \qquad (3.2.36)$$

Thus the constitutive equation (3.2.16) simplifies to

$$\begin{bmatrix} \sigma_4 \\ \sigma_5 \\ D_1 \\ D_2 \end{bmatrix} = \begin{bmatrix} c_{44} & 0 & 0 & e_{15} \\ 0 & c_{44} & e_{15} & 0 \\ 0 & e_{15} & -\kappa_{11} & 0 \\ e_{15} & 0 & 0 & -\kappa_{11} \end{bmatrix} \begin{bmatrix} \varepsilon_4 \\ \varepsilon_5 \\ -E_1 \\ -E_2 \end{bmatrix} \qquad (3.2.37)$$

and the governing equation (3.2.12) becomes

$$\begin{aligned}
c_{44}\nabla^2 u_3 + e_{15}\nabla^2 \phi + b_3 &= 0 \\
e_{15}\nabla^2 u_3 - \kappa_{11}\nabla^2 \phi + b_e &= 0
\end{aligned} \qquad (3.2.38)$$

where $\nabla^2 = (\ )_{,11} + (\ )_{,22}$ is the two-dimensional Laplacian operator.

3.3 Two classical solution approaches for piezoelectricity

For two-dimensional deformations in a general anisotropic piezoelectric material, in which u_i and ϕ depend on x_1 and x_2 (or x_3) only, there are two powerful solution procedures in the literature. One is Lekhnitskii's approach [30], which begins with equilibrated stress functions, followed by compatibility equations. This approach is discussed in Subsection 3.3.2. Another is Stroh's formalism [31], which begins with the displacements and electric potential, followed by equilibrium equations. The equivalence of these two formalisms has been discussed by Suo [32]. The details of Stroh's formulation are given below.

3.3.1   Solution with Stroh formalism

We begin by introducing the shorthand notation given by Barnett and Lothe [33], as it greatly simplifies the following writing. With this shorthand notation, the governing equation (3.2.12) and the constitutive relationship (3.2.16) can be rewritten as

$$\Pi_{iJ,i} + b_J = 0 \tag{3.3.1}$$

$$\Pi_{iJ} = E_{iJKm} U_{K,m} \tag{3.3.2}$$

where $b_4 = b_e$ ($J = 4$), and

$$\Pi_{iJ} = \begin{cases} \sigma_{ij}, & i,J = 1,\ 2,\ 3 \\ D_i, & J = 4 \quad i = 1,\ 2,\ 3 \end{cases} \tag{3.3.3}$$

$$U_K = \begin{cases} u_k, & K = 1,\ 2,\ 3 \\ \phi, & K = 4 \end{cases} \tag{3.3.4}$$

$$E_{iJKm} = \begin{cases} C_{ijkm}, & i,J,K,m = 1,\ 2,\ 3 \\ e_{mij}, & K = 4, \quad i,J,m = 1,\ 2,\ 3 \\ e_{ikm}, & J = 4, \quad i,K,m = 1,\ 2,\ 3 \\ -\kappa_{im}, & J = K = 4, \quad i,m = 1,\ 2,\ 3 \end{cases} \tag{3.3.5}$$

For two-dimensional deformations in which $U = [u_1\ u_2\ u_3\ \phi]^{\mathrm{T}}$ depends on x_1 and x_2 only, where the superscript T denotes the transpose, a general solution can be obtained by considering an arbitrary function of the form [33]

$$U = a f(z) \tag{3.3.6}$$

where $z = x_1 + p x_2$, p and a are determined by inserting Eq.(3.3.6) into Eq.(3.3.2), and later into Eq.(3.3.1). In the absence of any body force and free charge distribution, we have

$$\left[Q + p(R + R^{\mathrm{T}}) + p^2 T \right] a = 0 \tag{3.3.7}$$

where Q, R and T are 4×4 real matrices whose components are

$$Q_{IK} = E_{1IK1}, \quad R_{IK} = E_{1IK2}, \quad T_{IK} = E_{2IK2} \tag{3.3.8}$$

The stress and electric displacement (SED) obtained by substituting Eq.(3.3.6) into Eq.(3.3.2) can be written in terms of a SED function φ as

$$\Pi_{i1} = -\varphi_{i,2}, \quad \Pi_{i2} = \varphi_{i,1} \tag{3.3.9}$$

where

$$\varphi = b f(z) \tag{3.3.10}$$

$$b = (R^{\mathrm{T}} + p T)a = -p^{-1}(Q + p R)a \tag{3.3.11}$$

The second equality in Eq.(3.3.11) follows from Eq.(3.3.7). It suffices therefore to consider the SED function φ because the stresses σ_{ij} and the electric displacement D_i can be obtained by differentiation.

There are eight eigenvalues p from Eq.(3.3.7) which consists of four pairs of complex conjugates [33]. If $p_J, \boldsymbol{a}_J (J = 1, 2, \cdots, 8)$ are the eigenvalues and the associated eigenvectors, let

$$\text{Im } p_J > 0, \quad p_{J+4} = \overline{p}_J, \quad \boldsymbol{a}_{J+4} = \overline{\boldsymbol{a}}_J, \quad \boldsymbol{b}_{J+4} = \overline{\boldsymbol{b}}_J \quad J{=}1{\sim}4 \qquad (3.3.12)$$

where "Im" stands for the imaginary part of a complex number and the overbar denotes the complex conjugate. Assuming that p_J are distinct, the general solutions for $\boldsymbol{U}$ and φ obtained by superposing eight solutions of the form of Eq.(3.3.6) and Eq.(3.3.10) are

$$\boldsymbol{U} = \sum_{J=1}^{4} \left\{ \boldsymbol{a}_J f_J(z_J) + \overline{\boldsymbol{a}}_J f_{J+4}(\overline{z}_J) \right\} \qquad (3.3.13)$$

$$\varphi = \sum_{J=1}^{4} \left\{ \boldsymbol{b}_J f_J(z_J) + \overline{\boldsymbol{b}}_J f_{J+4}(\overline{z}_J) \right\} \qquad (3.3.14)$$

where $f_J(J = 1, 2, \cdots, 8)$ are arbitrary functions of their argument $z_J = x_1 + p_J x_2$. In most applications f_J assume the same functional form, so that we may write

$$f_J(z_J) = q_J f(z_J), \quad f_{J+4}(\overline{z}_J) = \overline{q}_J \overline{f}(\overline{z}_J), \quad J{=}1{\sim}4 \qquad (3.3.15)$$

where q_J are complex constants to be determined. Expressions (3.3.13) and (3.3.14) can then be written in a compact form

$$\boldsymbol{U} = 2\,\text{Re}\left\{ \boldsymbol{A}\boldsymbol{f}(z) \right\} = 2\,\text{Re}\left\{ \boldsymbol{A}\langle f(z_\alpha) \rangle \boldsymbol{q} \right\} \qquad (3.3.16)$$

$$\varphi = 2\,\text{Re}\left\{ \boldsymbol{B}\boldsymbol{f}(z) \right\} = 2\,\text{Re}\left\{ \boldsymbol{B}\langle f(z_\alpha) \rangle \boldsymbol{q} \right\} \qquad (3.3.17)$$

in which "Re" stands for the real part of a complex number, $\boldsymbol{f}(z){=}[f_1(z_1) \ \ f_2(z_2) \ \ f_3(z_3) \ \ f_4(z_4)]^{\text{T}}$, $\boldsymbol{A}, \boldsymbol{B}$ are 4×4 complex matrices defined by

$$\boldsymbol{A} = [\boldsymbol{a}_1 \ \boldsymbol{a}_2 \ \boldsymbol{a}_3 \ \boldsymbol{a}_4], \quad \boldsymbol{B} = [\boldsymbol{b}_1 \ \boldsymbol{b}_2 \ \boldsymbol{b}_3 \ \boldsymbol{b}_4] \qquad (3.3.18)$$

and $\langle f(z_\alpha) \rangle$ is a diagonal matrix

$$\langle f(z_\alpha) \rangle = \text{diag}\left[f(z_1) \ \ f(z_2) \ \ f(z_3) \ \ f(z_4) \right] \qquad (3.3.19)$$

For a given problem, it is clear that all that is required is to determine the unknown function $f(z_J)$ and the complex constant vector $\boldsymbol{q}$.

3.3.2 Solution with Lekhnitskii formalism

The mathematical method known as Lekhnitskii formalism was developed originally to solve two-dimensional problems in elastic anisotropic materials [30].

The evolution of the method and a number of extensions to electroelastic problems were described in [34-37]. In this section the Lekhnitskii formalism used in linear piezoelectricity is briefly summarized. For a complete derivation and discussion, the reader is referred to [30, 34-37].

Consider a two-dimensional piezoelectric plate where the material is transversely isotropic and coupling between in-plane stresses and in-plane electric fields takes place. For a Cartesian coordinate system $Oxyz$, choose the z-axis as the poling direction, and denote the coordinates x and z by x_1 and x_2 in order to generate a compacted notation. The plane strain constitutive equations are governed by Eq.(3.2.31) or Eq.(3.2.32), except that all indices 3 should be replaced by 2 here. That is [35]

$$\begin{bmatrix} \varepsilon_{11} \\ \varepsilon_{22} \\ 2\varepsilon_{12} \\ -E_1 \\ -E_2 \end{bmatrix} = \begin{bmatrix} f_{11} & f_{12} & 0 & 0 & p_{21} \\ f_{12} & f_{22} & 0 & 0 & p_{22} \\ 0 & 0 & f_{33} & p_{13} & 0 \\ 0 & 0 & p_{13} & -\beta_{11} & 0 \\ p_{31} & p_{33} & 0 & 0 & -\beta_{22} \end{bmatrix} \begin{bmatrix} \sigma_{11} \\ \sigma_{22} \\ \sigma_{12} \\ D_1 \\ D_2 \end{bmatrix} \qquad (3.3.20)$$

From the constitutive equations, we observe that D_2 produces normal strains ε_{11} and ε_{22}, while the stress component σ_{12} induces an electric field E_1, and σ_{11} and σ_{22} produce E_2. Equation (3.3.20) constitutes a system of five equations in ten unknowns. Additional equations are provided by elastic equilibrium and Gauss' law

$$\sigma_{11,1} + \sigma_{12,2} = 0, \quad \sigma_{12,1} + \sigma_{22,2} = 0, \quad D_{1,1} + D_{2,2} = 0 \qquad (3.3.21)$$

in which the absence of body forces and free electric volume charge has been assumed, and by one elastic and one electric compatibility condition

$$\varepsilon_{11,22} + \varepsilon_{22,11} - 2\varepsilon_{12,12} = 0, \quad E_{1,2} - E_{2,1} = 0 \qquad (3.3.22)$$

Having formulated the electroelastic problem, we seek a solution to Eqs.(3.3.20)~(3.3.22) subjected to a given loading and boundary condition. To this end, the well-known Lekhnitskii stress function F and induction function V satisfying the foregoing equilibrium equations are introduced as follows [34,35]:

$$\sigma_{11} = F_{,22}, \quad \sigma_{22} = F_{,11}, \quad \sigma_{12} = -F_{,12}, \quad D_1 = V_{,2}, \quad D_2 = -V_{,1} \qquad (3.3.23)$$

Inserting Eq.(3.3.23) into Eq.(3.3.20), and later into Eq.(3.3.22) leads to

$$L_4 F - L_3 V = 0, \quad L_3 F + L_2 V = 0 \qquad (3.3.24)$$

where

$$L_4 = f_{22}\frac{\partial^4}{\partial x_1^4} + f_{11}\frac{\partial^4}{\partial x_2^4} + (2f_{12} + f_{33})\frac{\partial^4}{\partial x_1^2 \partial x_2^2}$$

$$L_3 = p_{22}\frac{\partial^3}{\partial x_1^3} + (p_{21} + p_{13})\frac{\partial^3}{\partial x_1 \partial x_2^2}, \quad L_2 = \beta_{22}\frac{\partial^2}{\partial x_1^2} + \beta_{11}\frac{\partial^2}{\partial x_2^2}$$

(3.3.25)

Note that if the problem was purely mechanical, L_4 would be the only nonzero operator and its form would coincide with the plane anisotropic case discussed, among others, by Lekhnitskii [30]. To solve Eq.(3.3.24) we reduce the system to a single partial differential equation of order six in either F or V. Choosing F, we obtain

$$(L_4 L_2 + L_3 L_3)F = 0 \tag{3.3.26}$$

As discussed in [30] within the framework of anisotropic elasticity, Eq.(3.3.26) can be solved by assuming a solution of $F(z)$ such that

$$F(z) = F(x_1 + px_2), \quad p = \alpha + i\beta \tag{3.3.27}$$

where α and β are real numbers. By introducing Eq.(3.3.27) into Eq.(3.3.26), and using the chain rule of differentiation, an expression of the form $\{\}F^{(6)} = 0$ is obtained. A nontrivial solution follows by setting the characteristic equation (i.e., $L_4 L_2 + L_3 L_3$) equal to zero, namely

$$f_{11}\beta_{11}p^6 + (f_{11}\beta_{22} + f_{33}\beta_{11} + 2f_{12}\beta_{11} + p_{21}^2 + p_{13}^2 + 2p_{21}p_{13})p^4 +$$

$$(f_{22}\beta_{11} + 2f_{12}\beta_{22} + f_{33}\beta_{22} + 2p_{21}p_{22} + 2p_{13}p_{22})p^2 + f_{22}\beta_{22} + p_{22}^2 = 0 \tag{3.3.28}$$

Owing to the particular material symmetry of the piezoelectric material under investigation, the polynomial is expressed in terms of even powers of p. This allows us to solve Eq.(3.3.28) analytically, rendering

$$p_1 = i\beta_1, \ p_2 = \alpha_2 + i\beta_2, \ p_3 = -\alpha_2 + i\beta_2, \ p_4 = \bar{p}_1, \ p_5 = \bar{p}_2, \ p_6 = \bar{p}_3 \tag{3.3.29}$$

where β_1, α_2 and β_2 depend on the material constants. Once the roots $p_j, j=1, 2, 3$ are known, the solution for stress function F is written as

$$F(x_1, x_2) = 2\,\mathrm{Re}\sum_{j=1}^{3} F_j(z_j) \tag{3.3.30}$$

The next step is to find the function V using one of Eq.(3.3.24). If we consider $L_3 F = -L_2 V$, assuming solutions of the form $F(z_k)$ and $V(z_k)$, we have

$$V_k''(z_k) = \bar{\omega}_k(p_k)F_k'''(z_k) \tag{3.3.31}$$

where primes indicate differentiation with respect to related argument, and

$$\bar{\omega}_k(p_k) = -\frac{(p_{21} + p_{13})p_k^2 + p_{22}}{\beta_{11}p_k^2 + \beta_{22}} \tag{3.3.32}$$

Integration of Eq.(3.3.31) leads to

$$V_k(z_k) = \overline{\omega}_k(p_k)F_k'(z_k) + c_1 z_k + c_2 \tag{3.3.33}$$

It should be noted here that the arbitrary constants of integration could be set zero [34]. Thus, the solution for the induction function can be expressed as follows:

$$V(x_1, x_2) = 2\operatorname{Re}\sum_{j=1}^{3} V_j(z_j) = 2\operatorname{Re}\sum_{j=1}^{3} \overline{\omega}_j F_j'(z_j) \tag{3.3.34}$$

With the aid of Eq.(3.3.30) and Eq.(3.3.34) we can obtain expressions for the stress and electric displacement components. Using Eqs. (3.3.23), (3.3.30) and (3.3.34), we obtain

$$\begin{bmatrix} \sigma_{11} \\ \sigma_{22} \\ \sigma_{12} \end{bmatrix} = 2\operatorname{Re}\sum_{k=1}^{3} \begin{bmatrix} p_k^2 \\ 1 \\ -p_k \end{bmatrix} \Phi_k'(z_k), \quad \begin{bmatrix} D_1 \\ D_2 \end{bmatrix} = 2\operatorname{Re}\sum_{k=1}^{3} \begin{bmatrix} \overline{\omega}_k p_k \\ -\overline{\omega}_k \end{bmatrix} \Phi_k'(z_k) \tag{3.3.35}$$

where $\Phi_k(z_k) = F_k'(z_k)$.

Finally, using the constitutive equation (3.3.20) in conjunction with Eq.(3.3.35) allows us to find expressions for the strain and electric field. They are

$$\begin{bmatrix} \varepsilon_{11} \\ \varepsilon_{22} \\ 2\varepsilon_{12} \end{bmatrix} = 2\operatorname{Re}\sum_{k=1}^{3} \begin{bmatrix} p_k^* \\ q_k^* \\ r_k^* \end{bmatrix} \Phi_k'(z_k), \quad \begin{bmatrix} E_1 \\ E_2 \end{bmatrix} = -\sum_{k=1}^{3} \begin{bmatrix} t_k^* \\ v_k^* \end{bmatrix} \Phi_k'(z_k) \tag{3.3.36}$$

where

$$p_k^* = f_{11}p_k^2 + f_{12} - p_{21}\overline{\omega}_k, \ q_k^* = f_{12}p_k^2 + f_{22} - p_{22}\overline{\omega}_k, \ r_k^* = (p_{13}\overline{\omega}_k - f_{33})p_k$$
$$t_k^* = -(p_{13} + \beta_{11}\overline{\omega}_k)p_k, \ v_k^* = p_{21}p_k^2 + p_{22} + \beta_{22}\overline{\omega}_k \tag{3.3.37}$$

Substitution of Eq.(3.2.3) and Eq.(3.2.5) into Eq.(3.3.36), and then integration of the normal strains and the electric field $E_i = -\phi_{,i}$ produces

$$\begin{bmatrix} u_1 \\ u_2 \\ \phi \end{bmatrix} = 2\operatorname{Re}\sum_{k=1}^{3} \begin{bmatrix} p_k^* \\ q_k^* \\ t_k^* \end{bmatrix} \Phi_k(z_k) + \begin{bmatrix} \omega_0 x_2 + u_0 \\ -\omega_0 x_1 + v_0 \\ \phi_0 \end{bmatrix} \tag{3.3.38}$$

where the constants ω_0, u_0, v_0 represent rigid body displacements, and ϕ_0 is a reference potential.

Recapitulating, based on the procedure above, the plane strain piezoelectric problem is reduced to one of finding three complex potentials, Φ_1, Φ_2 and Φ_3, in some region Ω of the material. Each potential is a function of a different generalized complex variable $z_k = x_1 + p_k x_2$.

3.3.3 Some identities

In this subsection some identities of matrices are presented in order to provide a source for usage in later sections and chapters. To this end rewrite Eq.(3.3.11) in the form

$$\begin{bmatrix} -\boldsymbol{R}^{\mathrm{T}} & \boldsymbol{I} \\ -\boldsymbol{Q} & 0 \end{bmatrix} \begin{bmatrix} \boldsymbol{a} \\ \boldsymbol{b} \end{bmatrix} = p \begin{bmatrix} \boldsymbol{T} & 0 \\ \boldsymbol{R} & \boldsymbol{I} \end{bmatrix} \begin{bmatrix} \boldsymbol{a} \\ \boldsymbol{b} \end{bmatrix} \tag{3.3.39}$$

where $\boldsymbol{I}$ is the identity matrix. Since $\boldsymbol{T}^{-1}$ exists, we can reduce Eq.(3.3.39) to

$$\boldsymbol{N}\boldsymbol{\xi} = p\boldsymbol{\xi} \tag{3.3.40}$$

where

$$\boldsymbol{N} = \begin{bmatrix} \boldsymbol{N}_1 & \boldsymbol{N}_2 \\ \boldsymbol{N}_3 & \boldsymbol{N}_1^{\mathrm{T}} \end{bmatrix}, \quad \boldsymbol{\xi} = \begin{bmatrix} \boldsymbol{a} \\ \boldsymbol{b} \end{bmatrix} \tag{3.3.41}$$

$$\boldsymbol{N}_1 = -\boldsymbol{T}^{-1}\boldsymbol{R}^{\mathrm{T}}, \quad \boldsymbol{N}_2 = \boldsymbol{T}^{-1} = \boldsymbol{N}_2^{\mathrm{T}}, \quad \boldsymbol{N}_3 = \boldsymbol{R}\boldsymbol{T}^{-1}\boldsymbol{R}^{\mathrm{T}} - \boldsymbol{Q} = \boldsymbol{N}_3^{\mathrm{T}} \tag{3.3.42}$$

Eq.(3.3.40) is a standard eigenrelation in the eight-dimensional space. The vector $\boldsymbol{\xi}$ in Eq.(3.3.40) is a right eigenvector. The left eigenvector $\boldsymbol{\eta}$ is defined by

$$\boldsymbol{\eta}^{\mathrm{T}}\boldsymbol{N} = p\boldsymbol{\eta}^{\mathrm{T}}, \quad \boldsymbol{N}^{\mathrm{T}}\boldsymbol{\eta} = p\boldsymbol{\eta} \tag{3.3.43}$$

and can be shown to be [33]

$$\boldsymbol{\eta} = \begin{bmatrix} \boldsymbol{b} \\ \boldsymbol{a} \end{bmatrix} \tag{3.3.44}$$

Normalization of $\boldsymbol{\xi}_I$ and $\boldsymbol{\eta}_J$ (which are orthogonal to each other) gives

$$\boldsymbol{\eta}_J^{\mathrm{T}}\boldsymbol{\xi}_K = \delta_{JK} \tag{3.3.45}$$

where δ_{JK} is the Kronecker delta. Making use of Eqs.(3.3.11), (3.3.41), and (3.3.44), Eq.(3.3.45) can be written as

$$\begin{bmatrix} \boldsymbol{B}^{\mathrm{T}} & \boldsymbol{A}^{\mathrm{T}} \\ \overline{\boldsymbol{B}}^{\mathrm{T}} & \overline{\boldsymbol{A}}^{\mathrm{T}} \end{bmatrix} \begin{bmatrix} \boldsymbol{A} & \overline{\boldsymbol{A}} \\ \boldsymbol{B} & \overline{\boldsymbol{B}} \end{bmatrix} = \begin{bmatrix} \boldsymbol{I} & 0 \\ 0 & \boldsymbol{I} \end{bmatrix} \tag{3.3.46}$$

This is the orthogonality relation. The two matrices on the left-hand side of Eq.(3.3.46) are the inverse of each other. Their product commutes so that

$$\begin{bmatrix} \boldsymbol{A} & \overline{\boldsymbol{A}} \\ \boldsymbol{B} & \overline{\boldsymbol{B}} \end{bmatrix} \begin{bmatrix} \boldsymbol{B}^{\mathrm{T}} & \boldsymbol{A}^{\mathrm{T}} \\ \overline{\boldsymbol{B}}^{\mathrm{T}} & \overline{\boldsymbol{A}}^{\mathrm{T}} \end{bmatrix} = \begin{bmatrix} \boldsymbol{I} & 0 \\ 0 & \boldsymbol{I} \end{bmatrix} \tag{3.3.47}$$

This is the closure relation and is equivalent to

$$\boldsymbol{A}\boldsymbol{B}^{\mathrm{T}} + \overline{\boldsymbol{A}}\overline{\boldsymbol{B}}^{\mathrm{T}} = \boldsymbol{B}\boldsymbol{A}^{\mathrm{T}} + \overline{\boldsymbol{B}}\overline{\boldsymbol{A}}^{\mathrm{T}} = \boldsymbol{I}, \quad \boldsymbol{A}\boldsymbol{A}^{\mathrm{T}} + \overline{\boldsymbol{A}}\overline{\boldsymbol{A}}^{\mathrm{T}} = \boldsymbol{B}\boldsymbol{B}^{\mathrm{T}} + \overline{\boldsymbol{B}}\overline{\boldsymbol{B}}^{\mathrm{T}} = 0 \tag{3.3.48}$$

Equation (3.3.48) tells us that the real part of $\boldsymbol{A}\boldsymbol{B}^{\mathrm{T}}$ is $\boldsymbol{I}/2$, while $\boldsymbol{A}\boldsymbol{A}^{\mathrm{T}}$ and $\boldsymbol{B}\boldsymbol{B}^{\mathrm{T}}$ are purely imaginary. Hence, the three Barnett-Lothe tensors $\boldsymbol{S}, \boldsymbol{H}, \boldsymbol{L}$, defined by

$$\boldsymbol{S} = i(2\boldsymbol{A}\boldsymbol{B}^{\mathrm{T}} - \boldsymbol{I}), \quad \boldsymbol{H} = 2i\boldsymbol{A}\boldsymbol{A}^{\mathrm{T}}, \quad \boldsymbol{L} = -2i\boldsymbol{B}\boldsymbol{B}^{\mathrm{T}} \tag{3.3.49}$$

are real. It is clear that H and L are symmetric. It can be shown that they are positive definite and that SH, LS, $H^{-1}S$, SL^{-1} are anti-symmetric [38]. Moreover the real matrices S, H, L are not entirely independent. Indeed, they are related by

$$LS + S^{T}L = 0 \tag{3.3.50}$$

$$SL^{-1} + L^{-1}S^{T} = 0 \tag{3.3.51}$$

$$HS^{T} + SH = 0 \tag{3.3.52}$$

$$S^{T}H^{-1} + H^{-1}S = 0 \tag{3.3.53}$$

$$HL - SS = I \tag{3.3.54}$$

Identities (3.3.50), (3.3.52) and (3.3.54) can be verified by a direct substitution of S, H and L from Eq.(3.3.49) with the aid of Eq.(3.3.48). Identity (3.3.51) is obtained from identity (3.3.50) by pre-multiplying and post-multiplying by L^{-1}. Similarly, identity (3.3.53) is obtained from identity (3.3.52) by multiplying by H^{-1}.

A generalized form of Eq.(3.3.40) and Eq.(3.3.43), which is related to the coordinate transformation and is useful for the development of identities, is written as

$$N(\omega)\xi = p(\omega)\xi \tag{3.3.55}$$

$$N^{T}(\omega)\eta(\omega) = p(\omega)\eta(\omega) \tag{3.3.56}$$

where

$$\eta(\omega) = \left[\cos\omega + p(0)\sin\omega\right]\eta(0) \tag{3.3.57}$$

$$p(\omega) = \frac{p(0)\cos\omega - \sin\omega}{p(0)\sin\omega + \cos\omega} \tag{3.3.58}$$

$$N_{1}(\omega) = -T^{-1}(\omega)R^{T}(\omega), \quad N_{2}(\omega) = T^{-1}(\omega), \quad N_{3}(\omega) = -R(\omega)N_{1}(\omega) - Q(\omega) \tag{3.3.59}$$

with

$$Q_{JK}(\omega) = E_{iJKs}n_{i}n_{s}, \quad R_{JK}(\omega) = E_{iJKs}n_{i}m_{s}, \quad T_{JK}(\omega) = E_{iJKs}m_{i}m_{s} \tag{3.3.60}$$

$$n = [\cos\omega \quad \sin\omega \quad 0]^{T}, \quad m = [-\sin\omega \quad \cos\omega \quad 0]^{T} \tag{3.3.61}$$

In Eq.(3.3.61), n and m are two mutually orthogonal unit vectors embedded in the material as shown in Fig.3.1. The plane defined by n and m is the plane of interest and $t=n\times m$ is the unit normal to the plane. Note that ξ in Eq.(3.3.55) is independent of ω, as has been shown in [39]. When $\omega=0$, Eq.(3.3.55) reduces to Eq.(3.3.40). By using Eq.(3.3.40) and Eq.(3.3.55), the following identities can be obtained [38,39]

$$2AP(\omega)A^{T} = N_{2}(\omega) - i\left[N_{2}(\omega)S^{T} + N_{1}(\omega)H\right] \tag{3.3.62}$$

$$2AP(\omega)B^{T} = N_{1}(\omega) - i\left[N_{1}(\omega)S - N_{2}(\omega)L\right] \tag{3.3.63}$$

$$2BP(\omega)A^{\mathrm{T}} = N_1^{\mathrm{T}}(\omega) - i\left[N_1^{\mathrm{T}}(\omega)S^{\mathrm{T}} + N_3(\omega)H\right] \qquad (3.3.64)$$

$$2BP(\omega)B^{\mathrm{T}} = N_3(\omega) - i\left[N_3(\omega)S - N_1^{\mathrm{T}}(\omega)L\right] \qquad (3.3.65)$$

in which $P(\omega)$ is a diagonal matrix defined by

$$P(\omega) = \mathrm{diag}\left[p_1(\omega) \quad p_2(\omega) \quad p_3(\omega) \quad p_4(\omega)\right] \qquad (3.3.66)$$

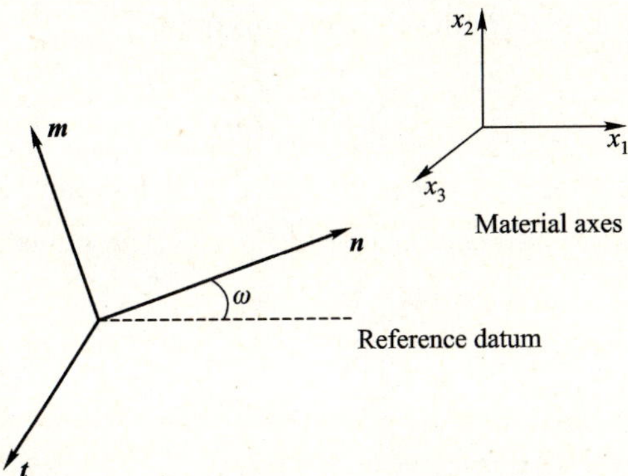

Fig.3.1   Mutually orthogonal unit vectors m, n and t used in analysis

Further, it has been shown in [38] that

$$S = \frac{1}{\pi}\int_0^{\pi} N_1(\omega)\mathrm{d}\omega, \quad H = \frac{1}{\pi}\int_0^{\pi} N_2(\omega)\mathrm{d}\omega, \quad L = -\frac{1}{\pi}\int_0^{\pi} N_3(\omega)\mathrm{d}\omega$$

$$(3.3.67)$$

Equation (3.3.67) provides an alternative to Eq.(3.3.49) for the Barnett-Lothe tensors S, H and L. In addition, for any integer number k, we have [40]

$$AP^k A^{\mathrm{T}} = (N_2^k M - iN_1^k)H/2 \qquad (3.3.68)$$

$$AP^k B^{\mathrm{T}} = (N_1^k M^{-1} + iN_2^k)L/2 \qquad (3.3.69)$$

$$BP^k A^{\mathrm{T}} = [(N_1^k)^{\mathrm{T}} M - iN_3^k]H/2 \qquad (3.3.70)$$

$$BP^k B^{\mathrm{T}} = [N_3^k M^{-1} + i(N_1^k)^{\mathrm{T}}]L/2 \qquad (3.3.71)$$

where

$$M = H^{-1}(I + iS) = (I - iS^{\mathrm{T}})H^{-1} \qquad (3.3.72)$$

3.4   Logarithmic singularity of crack-tip fields in homogeneous piezoelectricity

3.4.1   General solution for crack-tip fields

The singularity of stress and electric displacement near the tip in homogeneous piezoelectricity has been studied by Ting [41] for anisotropic elasticity, Qin and Yu [42] for electroelastic problems, and Yu and Qin [10] for thermo-electro-elastic problems. In this section we follow the results given in [42].

Consider a semi-infinite crack along the negative x-axis. The SED singularities at the tips of the crack can be determined by assuming the function f in Eq.(3.3.6) and Eq.(3.3.10) in the following form [43]

$$f(z_J) = \frac{z_J^{1-\eta}}{1-\eta} \tag{3.4.1}$$

where $\eta = a + ib$ is a complex constant with a and b being two real constants. Substituting Eq.(3.4.1) into Eq.(3.3.9) and Eq. (3.3.13) yields

$$U = 2\,\mathrm{Re}\left\{ A\left\langle z_\alpha^{1-\eta}\right\rangle \frac{q}{1-\eta} \right\} \tag{3.4.2}$$

$$\Pi_2 = 2\,\mathrm{Re}\left\{ B\left\langle z_\alpha^{-\eta}\right\rangle q \right\} \tag{3.4.3}$$

where $\Pi_2 = [\sigma_{21}\ \ \sigma_{22}\ \ \sigma_{23}\ \ D_2]^\mathrm{T}$. If we use the polar coordinate system (r, θ) originating at the crack-tip, the complex variable z_α becomes

$$z_\alpha = r(\cos\theta + p_\alpha \sin\theta) \tag{3.4.4}$$

We see that with the assumption of Eq.(3.4.1) the SED given by Eq.(3.4.3) is of the order $r^{-\eta}$. It is obvious that the SED is singular if the real part of η, i.e. a, is positive. For the potential energy to be bounded at the crack tip, we require that $a<1$. So we focus our attention on the interval $0< a<1$. Using the traction-charge free condition on the crack surfaces and noting that $z = r$ when $\theta = 0$ and $z = re^{\pm i\pi}$ when $\theta = \pm\pi$, we know that

$$\Pi_2(\pi) = -r^{-a}(r^{-ib}e^{-i\pi\eta}Bq + r^{ib}e^{i\pi\bar\eta}\overline{Bq}) = 0 \tag{3.4.5}$$

$$\Pi_2(-\pi) = -r^{-a}(r^{-ib}e^{i\pi\eta}Bq + r^{ib}e^{-i\pi\bar\eta}\overline{Bq}) = 0 \tag{3.4.6}$$

or in matrix form

$$X(\eta)Q = 0 \tag{3.4.7}$$

where $Q = [Bq\ \ \overline{Bq}]^\mathrm{T}$. To obtain a nontrivial solution for Q we should let the determinant of X vanish, i.e.

$$\|X\| = 0 \tag{3.4.8}$$

where the symbol $\|\cdot\|$ denotes the determinant, which leads to

$$b=0, \quad (1-e^{4i\pi a})^4 = 0 \tag{3.4.9}$$

The solution of Eq.(3.4.9) reads

$$a = \frac{1-n}{2}, \quad n = 0,\ 1,\ 2,\cdots \tag{3.4.10}$$

Hence, to satisfy $0<a<1$, we should take $n=0$, which is a fourfold root of Eq.(3.4.9). The elastic displacement and electric potential, U, and SED, Π_2, may now be written in their asymptotic forms by combining Eqs. (3.4.2), (3.4.3) and (3.4.10) as

$$U = 4r^{1/2} \operatorname{Re}\left[A\left\langle (\cos\theta + p_\alpha \sin\theta)^{1/2} \right\rangle q \right] \qquad (3.4.11)$$

$$\Pi_2 = 2r^{-1/2} \operatorname{Re}\left[B\left\langle (\cos\theta + p_\alpha \sin\theta)^{-1/2} \right\rangle q \right] \qquad (3.4.12)$$

3.4.2 Modified solution for *p* being a multiple root

The analyses presented so far tacitly assume that the eigenvalues p's are distinct. When one of the p's is a double root, one may or may not have four independent functions in Eq.(3.4.11) and Eq.(3.4.12), and a set of additional solutions are required [43]. It is not difficult to see that if Eq.(3.4.11) and Eq.(3.4.12) are the solutions corresponding to the double root p_i, so are [43]

$$U^{(2)} = 4r^{1/2} \operatorname{Re}\left\{ \frac{\mathrm{d}}{\mathrm{d}p_i}\left\{ A\left\langle (\cos\theta + p_\alpha \sin\theta)^{1/2} \right\rangle \right\} q \right\} \qquad (3.4.13)$$

$$\Pi_2^{(2)} = 2r^{-1/2} \operatorname{Re}\left\{ \frac{\mathrm{d}}{\mathrm{d}p_i}\left\{ B\left\langle (\cos\theta + p_\alpha \sin\theta)^{-1/2} \right\rangle \right\} q \right\} \qquad (3.4.14)$$

where $\mathrm{d}A/\mathrm{d}p_i$ and $\mathrm{d}B/\mathrm{d}p_i$ can be obtained by differentiating Eq.(3.3.7) and Eq.(3.3.11) with respect to p_i, that is

$$\frac{\mathrm{d}}{\mathrm{d}p_i}\{DA\} = 0 \qquad (3.4.15)$$

$$\frac{\mathrm{d}B}{\mathrm{d}p_i} = \frac{\mathrm{d}}{\mathrm{d}p_i}\left\{ (R^{\mathrm{T}} + pT)A \right\} \qquad (3.4.16)$$

where $D = Q + p(R + R^{\mathrm{T}}) + p^2 T$ is a 4×4 matrix. The new solutions (3.4.13) and (3.4.14) exist if the following equation holds true [44]:

$$\frac{\mathrm{d}^n}{\mathrm{d}p_i^n}\|D\|\Big|_{p=p_i} = 0, \qquad n = N - M \qquad (3.4.17)$$

where N and M are the order and rank of D, respectively. However, it is found that the order of singularity is not changed in the presence of the new solution (3.4.14).

3.4.3 Modified solution for η being a multiple root

If η is a multiple root of Eq.(3.4.8), the components of q may not be unique and one must find other independent solutions. For a root of multiplicity m, the new solutions are given by (taking Π_2 as an example)

$$\Pi_2^{(i)} = 2\,\mathrm{Re}\left\{\boldsymbol{B}[z_1^{*(i)} \quad z_2^{*(i)} \quad z_3^{*(i)} \quad z_4^{*(i)}]^{\mathrm{T}}\right\}, \quad i=1,\,2,\cdots,\,m-1 \quad (3.4.18)$$

where $z_\alpha^{*(i)} = z_\alpha^{-\eta}\left(-\ln z_\alpha + \dfrac{\partial}{\partial \eta}\right)^i q_\alpha$. Likewise, new solutions exist if

$$\frac{\mathrm{d}^n}{\mathrm{d}\eta^n}\|\boldsymbol{X}\|\Big|_{\eta=1/2} = 0, \quad n = 8 - M \tag{3.4.19}$$

holds true. Here M is the rank of matrix $\boldsymbol{X}$. Since $\eta=1/2$ is a fourfold root [see Eq.(3.4.9)], the SED singularities at the tip of a semi-infinite crack must occur in one of the following cases

$$\Pi_2(r) = \begin{cases} O(r^{-1/2}), & \text{only satisfying } \boldsymbol{XQ} = 0 \\[2mm] O(r^{-1/2}\ln r), & \text{satisfying } \mathrm{d}\boldsymbol{X}/\mathrm{d}\eta\big|_{\eta=1/2} = 0 \\[2mm] O(r^{-1/2}\ln^2 r), & \text{satisfying } \mathrm{d}^2\boldsymbol{X}/\mathrm{d}\eta^2\big|_{\eta=1/2} = 0 \\[2mm] O(r^{-1/2}\ln^3 r), & \text{satisfying } \mathrm{d}^3\boldsymbol{X}/\mathrm{d}\eta^3\big|_{\eta=1/2} = 0 \end{cases} \tag{3.4.20}$$

For a semi-infinite crack in an anisotropic piezoelectric medium, it is therefore shown that both stress and electric displacement at the crack tip may be in the order of $r^{-1/2}$, or $r^{-1/2}\ln r$, $r^{-1/2}\ln^2 r$, $r^{-1/2}\ln^3 r$, as $r \to 0$, where r is the distance from crack tip to field point, depending on which boundary conditions are satisfied.

3.5   Trefftz finite element method for piezoelectricity

The Hybrid-Trefftz (HT) finite element (FE) model was originally developed in 1977 for analysis of the effect of mesh distortion on thin plate elements [45]. During the following three decades, the potential of Trefftz finite elements for the solution of different types of applied science and engineering problems was recognised. Over the years, the HT finite element method (FEM) has become increasingly popular as an efficient numerical tool in computational mechanics and has been widely used in the analysis of plane elasticity, thin and thick plate bending, Poisson's equation, shell, heat conduction, and piezoelectric materials. Detailed discussion of the development in this area can be found in [46]. In contrast to conventional FEM, the class of finite elements associated with the Trefftz method is based on a hybrid method which includes the use of an auxiliary inter-element displacement or traction frame to link the internal displacement fields of the elements. Such internal fields, chosen so as to a priori satisfy the governing differential equations, have conveniently been represented as the sum of a particular integral of non-homogeneous equations and a suitably truncated Trefftz complete set of regular homogeneous solutions multiplied by undetermined coefficients. Inter-element continuity is enforced by using a modified variational principle together with an independent frame field defined on each element boundary. The element formulation, during which the internal parameters are eliminated at the element level, in the end leads to the standard force-displacement relationship, with a symmetric positive definite stiffness matrix. Clearly, whereas the conventional FE formulation may be assimilated to a particular form of the Rayleigh-Ritz method, the HT FE approach has a close relationship with the Trefftz method [46]. This section addresses applications of the Trefftz FEM to piezoelectric materials. The presentation below follows the developments appearing in [47,48].

3.5.1   Basic field equations and boundary conditions

Consider a linear piezoelectric material, in which the differential governing equations in the Cartesian coordinates x_i (i=1, 2, 3) are given by

$$\sigma_{ij,j} + b_i = 0, \qquad D_{i,i} + b_e = 0, \qquad \text{in } \Omega \tag{3.5.1}$$

where Ω is the solution domain and the Einstein summation convention over repeated indices is used. For an anisotropic piezoelectric material, the constitutive relation is

$$\varepsilon_{ij} = -\frac{\partial H(\boldsymbol{\sigma}, \boldsymbol{D})}{\partial \sigma_{ij}} = s_{ijkl}^{D} \sigma_{kl} + g_{kij} D_k, \qquad E_i = \frac{\partial H(\boldsymbol{\sigma}, \boldsymbol{D})}{\partial D_i} = -g_{ikl} \sigma_{kl} + \lambda_{ik}^{\sigma} D_k$$

$$\tag{3.5.2}$$

for $(\boldsymbol{\sigma}, \boldsymbol{D})$ as basic variables,

$$\sigma_{ij} = \frac{\partial H(\boldsymbol{\varepsilon}, \boldsymbol{E})}{\partial \varepsilon_{ij}} = c_{ijkl}^{E} \varepsilon_{kl} - e_{kij} E_k, \qquad D_i = -\frac{\partial H(\boldsymbol{\varepsilon}, \boldsymbol{E})}{\partial E_i} = e_{ikl} \varepsilon_{kl} + \kappa_{ik}^{\varepsilon} E_k$$

$$\tag{3.5.3}$$

for $(\boldsymbol{\varepsilon}, \boldsymbol{E})$ as basic variables,

$$\sigma_{ij} = \frac{\partial H(\boldsymbol{\varepsilon}, \boldsymbol{D})}{\partial \sigma_{ij}} = c_{ijkl}^{D} \varepsilon_{kl} + h_{kij} D_k, \qquad E_i = \frac{\partial H(\boldsymbol{\varepsilon}, \boldsymbol{D})}{\partial D_i} = h_{ikl} \varepsilon_{kl} + \lambda_{ik}^{\varepsilon} D_k$$

$$\tag{3.5.4}$$

for (ε, D) as basic variables, and

$$\varepsilon_{ij} = -\frac{\partial H(\sigma, E)}{\partial \sigma_{ij}} = s^{E}_{ijkl}\sigma_{kl} + d_{kij}D_{k}, \qquad D_{i} = -\frac{\partial H(\sigma, E)}{\partial E_{i}} = d_{ikl}\sigma_{kl} + \kappa^{\sigma}_{ik}E_{k}$$

(3.5.5)

for (σ, E) as basic variables, with

$$H(\sigma, D) = -\frac{1}{2}s^{D}_{ijkl}\sigma_{ij}\sigma_{kl} + \frac{1}{2}\lambda^{\sigma}_{ij}D_{i}D_{j} - g_{kij}\sigma_{ij}D_{k} \qquad (3.5.6)$$

$$H(\varepsilon, E) = \frac{1}{2}c^{E}_{ijkl}\varepsilon_{ij}\varepsilon_{kl} - \frac{1}{2}\kappa^{\varepsilon}_{ij}E_{i}E_{j} - e_{kij}\varepsilon_{ij}E_{k} \qquad (3.5.7)$$

$$H(\varepsilon, D) = \frac{1}{2}c^{D}_{ijkl}\varepsilon_{ij}\varepsilon_{kl} + \frac{1}{2}\lambda^{\varepsilon}_{ij}D_{i}D_{j} + h_{kij}\varepsilon_{ij}D_{k} \qquad (3.5.8)$$

$$H(\sigma, E) = -\frac{1}{2}s^{E}_{ijkl}\sigma_{ij}\sigma_{kl} - \frac{1}{2}\kappa^{\sigma}_{ij}E_{i}E_{j} - d_{kij}\sigma_{ij}E_{k} \qquad (3.5.9)$$

where c^{E}_{ijkl}, c^{D}_{ijkl} and s^{E}_{ijkl}, s^{D}_{ijkl} are the stiffness and compliance coefficient tensor for $E=0$ or $D=0$, κ^{σ}_{ij}, $\kappa^{\varepsilon}_{ij}$ and λ^{σ}_{ij}, $\lambda^{\varepsilon}_{ij}$ are the permittivity matrix and the conversion of the permittivity constant matrix for $\sigma=0$ or $\varepsilon=0$.

The boundary conditions of the electroelastic problem are defined by

$$u_{i} = \bar{u}_{i}, \qquad\qquad\qquad \text{on } \Gamma_{u} \qquad (3.5.10)$$

$$t_{i} = \sigma_{ij}n_{j} = \bar{t}_{i}, \qquad\qquad \text{on } \Gamma_{t} \qquad (3.5.11)$$

$$D_{n} = D_{i}n_{i} = -\bar{q}_{n} = \bar{D}_{n}, \qquad \text{on } \Gamma_{D} \qquad (3.5.12)$$

$$\phi = \bar{\phi}, \qquad\qquad\qquad\quad \text{on } \Gamma_{\phi} \qquad (3.5.13)$$

where $\bar{u}_{i}$, $\bar{t}_{i}$, $\bar{q}_{n}$ and $\bar{\phi}$ are, respectively, prescribed boundary displacement, traction vector, surface charge and electric potential, an overhead bar denotes prescribed value, $\Gamma = \Gamma_{u} + \Gamma_{t} = \Gamma_{D} + \Gamma_{\phi}$ is the boundary of the solution domain Ω.

Moreover, in the Trefftz FE form, Eqs. (3.5.1)~(3.5.13) should be completed by the following inter-element continuity requirements:

$$u_{ie} = u_{if}, \qquad \phi_{e} = \phi_{f}, \qquad \text{on } \Gamma_{e}\bigcap\Gamma_{f}, \text{ conformity} \quad (3.5.14)$$

$$t_{ie} + t_{if} = 0, \qquad D_{ne} + D_{nf} = 0, \quad \text{on } \Gamma_{e}\bigcap\Gamma_{f}, \text{ reciprocity} \quad (3.5.15)$$

where "e" and "f" stand for any two neighboring elements. Eqs.(3.5.1)~(3.5.15) are taken as the basis to establish the modified variational principle for Trefftz FE analysis of piezoelectric materials.

3.5.2   Assumed displacement and electric potential fields

The main idea of the HT FEM is to establish a FE formulation whereby the intra-element continuity is enforced on a non-conforming internal displacement field chosen so as to a priori satisfy the governing differential equation of the problem under consideration [46]. In other words, as an obvious alternative to the Rayleigh-Ritz method as a basis for a FE formulation, the model here is based on the method of Trefftz [49]. With this method the solution domain Ω is subdivided into elements, and over each element "e," the assumed intra-element fields are

$$U = \begin{bmatrix} u_1 \\ u_2 \\ u_3 \\ \phi \end{bmatrix} = \begin{bmatrix} \breve{u}_1 \\ \breve{u}_2 \\ \breve{u}_3 \\ \breve{\phi} \end{bmatrix} + \begin{bmatrix} N_1 \\ N_2 \\ N_3 \\ N_4 \end{bmatrix} c = \breve{u} + \sum_{j=1} N_j c_j = \breve{u} + Nc \qquad (3.5.16)$$

where c_j stands for undetermined coefficient, and $\breve{u} = [\breve{u}_1 \ \ \breve{u}_2 \ \ \breve{u}_3 \ \ \breve{\phi}]^{\mathrm{T}}$ and N are known functions. If the governing differential equation (3.5.1) is rewritten in a general form

$$\mathcal{R} u(x) + b(x) = 0 \qquad (x \in \Omega_e) \qquad (3.5.17)$$

where $\mathcal{R}$ stands for the differential operator matrix for Eq.(3.5.1), x for the position vector, $b = [b_1 \ \ b_2 \ \ b_3 \ \ b_e]^{\mathrm{T}}$ for the known right-hand side term, and Ω_e stands for the eth element sub-domain, then $\breve{u} = \breve{u}(x)$ and $N = N(x)$ in Eq.(3.5.16) must be chosen such that

$$\mathcal{R}\breve{u} + b = 0 \quad \text{and} \quad \mathcal{R}N = 0 \qquad (3.5.18)$$

everywhere in Ω_e. A complete system of homogeneous solutions N_j can be generated by way of the solution in Stroh formalism

$$u = 2\,\mathrm{Re}\left\{ A \langle f(z_\alpha) \rangle c \right\} \qquad (3.5.19)$$

where $\langle f(z_\alpha) \rangle = \mathrm{diag}[f(z_1) \ \ f(z_2) \ \ f(z_3) \ \ f(z_4)]$ is a diagonal 4×4 matrix, while $f(z_i)$ is an arbitrary function with argument $z_i = x_1 + p_i x_2$. p_i (i=1~4) are the material eigenvalues. Of particular interest is a complete set of polynomial solutions which may be generated by setting in Eq.(3.5.19) in turn

$$\left. \begin{aligned} f(z_\alpha) &= z_\alpha^k \\ f(z_\alpha) &= i z_\alpha^k \end{aligned} \right\}, \quad k = 1, 2, \cdots \qquad (3.5.20)$$

where $i = \sqrt{-1}$. This leads, for N_j of Eq.(3.5.16), to the following sequence

$$N_{2j} = 2\,\mathrm{Re}\left\{ A \langle z_\alpha^j \rangle \right\}, \quad N_{2j+1} = 2\,\mathrm{Re}\left\{ A \langle i z_\alpha^j \rangle \right\} \qquad (3.5.21)$$

The unknown coefficient c in Eq.(3.5.19) can be written as

$$c = [c_1 \ \ c_2 \ \ \cdots \ \ c_m]^{\mathrm{T}} \qquad (3.5.22)$$

in which m is the dimension of vector c. The choice of m has been discussed in [46]. For the reader's convenience, we briefly describe the basic rule for determining m. It is important to choose the proper number m of trial functions N_j for the Trefftz element with the hybrid technique. The basic rule used to prevent spurious energy modes is analogous to that in the hybrid-stress model. The necessary (but not sufficient) condition for the matrix H, which is later defined by Eq.(3.5.47) in Subsection 3.5.4, to have full rank is stated as [46]

$$m_{\min} = N_{DOF} - N_{RIG} \qquad (3.5.23)$$

where N_{DOF} and N_{RIG} are numbers of nodal degrees of freedom of the element under consideration and of the discarded rigid body motion terms, or more generally the number of zero eigenvalues. Although the use of the minimum number $m = N_{DOF} - N_{RIG}$ of flux mode terms in Eq.(3.5.23) does not always guarantee a stiffness matrix with full rank, full rank may always be achieved by suitably augmenting m. The optimal value of m for a given type of element should be found by numerical experimentation.

The unknown coefficient c in Eq.(3.5.19) may be calculated from the conditions on the external boundary and/or the continuity conditions on the inter-element boundary. Thus various Trefftz element models can be obtained by using different approaches to enforce these conditions. In the majority of cases a hybrid technique is used, whereby the elements are linked through an auxiliary conforming displacement frame which has the same form as in the conventional FE method. This means that, in the Trefftz FE approach, a conforming electric potential and displacement (EPD) field should be independently defined on the element boundary to enforce the field continuity between elements and also to link the coefficient c, appearing in Eq.(3.5.19), with nodal EPD d. The frame is defined as

$$\tilde{u}(x) = \begin{bmatrix} \tilde{u}_1 \\ \tilde{u}_2 \\ \tilde{u}_3 \\ \tilde{\phi} \end{bmatrix} = \begin{bmatrix} \tilde{N}_1 \\ \tilde{N}_2 \\ \tilde{N}_3 \\ \tilde{N}_4 \end{bmatrix} d = \tilde{N}d, \qquad x \in \Gamma_e \qquad (3.5.24)$$

where the symbol "~" is used to specify that the field is defined on the element boundary only, $d=d(c)$ stands for the vector of the nodal displacements which are the final unknowns of the problem, Γ_e represents the boundary of element e, and $\tilde{N}$ is a matrix of the corresponding shape functions which are the same as those in conventional FE formulation. For example, along the side *A-O-B* of a particular element (see Fig.3.2), a simple interpolation of the frame displacement and electric potential can be given in the form

$$\tilde{u}(x) = \begin{bmatrix} \tilde{u}_1 \\ \tilde{u}_2 \\ \tilde{u}_3 \\ \tilde{\phi} \end{bmatrix} = \begin{bmatrix} \tilde{N}_A & \tilde{N}_B \end{bmatrix} \begin{bmatrix} d_A \\ d_B \end{bmatrix}, \qquad x \in \Gamma_e \qquad (3.5.25)$$

where

$$\tilde{N}_A = \text{diag}[\tilde{N}_1 \ \tilde{N}_1 \ \tilde{N}_1 \ \tilde{N}_1], \qquad N_B = \text{diag}[\tilde{N}_2 \ \tilde{N}_2 \ \tilde{N}_2 \ \tilde{N}_2] \qquad (3.5.26)$$

$$d_A = [u_{1A} \ u_{2A} \ u_{3A} \ \phi_A]^{\mathrm{T}}, \qquad d_B = [u_{1B} \ u_{2B} \ u_{3B} \ \phi_B]^{\mathrm{T}} \qquad (3.5.27)$$

with

$$\tilde{N}_1 = \frac{1-\xi}{2}, \qquad \tilde{N}_2 = \frac{1+\xi}{2} \qquad (3.5.28)$$

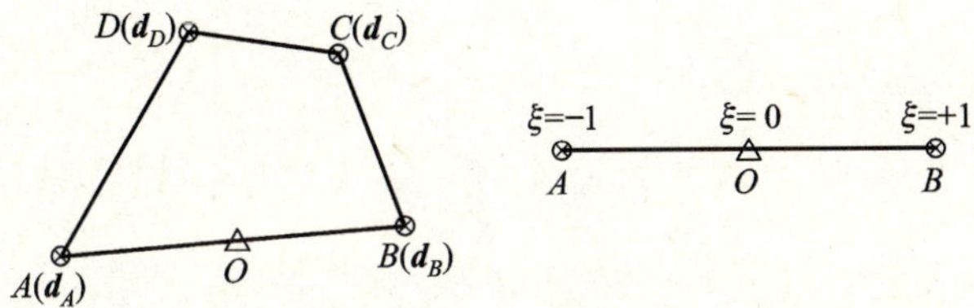

Fig.3.2   A quadrilateral element generalized two-dimensional problem

Using the above definitions, the generalized boundary forces and electric displacements can be derived from Eqs.(3.5.11), (3.5.12) and (3.5.16), and denoted

$$T = \begin{bmatrix} t_1 \\ t_2 \\ t_3 \\ D_n \end{bmatrix} = \begin{bmatrix} \sigma_{1j}n_j \\ \sigma_{2j}n_j \\ \sigma_{3j}n_j \\ D_j n_j \end{bmatrix} = \begin{bmatrix} \breve{t}_1 \\ \breve{t}_2 \\ \breve{t}_3 \\ \breve{D}_n \end{bmatrix} + \begin{bmatrix} Q_1 \\ Q_2 \\ Q_3 \\ Q_4 \end{bmatrix} c = \breve{T} + Qc \qquad (3.5.29)$$

where $\breve{t}_i$ and $\breve{D}_n$ are derived from $\breve{u}$.

3.5.3   Variational principles

The Trefftz FE equation for piezoelectric materials can be established by the variational approach [46]. Since stationary conditions of the traditional potential and complementary variational functional cannot satisfy the inter-element continuity condition which is required in Trefftz FE analysis, some new variational functionals need to be developed. For this purpose, we present the following two modified variational functionals suitable for Trefftz FE analysis:

$$\Theta_m^{\sigma D} = \sum_e \Theta_{me}^{\sigma D} = \sum_e \{\Theta_e^{\sigma D} - \int_{\Gamma_{De}} (\bar{D}_n - D_n)\tilde{\phi}\mathrm{d}s - \int_{\Gamma_{te}} (\bar{t}_i - t_i)\tilde{u}_i\mathrm{d}s +$$

$$\int_{\Gamma_{Ie}} (D_n\tilde{\phi} + t_i\tilde{u}_i)\mathrm{d}s\} \tag{3.5.30}$$

$$\Theta_m^{\varepsilon E} = \sum_e \Theta_{me}^{\varepsilon E} = \sum_e \left\{\Theta_e^{\varepsilon E} + \int_{\Gamma_{\phi e}} (\bar{\phi} - \phi)\tilde{D}_n\mathrm{d}s + \int_{\Gamma_{ue}} (\bar{u}_i - u_i)\tilde{t}_i\mathrm{d}s - \right.$$

$$\left. 2\int_{\Gamma_{te}} \tilde{u}_i t_i\mathrm{d}s - 2\int_{\Gamma_{De}} \tilde{\phi}D_n\mathrm{d}s - \int_{\Gamma_{Ie}} (\tilde{\phi}D_n + \tilde{u}_i t_i\mathrm{d}s\right\} \tag{3.5.31}$$

$$\Theta_m^{\varepsilon D} = \sum_e \Theta_{me}^{\varepsilon D} = \sum_e \left\{\Theta_e^{\varepsilon D} - \int_{\Gamma_{De}} (\bar{D}_n - D_n)\tilde{\phi}\mathrm{d}s + \int_{\Gamma_{ue}} (\bar{u}_i - u_i)\tilde{t}_i\mathrm{d}s - \right.$$

$$\left. 2\int_{\Gamma_{te}} \tilde{u}_i t_i\mathrm{d}s + \int_{\Gamma_{Ie}} (D_n\tilde{\phi} - t_i\tilde{u}_i)\mathrm{d}s\right\} \tag{3.5.32}$$

$$\Theta_m^{\sigma E} = \sum_e \Theta_{me}^{\sigma E} = \sum_e \left\{\Theta_e^{\sigma E} + \int_{\Gamma_{\phi e}} (\bar{\phi} - \phi)\tilde{D}_n\mathrm{d}s - \int_{\Gamma_{te}} (\bar{t}_i - t_i)\tilde{u}_i\mathrm{d}s - \right.$$

$$\left. 2\int_{\Gamma_{De}} D_n\tilde{\phi}\mathrm{d}s - \int_{\Gamma_{Ie}} (D_n\tilde{\phi} - t_i\tilde{u}_i)\mathrm{d}s\right\} \tag{3.5.33}$$

where

$$\Theta_e^{\sigma D} = \iint_{\Omega_e} H(\boldsymbol{\sigma}, \boldsymbol{D})\mathrm{d}\Omega + \int_{\Gamma_{ue}} t_i\bar{u}_i\mathrm{d}s + \int_{\Gamma_{\phi e}} D_n\bar{\phi}\mathrm{d}s \tag{3.5.34}$$

$$\Theta_e^{\varepsilon E} = \iint_{\Omega_e} [H(\boldsymbol{\varepsilon}, \boldsymbol{E}) - b_i u_i - b_e\phi]\mathrm{d}\Omega + \int_{\Gamma_{te}} \bar{t}_i\tilde{u}_i\mathrm{d}s + \int_{\Gamma_{De}} \bar{D}_n\tilde{\phi}\mathrm{d}s \tag{3.5.35}$$

$$\Theta_e^{\varepsilon D} = \iint_{\Omega_e} [H(\boldsymbol{\varepsilon}, \boldsymbol{D}) - b_i u_i]\mathrm{d}\Omega + \int_{\Gamma_{te}} \bar{t}_i\tilde{u}_i\mathrm{d}s + \int_{\Gamma_{\phi e}} D_n\bar{\phi}\mathrm{d}s \tag{3.5.36}$$

$$\Theta_e^{\sigma E} = \iint_{\Omega_e} [H(\boldsymbol{\sigma}, \boldsymbol{E}) - b_e\phi]\mathrm{d}\Omega + \int_{\Gamma_{ue}} t_i\bar{u}_i\mathrm{d}s + \int_{\Gamma_{De}} \bar{D}_n\tilde{\phi}\mathrm{d}s \tag{3.5.37}$$

The boundary Γ_e of a particular element consists of the following parts:

$$\Gamma_e = \Gamma_{ue} \cup \Gamma_{te} \cup \Gamma_{Ie} = \Gamma_{\phi e} \cup \Gamma_{De} \cup \Gamma_{Ie} \tag{3.5.38}$$

where

$$\Gamma_{ue} = \Gamma_u \cap \Gamma_e, \quad \Gamma_{te} = \Gamma_t \cap \Gamma_e, \quad \Gamma_{\phi e} = \Gamma_\phi \cap \Gamma_e, \quad \Gamma_{De} = \Gamma_D \cap \Gamma_e$$

$$\tag{3.5.39}$$

and Γ_{Ie} is the inter-element boundary of the element "e". We now show that the stationary condition of any one functional in Eqs.(3.5.30)~(3.5.33) leads to Eqs.(3.5.10)~(3.5.15), $u_i = \tilde{u}_i$ (on Γ_t) and $\phi = \tilde{\phi}$ (on Γ_D), and present the theorem on the existence of extremum of the functional, which ensures that an approximate solution can converge to the exact one. Taking $\Theta_m^{\sigma D}$ as an example, we have the following two statements:

(1) Modified complementary principle.

$$\delta\Theta_m^{\sigma D} = 0 \Rightarrow (3.5.10) \sim (3.5.15), \quad u_i = \tilde{u}_i \ (\text{on } \Gamma_t) \ \text{ and } \ \phi = \tilde{\phi} \ (\text{on } \Gamma_D)$$

$$(3.5.40)$$

where δ stands for the variation symbol.

(2) Theorem on the existence of extremum.

If the expression

$$\iint_\Omega \delta^2 H(\boldsymbol{\sigma},\boldsymbol{D})\mathrm{d}\Omega + \int_{\Gamma_t} \delta t_i \delta\tilde{u}_i \mathrm{d}s + \int_{\Gamma_D} \delta D_n \delta\tilde{\phi}\mathrm{d}s + \sum_e \int_{\Gamma_{Ie}} (\delta\tilde{\phi}\delta D_n + \delta\tilde{u}_i \delta t_i)\mathrm{d}s$$

$$(3.5.41)$$

is uniformly positive (or negative) in the neighborhood of $\boldsymbol{U}_0$, where $\boldsymbol{U}_0$ is such a value that $\Theta_m^{\sigma D}(\boldsymbol{U}_0) = (\Theta_m^{\sigma D})_0$, and where $(\Theta_m^{\sigma D})_0$ stands for the stationary value of $\Theta_m^{\sigma D}$, we have

$$\Theta_m^{\sigma D} \geqslant (\Theta_m^{\sigma D})_0 \ [\text{or } \Theta_m^{\sigma D} \leqslant (\Theta_m^{\sigma D})_0] \qquad (3.5.42)$$

in which the relation that $\tilde{\boldsymbol{u}}_e = \tilde{\boldsymbol{u}}_f$ is identical on $\Gamma_e \cap \Gamma_f$ has been used.

Proof: First, we derive the stationary conditions of functional (3.5.30). To this end, performing variation of $\Theta_m^{\sigma D}$ and noting that Eq.(3.5.1) holds true a priori by the previous assumption, we obtain

$$\delta\Theta_m^{\sigma D} = \int_{\Gamma_u} (\overline{u}_i - u_i)\delta t_i \mathrm{d}s + \int_{\Gamma_\phi} (\overline{\phi} - \phi)\delta D_n \mathrm{d}s -$$

$$\int_{\Gamma_t} \left[(\overline{t}_i - t_i)\delta\tilde{u}_i - (\tilde{u}_i - u_i)\delta t_i \right]\mathrm{d}s - \int_{\Gamma_D} \left[(\overline{D}_n - D_n)\delta\tilde{\phi} - (\tilde{\phi} - \phi)\delta D_n \right]\mathrm{d}s +$$

$$\sum_e \int_{\Gamma_{Ie}} \left[(\tilde{u}_i - u_i)\delta t_i + (\tilde{\phi} - \phi)\delta D_n + t_i \delta\tilde{u}_i + D_n \delta\tilde{\phi} \right]\mathrm{d}s \qquad (3.5.43)$$

Therefore, the Euler equations for expression (3.5.43) are Eqs. (3.5.10)~(3.5.15), $u_i = \tilde{u}_i$ (on Γ_t), and $\phi = \tilde{\phi}$ (on Γ_D), as the quantities δt_i, δu_i, $\delta\phi$, δD_n, $\delta\tilde{u}_i$ and $\delta\tilde{\phi}$ may be arbitrary. The principle (3.5.40) has thus been proved. This indicates that the stationary condition of the functional satisfies the required boundary and inter-element continuity equations and can thus be used for deriving Trefftz FE formulation.

As for the proof of the theorem on the existence of extremum, we may complete it by way of the so-called "second variational approach" [50]. In doing this, performing variation of $\delta\Theta_m^{\sigma D}$ and using the constrained conditions (3.5.1), we find

$$\delta^2 \Theta_m^{\sigma D} = \iint_\Omega \delta^2 H(\boldsymbol{\sigma}, \boldsymbol{D}) \mathrm{d}\Omega + \int_{\Gamma_t} \delta t_i \delta \tilde{u}_i \mathrm{d}s +$$

$$\int_{\Gamma_D} \delta D_n \delta \tilde{\phi} \mathrm{d}s + \sum_e \int_{\Gamma_{Ie}} (\delta \tilde{\phi} \delta D_n + \delta \tilde{u}_i \delta t_i) \mathrm{d}s \qquad (3.5.44)$$

Therefore the theorem has been proved from the sufficient condition of the existence of a local extreme of a functional [50]. This completes the proof. The functional given in Eqs.(3.5.31)~(3.5.33) can be stated and proved similarly. We omit those details for the sake of conciseness.

3.5.4   Elemental stiffness matrix

The element matrix equation can be generated by setting $\delta \Theta_{me}^{xy} = 0$. To simplify the derivation, we first transform all domain integrals in Eq.(3.5.30) into boundary ones. In fact, by reason of the solution properties of the intra-element trial functions, the functional $\Theta_{me}^{\sigma D}$ can be simplified to

$$\Theta_{me}^{\sigma D} = -\frac{1}{2} \int_{\Gamma_e} (t_i u_i + D_n \phi) \mathrm{d}s - \frac{1}{2} \int_\Omega (\bar{b}_i u_i + \bar{q}_b \phi) \mathrm{d}\Omega - \int_{\Gamma_{De}} (\bar{D}_n - D_n) \tilde{\phi} \mathrm{d}s -$$

$$\int_{\Gamma_{te}} (\bar{t}_i - t_i) \tilde{u}_i \mathrm{d}s + \int_{\Gamma_{Ie}} (D_n \tilde{\phi} + t_i \tilde{u}_i) \mathrm{d}s + \int_{\Gamma_{ue}} t_i \bar{u}_i \mathrm{d}s + \int_{\Gamma_{\phi e}} D_n \bar{\phi} \mathrm{d}s$$

$$(3.5.45)$$

Substituting the expressions given in Eqs.(3.5.16), (3.5.24), and (3.5.29) into (3.5.45) produces

$$\Theta_{me}^{\sigma D} = -\frac{1}{2} \boldsymbol{c}^\mathrm{T} \boldsymbol{H} \boldsymbol{c} + \boldsymbol{c}^\mathrm{T} \boldsymbol{S} \boldsymbol{d} + \boldsymbol{c}^\mathrm{T} \boldsymbol{r}_1 + \boldsymbol{d}^\mathrm{T} \boldsymbol{r}_2 + \text{terms without } \boldsymbol{c} \text{ or } \boldsymbol{d} \qquad (3.5.46)$$

in which the matrices $\boldsymbol{H}$, $\boldsymbol{S}$ and the vectors $\boldsymbol{r}_1$, $\boldsymbol{r}_2$ are defined by

$$\boldsymbol{H} = \int_{\Gamma_e} \boldsymbol{Q}^\mathrm{T} \boldsymbol{N} \mathrm{d}s \qquad (3.5.47)$$

$$\boldsymbol{S} = \int_{\Gamma_{De}} \boldsymbol{Q}_4^\mathrm{T} \tilde{\boldsymbol{N}}_4 \mathrm{d}s + \int_{\Gamma_{te}} \begin{bmatrix} \boldsymbol{Q}_1 \\ \boldsymbol{Q}_2 \\ \boldsymbol{Q}_3 \end{bmatrix}^\mathrm{T} \begin{bmatrix} \tilde{\boldsymbol{N}}_1 \\ \tilde{\boldsymbol{N}}_2 \\ \tilde{\boldsymbol{N}}_3 \end{bmatrix} \mathrm{d}s + \int_{\Gamma_{Ie}} \boldsymbol{Q}^\mathrm{T} \tilde{\boldsymbol{N}} \mathrm{d}s \qquad (3.5.48)$$

$$\boldsymbol{r}_1 = -\frac{1}{2} \int_{\Gamma_e} (\boldsymbol{N}^\mathrm{T} \breve{\boldsymbol{T}} + \boldsymbol{Q}^\mathrm{T} \breve{\boldsymbol{u}}) \mathrm{d}s - \frac{1}{2} \int_\Omega \boldsymbol{N}^\mathrm{T} \bar{\boldsymbol{b}} \mathrm{d}\Omega +$$

$$\int_{\Gamma_{\phi e}} \boldsymbol{Q}_4^\mathrm{T} \bar{\phi} \mathrm{d}s + \int_{\Gamma_{ue}} \begin{bmatrix} \boldsymbol{Q}_1 \\ \boldsymbol{Q}_2 \\ \boldsymbol{Q}_3 \end{bmatrix}^\mathrm{T} \begin{bmatrix} \bar{u}_1 \\ \bar{u}_2 \\ \bar{u}_3 \end{bmatrix} \mathrm{d}s \qquad (3.5.49)$$

$$r_2 = \int_{\Gamma_{De}} \tilde{N}_4^{\mathrm{T}}(\breve{D}_n - \bar{D}_n)\mathrm{d}s + \int_{\Gamma_e} \tilde{N}^{\mathrm{T}}\breve{T}\mathrm{d}s + \int_{\Gamma_{te}} \begin{bmatrix} \tilde{N}_1 \\ \tilde{N}_2 \\ \tilde{N}_3 \end{bmatrix}^{\mathrm{T}} \left(\begin{bmatrix} \breve{t}_1 \\ \breve{t}_2 \\ \breve{t}_3 \end{bmatrix} - \begin{bmatrix} \bar{t}_1 \\ \bar{t}_2 \\ \bar{t}_3 \end{bmatrix} \right) \mathrm{d}s \quad (3.5.50)$$

To enforce inter-element continuity on the common element boundary, the unknown vector c should be expressed in terms of nodal DOF d. An optional relationship between c and d in the sense of variation can be obtained from

$$\frac{\partial \Theta_{me}^{\sigma D}}{\partial c^{\mathrm{T}}} = -Hc + Sd + r_1 = 0 \quad (3.5.51)$$

This leads to

$$c = Gd + g \quad (3.5.52)$$

where $G = H^{-1}S$ and $g = H^{-1}r_1$, and then straightforwardly yields the expression of $\Theta_{me}^{\sigma D}$ only in terms of d and other known matrices

$$\Theta_{me}^{\sigma D} = \frac{1}{2}d^{\mathrm{T}}G^{\mathrm{T}}HGd + d^{\mathrm{T}}(G^{\mathrm{T}}Hg + r_2) + \text{terms without } d \quad (3.5.53)$$

Therefore, the element stiffness matrix equation can be obtained by taking the vanishing variation of the functional $\Theta_{me}^{\sigma D}$ as

$$\frac{\partial \Theta_{me}^{\sigma D}}{\partial d^{\mathrm{T}}} = 0 \Rightarrow \quad Kd = P \quad (3.5.54)$$

where $K=G^{\mathrm{T}}HG$ and $P=-G^{\mathrm{T}}Hg-r_2$ are, respectively, the element stiffness matrix and the equivalent nodal flow vector. The expression (3.5.54) is the elemental stiffness-matrix equation for Trefftz FE analysis.

3.5.5 Application to anti-plane problem

The formulation presented in Subsection 3.5.4 is for a general three-dimensional piezoelectric solid. To show typical applications of the above FE model, let us consider an anti-plane crack problem.

In the case of anti-plane shear deformation involving only out-of-plane displacement u_3 and in-plane electric fields, and these variables depends on x_1 and x_2 as defined in Eq.(3.2.36), the constitutive relation and equilibrium equation are governed by Eq.(3.2.37) and Eq.(3.2.38), respectively. When the coordinate system (x,y,z), rather than (x_1, x_2, x_3), is used, Eq.(3.3.27) and Eq.(3.3.28) are rewritten as

$$c_{44}\nabla^2 u_z + e_{15}\nabla^2 \phi = 0, \quad e_{15}\nabla^2 u_z - \kappa_{11}\nabla^2 \phi = 0 \quad (3.5.55)$$

$$\begin{bmatrix} \sigma_{xz} \\ \sigma_{yz} \\ D_x \\ D_y \end{bmatrix} = \begin{bmatrix} c_{44} & 0 & -e_{15} & 0 \\ 0 & c_{44} & 0 & -e_{15} \\ e_{15} & 0 & \kappa_{11} & 0 \\ 0 & e_{15} & 0 & \kappa_{11} \end{bmatrix} \begin{bmatrix} \gamma_{xz} \\ \gamma_{yz} \\ E_x \\ E_y \end{bmatrix} \tag{3.5.56}$$

or inversely

$$\begin{bmatrix} \gamma_{xz} \\ \gamma_{yz} \\ E_x \\ E_y \end{bmatrix} = \begin{bmatrix} s_{44} & 0 & g_{15} & 0 \\ 0 & s_{44} & 0 & g_{15} \\ -g_{15} & 0 & \lambda_{11} & 0 \\ 0 & -g_{15} & 0 & \lambda_{11} \end{bmatrix} \begin{bmatrix} \sigma_{xz} \\ \sigma_{yz} \\ D_x \\ D_y \end{bmatrix} \tag{3.5.57}$$

where γ_{xz}, γ_{yz} and E_x, E_y are, respectively, shear strains and electric fields given by

$$\gamma_{xz} = \frac{\partial u_z}{\partial x}, \quad \gamma_{yz} = \frac{\partial u_z}{\partial y}, \quad E_x = -\frac{\partial \phi}{\partial x}, \quad E_y = -\frac{\partial \phi}{\partial y} \tag{3.5.58}$$

The constants s_{44}, g_{15} and λ_{11} are defined by the relations

$$s_{44} = \frac{\kappa_{11}}{\Delta}, \quad g_{15} = \frac{e_{15}}{\Delta}, \quad \lambda_{11} = \frac{c_{44}}{\Delta}, \quad \Delta = c_{44}\kappa_{11} + e_{15}^2 \tag{3.5.59}$$

The boundary conditions of the anti-plane problem are given by

$$u_z = \overline{u}_z, \qquad\qquad \text{on } \Gamma_u \tag{3.5.60}$$

$$t = \sigma_{3j} n_j = \overline{t}, \qquad\qquad \text{on } \Gamma_t \tag{3.5.61}$$

$$D_n = D_i n_i = -\overline{q}_n = \overline{D}_n, \qquad \text{on } \Gamma_D \tag{3.5.62}$$

$$\phi = \overline{\phi}, \qquad\qquad \text{on } \Gamma_\phi \tag{3.5.63}$$

where $\overline{u}$, $\overline{t}$, $\overline{q}_n$ and $\overline{\phi}$ are, respectively, prescribed boundary displacement, traction vector, surface charge and electric potential, an overhead bar denotes prescribed value, $\Gamma = \Gamma_u + \Gamma_t = \Gamma_D + \Gamma_\phi$ is the boundary of the solution domain Ω.

In the Trefftz FE form, Eqs.(3.5.55)~(3.5.63) should be completed by the following inter-element continuity requirements:

$$u_{ze} = u_{zf}, \qquad \phi_e = \phi_f, \qquad\qquad \text{on } \Gamma_e \cap \Gamma_f \tag{3.5.64}$$

$$t_e + t_f = 0, \qquad D_{ne} + D_{nf} = 0, \qquad \text{on } \Gamma_e \cap \Gamma_f \tag{3.5.65}$$

where "e" and "f" stand for any two neighbouring elements.

It is obvious from Eq.(3.5.55) that it requires

$$c_{44}\kappa_{11} + e_{15}^2 \neq 0 \tag{3.5.66}$$

to have non-trivial solutions for the out-of-plane displacement and in-plane electric fields. This results in

$$\nabla^2 u_z = 0, \qquad \nabla^2 \phi = 0 \qquad (3.5.67)$$

(1) Trefftz functions. It is well known that the solutions of the Laplace equation (3.5.67) may be found using the method of variable separation. By this method, the Trefftz functions are obtained as [51]

$$u_z(r,\theta) = \sum_{m=0}^{\infty} r^m (a_m \cos m\theta + b_m \sin m\theta) \qquad (3.5.68)$$

$$\phi(r,\theta) = \sum_{m=0}^{\infty} r^m (c_m \cos m\theta + d_m \sin m\theta) \qquad (3.5.69)$$

for a bounded region and

$$u_z(r,\theta) = a_0^* + a_0 \ln r + \sum_{m=1}^{\infty} r^{-m} (a_m \cos m\theta + b_m \sin m\theta) \qquad (3.5.70)$$

$$\phi(r,\theta) = c_0^* + c_0 \ln r + \sum_{m=1}^{\infty} r^{-m} (c_m \cos m\theta + d_m \sin m\theta) \qquad (3.5.71)$$

for an unbounded region, where r and θ are a pair of polar coordinates. Thus, the associated Trefftz complete sets of Eqs.(3.5.68)~(3.5.71) can be expressed in the form

$$T = \{1,\ r^m \cos m\theta,\ r^m \sin m\theta\} = \{T_i\} \qquad (3.5.72)$$

$$T = \{1,\ \ln r,\ r^{-m} \cos m\theta,\ r^{-m} \sin m\theta\} = \{T_i\} \qquad (3.5.73)$$

(2) Assumed fields. To perform FE analysis, the solution domain Ω is divided into elements, and over each element "e" two independent fields are assumed in the following way:

(a) The non-conforming intra-element field is expressed by

$$u = \begin{bmatrix} u_z \\ \phi \end{bmatrix} = \sum_{j=1}^{m} \begin{bmatrix} N_{1j} & 0 \\ 0 & N_{2j} \end{bmatrix} \begin{bmatrix} c_{uj} \\ c_{\phi j} \end{bmatrix} = \begin{bmatrix} N_1 & 0 \\ 0 & N_2 \end{bmatrix} c = Nc \qquad (3.5.74)$$

where c is a vector of undetermined coefficient, N_i are taken from the components of the series (3.5.68)~(3.5.71).

(b) An auxiliary conforming field

$$\tilde{u} = \begin{bmatrix} \tilde{u}_z \\ \tilde{\phi} \end{bmatrix} = \begin{bmatrix} \tilde{N}_1 & 0 \\ 0 & \tilde{N}_2 \end{bmatrix} \begin{bmatrix} d_u \\ d_\phi \end{bmatrix} + \begin{bmatrix} \tilde{N}_{1c} & 0 \\ 0 & \tilde{N}_{2c} \end{bmatrix} \begin{bmatrix} d_{uc} \\ d_{\phi c} \end{bmatrix} = \tilde{N}d + \tilde{N}_c d_c \qquad (3.5.75)$$

is independently assumed along the element boundary in terms of nodal DOF

$d = \begin{bmatrix} d_u & d_\phi \end{bmatrix}^{\mathrm{T}}$ and $d_c = \begin{bmatrix} d_{uc} & d_{\phi c} \end{bmatrix}^{\mathrm{T}}$, where $\tilde{N}$ represents the conventional finite element interpolating functions and $\tilde{N}_{1c}$, $\tilde{N}_{2c}$ are given in Eq.(3.5.75) above. For example, in a simple interpolation of the frame field on the side 1-C-2 of a particular element (Fig.3.3), the frame functions are defined in the following way:

$$\left. \begin{aligned} \tilde{u}_{z12} &= \tilde{N}_1 u_{z1} + \tilde{N}_2 u_{z2} + \sum_{J=1}^{M_u} \xi^{J-1}(1-\xi^2)u_{zCJ} \\ \tilde{\phi}_{12} &= \tilde{N}_1 \phi_1 + \tilde{N}_2 \phi_2 + \sum_{J=1}^{M_\phi} \xi^{J-1}(1-\xi^2)\phi_{CJ} \end{aligned} \right\} \tag{3.5.76}$$

where u_{zCJ} and ϕ_{CJ} are shown in Fig.3.3, and

$$\tilde{N}_1 = \frac{1-\xi}{2}, \quad \tilde{N}_2 = \frac{1+\xi}{2} \tag{3.5.77}$$

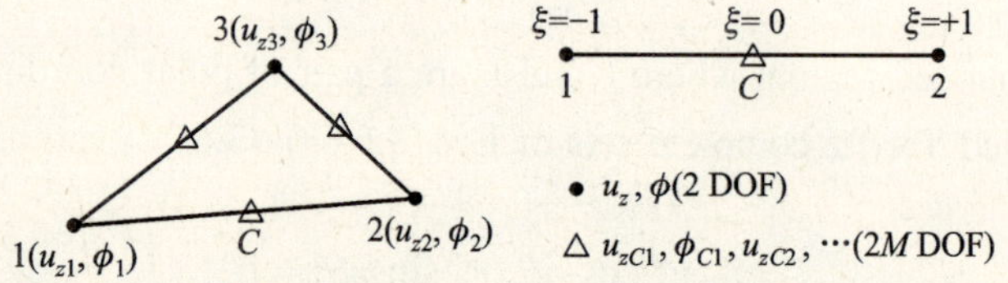

Fig.3.3   Geometry of a triangular element

Using the above definitions, the generalized boundary forces and electric displacements can be derived from Eqs.(3.5.61), (3.5.62) and (3.5.74), denoting

$$T = \begin{bmatrix} t \\ D_n \end{bmatrix} = \begin{bmatrix} \sigma_{3j} n_j \\ D_j n_j \end{bmatrix} = \begin{bmatrix} Q_1 \\ Q_2 \end{bmatrix} c = Qc \tag{3.5.78}$$

(3) Special element containing angular corner. It is well known that singularities induced by local defects such as angular corners, cracks, and so on, can be accurately accounted for in the conventional FE model by way of appropriate local refinement of the element mesh. However, an important feature of the HT FEM is that such problems can be far more efficiently handled by the use of special purpose functions [46]. Elements containing local defects (see Fig.3.4) are treated by simply replacing the standard regular functions N in Eq.(3.5.74) by appropriate special purpose functions. One common characteristic of such trial functions is that it is not only the governing differential equations, which here are Laplace equations, that are satisfied exactly, but also some prescribed boundary

conditions at a particular portion Γ_{eS} (see Fig.3.4) of the element boundary. This enables various singularities to be specifically taken into account without troublesome mesh refinement. Since the whole element formulation remains unchanged [except that now the frame function $\tilde{u}$ in Eq.(3.5.75) is defined and the boundary integration is performed only at the portion Γ_{e*} of the element boundary $\Gamma_e = \Gamma_{e*} + \Gamma_{eS}$, see Fig.3.4] [46], all that is needed to implement the elements containing such special trial functions is to provide the element subroutine of the standard, regular elements with a library of various optional sets of special purpose functions.

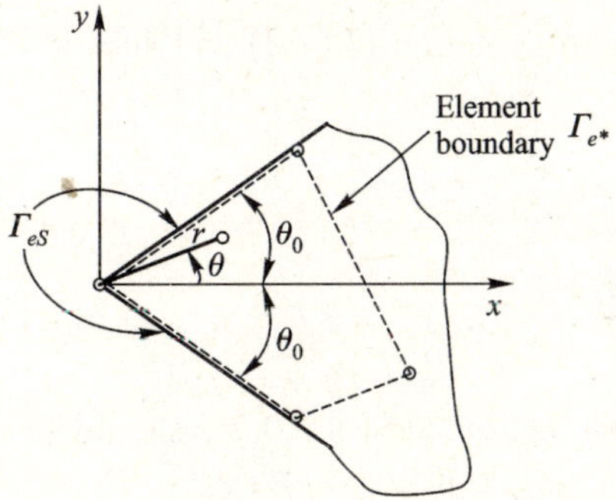

Fig. 3.4   Special element containing a singular corner

In this section we show how special purpose functions can be constructed to satisfy both the Laplace equation (3.5.67) and the traction-free boundary conditions on angular corner faces (Fig.3.4). The derivation of such functions is based on the general solution of the two-dimensional Laplace equation

$$u_z(r,\theta) = a_0 + \sum_{n=1}^{\infty}(a_n r^{\lambda_n} + b_n r^{-\lambda_n})\cos(\lambda_n\theta) + \sum_{n=1}^{\infty}(d_n r^{\lambda_n} + e_n r^{-\lambda_n})\sin(\lambda_n\theta)$$

(3.5.79)

$$\phi(r,\theta) = e_0 + \sum_{n=1}^{\infty}(e_n r^{\lambda_n} + f_n r^{-\lambda_n})\cos(\lambda_n\theta) + \sum_{n=1}^{\infty}(g_n r^{\lambda_n} + h_n r^{-\lambda_n})\sin(\lambda_n\theta)$$

(3.5.80)

Appropriate trial functions for a singular corner element are obtained by considering an infinite wedge (Fig.3.4) with particular boundary conditions prescribed along the sides $\theta=\pm\theta_0$ forming the angular corner. The boundary conditions on the upper and lower surfaces of the wedge are free of surface traction and surface charge

$$\sigma_{r\theta} = c_{44}\frac{\partial u_z}{r\partial\theta} + e_{15}\frac{\partial\phi}{r\partial\theta} = 0, \quad D_\theta = e_{15}\frac{\partial u_z}{r\partial\theta} - \kappa_{11}\frac{\partial\phi}{r\partial\theta} = 0 \tag{3.5.81}$$

This leads to

$$\frac{\partial u_z}{\partial\theta} = 0, \qquad \frac{\partial\phi}{\partial\theta} = 0, \qquad \theta = \pm\theta_0 \tag{3.5.82}$$

To solve this problem, we rewrite the general solution (3.5.79) as

$$u_z(r,\theta) = a_0 + \sum_{n=1}^{\infty}(a_n r^{\lambda_n} + b_n r^{-\lambda_n})\cos(\lambda_n\theta) + \sum_{n=1}^{\infty}(d_n r^{\beta_n} + e_n r^{-\beta_n})\sin(\beta_n\theta)$$

$$\tag{3.5.83}$$

where λ_n and β_n are two sets of constants which are assumed to be greater than zero. Differentiating solution (3.5.83) and substituting it into Eq.(3.5.82) yields

$$\frac{\partial u_z}{\partial\theta}\bigg|_{\theta=\pm\theta_0} = -\sum_{n=1}^{\infty}\lambda_n(a_n r^{\lambda_n} + b_n r^{-\lambda_n})\sin(\pm\lambda_n\theta_0) +$$

$$\sum_{n=1}^{\infty}\beta_n(d_n r^{\beta_n} + e_n b^{-\beta_n})\cos(\pm\beta_n\theta_0) = 0 \tag{3.5.84}$$

Since the solution must be limited for $r=0$, we should specify

$$b_n = e_n = 0 \tag{3.5.85}$$

From Eq.(3.5.84) it can be deduced that

$$\sin(\pm\lambda_n\theta_0) = 0, \qquad \cos(\pm\beta_n\theta_0) = 0 \tag{3.5.86}$$

leading to

$$\lambda_n\theta_0 = n\pi, \qquad n=1,2,3,\cdots \tag{3.5.87}$$

$$2\beta_n\theta_0 = n\pi, \qquad n=1,3,5,\cdots \tag{3.5.88}$$

Thus, for an element containing an edge crack (in this case $\theta_0 = \pi$), the solution can be written in the form

$$u_z(r,\theta) = a_0 + \sum_{n=1}^{\infty}a_n r^n \cos(n\theta) + \sum_{n=1,3,5}^{\infty} d_n r^{\frac{n}{2}}\sin(\frac{n}{2}\theta) \tag{3.5.89}$$

With the solution (3.5.89), the internal function defined in Eq.(3.5.74) can be taken as

$$N_{2n-1} = r^n \cos(n\theta), \quad N_{2n} = r^{\frac{2n-1}{2}}\sin\left(\frac{2n-1}{2}\theta\right), \quad n=1,2,3,\cdots \tag{3.5.90}$$

It is obvious that the displacement function (3.5.89) includes the term proportional to $r^{1/2}$, whose derivative is singular at the crack tip. The solution for the second equation of (3.5.82) can be obtained similarly.

(4) Variational principle. For the boundary value problem described by Eqs.(3.5.55)~(3.5.67), the corresponding dual variational functional is constructed in the form

$$\Theta_m^{\sigma D} = \sum_e \Theta_{me}^{\sigma D} = \sum_e \left\{ \Theta_e^{\sigma D} - \int_{\Gamma_{De}} (\bar{D}_n - D_n)\tilde{\phi}\,\mathrm{d}s - \int_{\Gamma_{te}} (\bar{t}-t)\tilde{u}_z\,\mathrm{d}s + \right.$$

$$\left. \int_{\Gamma_{Ie}} (D_n\tilde{\phi} + t\tilde{u}_z)\,\mathrm{d}s \right\} \tag{3.5.91}$$

$$\Theta_m^{\varepsilon E} = \sum_e \Theta_{me}^{\varepsilon E} = \sum_e \left\{ \Theta_e^{\varepsilon E} + \int_{\Gamma_{\phi e}} (\bar{\phi} - \phi)\tilde{D}_n\,\mathrm{d}s + \int_{\Gamma_{ue}} (\bar{u}_z - u_z)\tilde{t}\,\mathrm{d}s - \right.$$

$$\left. 2\int_{\Gamma_{te}} \tilde{u}_z t\,\mathrm{d}s - 2\int_{\Gamma_{De}} \tilde{\phi} D_n\,\mathrm{d}s - \int_{\Gamma_{Ie}} (\tilde{\phi} D_n + \tilde{u}_z t\,\mathrm{d}s \right\} \tag{3.5.92}$$

where

$$\Theta_e^{\sigma D} = \iint_{\Omega_e} H(\sigma_{ij}, D_k)\,\mathrm{d}\Omega + \int_{\Gamma_{ue}} t\bar{u}_z\,\mathrm{d}s + \int_{\Gamma_{\phi e}} D_n\bar{\phi}\,\mathrm{d}s \tag{3.5.93}$$

$$\Theta_e^{\varepsilon E} = \iint_{\Omega_e} H(\varepsilon_{ij}, E_k)\,\mathrm{d}\Omega + \int_{\Gamma_{te}} \bar{t}\tilde{u}_z\,\mathrm{d}s + \int_{\Gamma_{De}} \bar{D}_n\tilde{\phi}\,\mathrm{d}s \tag{3.5.94}$$

$$H(\sigma_{ij}, D_k) = -\frac{1}{2}s_{44}(\sigma_{xz}^2 + \sigma_{yz}^2) - g_{15}\sigma_{xz}D_x - g_{15}\sigma_{yz}D_y + \frac{1}{2}\lambda_{11}(D_x^2 + D_y^2)$$

$$\tag{3.5.95}$$

$$H(\gamma_{ij}, E_k) = \frac{1}{2}c_{44}(\gamma_{xz}^2 + \gamma_{yz}^2) - e_{15}\gamma_{xz}E_x - e_{15}\gamma_{yz}E_y - \frac{1}{2}\kappa_{11}(E_x^2 + E_y^2) \tag{3.5.96}$$

The boundary Γ_e of a particular element consists of the following parts:

$$\Gamma_e = \Gamma_{ue} \bigcup \Gamma_{te} \bigcup \Gamma_{Ie} = \Gamma_{\phi e} \bigcup \Gamma_{De} \bigcup \Gamma_{Ie} \tag{3.5.97}$$

where

$$\Gamma_{ue} = \Gamma_u \bigcap \Gamma_e, \quad \Gamma_{te} = \Gamma_t \bigcap \Gamma_e, \quad \Gamma_{\phi e} = \Gamma_\phi \bigcap \Gamma_e, \quad \Gamma_{De} = \Gamma_D \bigcap \Gamma_e$$

$$\tag{3.5.98}$$

and Γ_{Ie} is the inter-element boundary of the element "e".

(5) Generation of element matrix. Similar to the treatment of Eq. (3.5.45), the domain integral in Eq.(3.5.93) is converted into a boundary integral by use of solution properties of the intra-element trial functions, for which the functional (3.5.91) is rewritten as

$$\Pi_{me}^{\sigma D} = -\int_{\Gamma_{De}} (\bar{D}_n - D_n)\tilde{\phi}\,\mathrm{d}s - \int_{\Gamma_{te}} (\bar{t}-t)\tilde{u}_z\,\mathrm{d}s + \int_{\Gamma_{Ie}} (D_n\tilde{\phi} + t\tilde{u}_z)\,\mathrm{d}s\} -$$

$$\frac{1}{2}\int_{\Gamma_e} (tu_z + D_n\phi)\,\mathrm{d}s + \int_{\Gamma_{ue}} t\bar{u}_z\,\mathrm{d}s + \int_{\Gamma_{\phi e}} D_n\bar{\phi}\,\mathrm{d}s \tag{3.5.99}$$

Substituting the expressions given in Eqs.(3.5.74),(3.5.75) and (3.5.78) into (3.5.99) produces

$$\Pi_{me}^{\sigma D} = -\frac{1}{2}\boldsymbol{c}^{\mathrm{T}}\boldsymbol{Hc} + \boldsymbol{c}^{\mathrm{T}}\boldsymbol{Sd} + \boldsymbol{c}^{\mathrm{T}}\boldsymbol{r}_1 + \boldsymbol{d}^{\mathrm{T}}\boldsymbol{r}_2 + \text{terms without } \boldsymbol{c} \text{ or } \boldsymbol{d} \qquad (3.5.100)$$

in which the matrices $\boldsymbol{H}$, $\boldsymbol{S}$ and the vectors $\boldsymbol{r}_1$, $\boldsymbol{r}_2$ are defined by

$$\boldsymbol{H} = \int_{\Gamma_e} \boldsymbol{Q}^{\mathrm{T}}\boldsymbol{N}\mathrm{d}s \qquad (3.5.101)$$

$$\boldsymbol{S} = \int_{\Gamma_{De}} \boldsymbol{Q}_2^{\mathrm{T}}\tilde{\boldsymbol{N}}_2\mathrm{d}s + \int_{\Gamma_{te}} \boldsymbol{Q}_1^{\mathrm{T}}\tilde{\boldsymbol{N}}_1\mathrm{d}s + \int_{\Gamma_{Ie}} \boldsymbol{Q}^{\mathrm{T}}\tilde{\boldsymbol{N}}\mathrm{d}s \qquad (3.5.102)$$

$$\boldsymbol{r}_1 = \int_{\Gamma_{\phi e}} \boldsymbol{Q}_2^{\mathrm{T}}\overline{\phi}\,\mathrm{d}s + \int_{\Gamma_{ue}} \boldsymbol{Q}_1^{\mathrm{T}}\overline{u}_z\mathrm{d}s \qquad (3.5.103)$$

$$\boldsymbol{r}_2 = -\int_{\Gamma_{De}} \tilde{\boldsymbol{N}}_2^{\mathrm{T}}\overline{D}_n\mathrm{d}s - \int_{\Gamma_{te}} \tilde{\boldsymbol{N}}_1^{\mathrm{T}}\overline{t}\mathrm{d}s \qquad (3.5.104)$$

The remaining derivation and the resulting equations are in the same form as in Eqs.(3.5.51)~(3.5.54).

3.5.6 Numerical examples

As a numerical illustration of the finite element formulation presented in this section, an example of a piezoelectric prism subjected to simple tension is considered (see Fig.3.5). This example was taken from [52] for a PZT-4 ceramic prism subject to a tension $P=10$ Nm^{-2} in the y-direction. The properties of the material are given as follows:

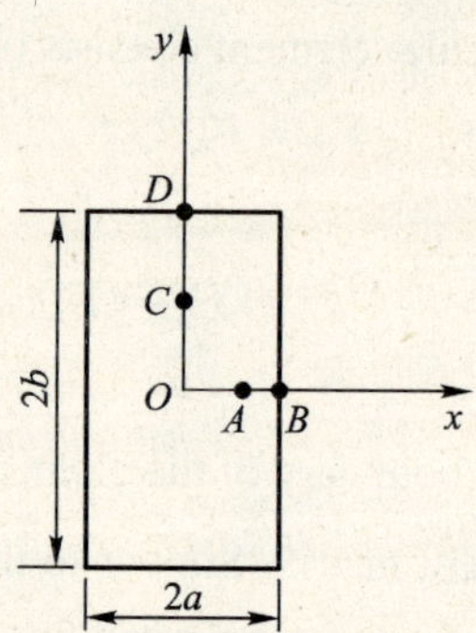

Fig.3.5 Geometry of the piezoelectric prism

$$c_{1111} = 12.6\times10^{10}\ \mathrm{Nm}^{-2},\quad c_{1122} = 7.78\times10^{10}\ \mathrm{Nm}^{-2},\quad c_{1133} = 7.43\times10^{10}\ \mathrm{Nm}^{-2}$$

$$c_{3333} = 11.5\times10^{10}\ \mathrm{Nm}^{-2},\quad c_{3232} = 2.56\times10^{10}\ \mathrm{Nm}^{-2},\quad e_{131} = 12.7\ \mathrm{Cm}^{-2}$$

$$e_{311} = -5.2\ \mathrm{Cm}^{-2},\quad e_{333} = 15.1\ \mathrm{Cm}^{-2},\quad \kappa_{11} = 730\kappa_0,\quad \kappa_{33} = 635\kappa_0$$

where $\kappa_0 = 8.854\times10^{-12}\ \mathrm{C}^2/\mathrm{Nm}^2$. The boundary conditions of the prism are

$$\sigma_{yy} = P,\quad \sigma_{xy} = D_y = 0, \text{ on edges } y = \pm b$$

$$\sigma_{xx} = \sigma_{xy} = D_x = 0 \text{ , on edges } x = \pm a$$

where a=3 m, b=10 m. Owing to the symmetry about load, boundary conditions and geometry, only one quadrant of the prism is modeled by 10 (x-direction) × 20 (y-direction) elements in the HT FEM analysis. Table 3.2 lists the displacements and electric potential at points A, B, C, and D using the present method and comparison is made with analytical results. It is shown that the TFEM results are in good agreement with the analytical ones [52].

Table 3.2 u_1, u_2, and ϕ of TFEM results and comparison with exact solution

Point		$A(2,0)$	$B(3,0)$	$C(0,5)$	$D(0,10)$
TFEM	$u_1/(10^{10}\text{m})$	−0.9674	−1.4510	0	0
	$u_2/(10^{9}\text{m})$	0	0	0.5009	1.0016
	$\phi(V)$	0	0	0.6890	1.3779
Exact [52]	$u_1/(10^{10}\text{m})$	−0.9672	−1.4508	0	0
	$u_2/(10^{9}\text{m})$	0	0	0.5006	1.0011
	$\phi(V)$	0	0	0.6888	1.3775

3.6   Theory of coupled thermo-piezoelectricity

In the previous sections of this chapter we described various problems of piezoelectric materials without considering thermal effects. In this section, an extension to include the thermal effect is presented. We begin with a discussion of the general theory of thermo-piezoelectricity, followed by an introduction of the uniqueness of the thermo-electro-elastic solution. The presentation focuses on the developments in [1,2].

3.6.1   Basic equations

The equations of the classical, linear theory of piezoelectricity, including the coupling among deformation, temperature, and electric field, were derived by Mindlin [1]. The coupling problem under consideration consists of determining the stress $\sigma_{ij}(x, t)$, electric displacement $D_i(x, t)$, elastic displacement $u_i(x, t)$ temperature $T(x,t)$ and electric potential $\phi(x,t)$ for $x \in \Omega$ and $t>0$.

In the region Ω and for $t>0$ without body force and free charge, the following equations should be satisfied:

(1) Divergence equations.

$$\sigma_{ij,j} = \rho \ddot{u}_i, \qquad D_{i,i} = 0, \qquad h_{i,i} = -T_0 \Delta s \qquad (3.6.1)$$

where ρ is the mass density. Using the notation introduced in Section 3.3 and considering a steady-state problem, Eq.(3.6.1) can be rewritten in a simple form

$$E_{iJKm} U_{K,mi} = \lambda_{iJ} T_{,i}, \qquad k_{ij} T_{,ij} = 0 \qquad (3.6.2)$$

where λ_{iJ} $(J=1,2,3)$ represent the thermal-stress constants, $\lambda_{i4} = p_i$ stands for the pyroelectric coefficient.

(2) Gradient equations.

$$\varepsilon_{ij} = \frac{1}{2}(u_{i,j} + u_{j,i}), \qquad E_i = -\phi_{,i}, \qquad h_i = -k_{ij} T_{,j} \qquad (3.6.3)$$

where k_{ij} is the heat conduction coefficient.

(3) Constitutive equations.

$$s = -\left[\frac{\partial g}{\partial T}\right]_{\varepsilon,E}, \qquad \sigma_{ij} = \left[\frac{\partial g}{\partial \varepsilon_{ij}}\right]_{T,E}, \qquad D_m = -\left[\frac{\partial g}{\partial E_m}\right]_{\varepsilon,T} \qquad (3.6.4)$$

where g is the "electric Gibbs function" defined by

$$g = \frac{1}{2} c_{ijkl} \varepsilon_{ij} \varepsilon_{kl} - \frac{1}{2} \kappa_{ij} E_i E_j - \frac{\rho C_v}{2T_0} T^2 - e_{ijk} E_i \varepsilon_{jk} - p_m T E_m - \lambda_{ij} T \varepsilon_{ij} \qquad (3.6.5)$$

Eqs.(3.6.1)~(3.6.4) comprise the 27 equations of linear thermo-piezoelectricity governing the 27 dependent variables u_i, σ_{ij}, ε_{ij}, D_i, E_i, ϕ, h_i, s, T. From Eq.(3.6.4) and Eq.(3.6.5), we find

$$s = \alpha T + \lambda_{ij} \varepsilon_{ij} + p_m E_m$$

$$\sigma_{ij} = -\lambda_{ij} T + c_{ijkl} \varepsilon_{kl} - e_{mij} E_m \qquad (3.6.6)$$

$$D_n = p_n T + e_{nij} \varepsilon_{ij} + \kappa_{mn} E_m$$

where $\alpha = \rho C_v T_0^{-1}$.

By successive substitution, the 27 equations may be reduced to five on u_i, ϕ and T

$$c_{ijkl} u_{k,li} + e_{kij} \phi_{,ki} - \lambda_{ij} T_{,i} = \rho \ddot{u}_j$$

$$e_{kij} u_{i,jk} - \kappa_{ij} \phi_{,ij} + p_i T_{,i} = 0 \qquad (3.6.7)$$

$$\lambda_{ij} \dot{u}_{i,j} - p_i \dot{\phi}_{,i} + \alpha \dot{T} = \frac{k_{ij} T_{,ij}}{T_0}$$

These equations should be completed with boundary and initial conditions. The following quantities may be assigned at the surface Γ of the body Ω:

(1) Displacement or surface traction.

$$u_i = \bar{u}_i(x, t) \quad (\text{on } \Gamma_1), \quad \sigma_{ij}n_j = \bar{t}_i(x, t) \quad (\text{on } \Gamma_2) \tag{3.6.8}$$

where $\bar{u}_i$ and $\bar{t}_i$ are known functions, and $\Gamma = \Gamma_1 \cup \Gamma_2$, $\Gamma_1 \cap \Gamma_2 = 0$.

(2) Electric potential or electric displacement.

$$\phi = \bar{\phi}(x, t) \quad (\text{on } \Gamma_3), \quad D_j n_j = -\bar{q}_s(x, t) \quad (\text{on } \Gamma_4) \tag{3.6.9}$$

where $\bar{\phi}$ and $\bar{q}_s$ are known functions, and $\Gamma = \Gamma_3 \cup \Gamma_4$, $\Gamma_3 \cap \Gamma_4 = 0$.

(3) Temperature or heat flux.

$$T = \bar{T}(x, t) \quad (\text{on } \Gamma_5), \quad -k_{ij}T_{,j}n_i = -\bar{h}_n(x, t) \quad (\text{on } \Gamma_6) \tag{3.6.10}$$

where $\bar{T}$ and $\bar{h}_n$ are known functions, and $\Gamma = \Gamma_5 \cup \Gamma_6$, $\Gamma_5 \cap \Gamma_6 = 0$.

(4) Initial conditions.

$$u_i(x, 0) = f_i(x), \quad \dot{u}_i(x, 0) = g_i(x), \quad T(x, 0) = T_0(x). \tag{3.6.11}$$

where f_i, g_i and T_0 are known functions.

3.6.2 Uniqueness of the solution

The uniqueness theorem for the differential equations of thermo-piezo-electricity can be established by way of the principle of virtual work. The energy functional used for this purpose is as follows:

$$-\int_\Omega \rho \ddot{u}_i \delta u_i \, d\Omega + \int_\Gamma T_i \delta u_i \, d\Gamma = \int_\Omega \sigma_{ij} \delta \varepsilon_{ij} \, d\Omega \tag{3.6.12}$$

in which the virtual increments have been replaced by the real increments

$$\delta u_i = \frac{\partial u_i}{\partial t} dt = \dot{u}_i dt, \quad \delta \varepsilon_{ij} = \frac{\partial \varepsilon_{ij}}{\partial t} dt = \dot{\varepsilon}_{ij} dt \tag{3.6.13}$$

Thus, we obtain the fundamental energy equation

$$-\int_\Omega \rho \ddot{u}_i \dot{u}_i \, d\Omega + \int_\Gamma T_i \dot{u}_i \, d\Gamma = \int_\Omega \sigma_{ij} \dot{\varepsilon}_{ij} \, d\Omega \tag{3.6.14}$$

into which we introduce the constitutive relations

$$\sigma_{ij} = -\lambda_{ij}T + c_{ijkl}\varepsilon_{kl} - e_{mij}E_m \tag{3.6.15}$$

Hence

$$\frac{d}{dt}(K + W) = \int_\Gamma T_i \dot{u}_i \, d\Gamma + \int_\Omega (\lambda_{ij}T + e_{mij}E_m)\dot{\varepsilon}_{ij} \, d\Omega \tag{3.6.16}$$

where K is the kinetic energy and W the work of deformation

$$K = \frac{\rho}{2}\int_\Omega \dot{u}_i \dot{u}_i \, d\Omega, \quad W = \frac{1}{2}\int_\Omega c_{ijkl}\varepsilon_{ij}\varepsilon_{kl} \, d\Omega \tag{3.6.17}$$

To eliminate the term $\int_\Omega \lambda_{ij}\dot{\varepsilon}_{ij}T \, d\Omega$, we consider the heat conduction equation

$$\frac{k_{ij}T_{,ij}}{T_0} - \alpha\dot{T} = \lambda_{ij}\dot{\varepsilon}_{ij} + \rho_i\dot{E}_i \tag{3.6.18}$$

Multiplying by T and integrating over the region Ω, after simple transformation we obtain

$$\int_{\Omega}\lambda_{ij}T\dot{\varepsilon}_{ij}\mathrm{d}\Omega = \frac{k_{ij}}{T_0}\int_{\Gamma}TT_{,j}n_i\mathrm{d}\Gamma - \rho_m\int_{\Omega}T\dot{E}_m\mathrm{d}\Omega - \frac{\mathrm{d}\mathcal{R}}{\mathrm{d}t} - \chi_{\theta} \tag{3.6.19}$$

where

$$\mathcal{R} = \frac{\alpha}{2}\int_{\Omega}T^2\mathrm{d}\Omega, \qquad \chi_{\theta} = \frac{k_{ij}}{T_0}\int_{\Omega}T_{,i}T_{,j}\mathrm{d}\Omega \tag{3.6.20}$$

Substituting Eq.(3.6.19) into Eq.(3.6.16) yields

$$\frac{\mathrm{d}}{\mathrm{d}t}(K + W + \mathcal{R}) + \chi_{\theta} = \int_{\Gamma}\left(T_i\dot{u}_i + \frac{k_{ij}}{T_0}TT_{,j}n_i\right)\mathrm{d}\Gamma + \int_{\Omega}(e_{mij}E_m\dot{\varepsilon}_{ij} - \rho_i\dot{E}_iT)\mathrm{d}\Omega$$

$$\tag{3.6.21}$$

To eliminate the term $\int_{\Omega}e_{mij}\dot{\varepsilon}_{ij}E_m\mathrm{d}\Omega$ in Eq.(3.6.21), we make use of the constitutive relations

$$D_n = \rho_nT + e_{nij}\varepsilon_{ij} + \kappa_{mn}E_m \tag{3.6.22}$$

Finally, we make use of the equation of the electric field $\dot{D}_{i,i} = 0$. Multiplying the equation by ϕ and integrating over the region Ω, we obtain

$$\int_{\Gamma}\dot{D}_in_i\phi\mathrm{d}\Gamma + \int_{\Omega}\dot{D}_mE_m\mathrm{d}\Omega = 0 \tag{3.6.23}$$

Using relation (3.6.22), after simple transformations we obtain

$$\int_{\Omega}(e_{mij}E_m\dot{\varepsilon}_{ij} - \rho_i\dot{E}_iT)\mathrm{d}\Omega = -\int_{\Gamma}\dot{D}_mn_m\phi\mathrm{d}\Gamma - \frac{\mathrm{d}}{\mathrm{d}t}\left(\aleph - \rho_k\int_{\Omega}TE_k\mathrm{d}\Omega\right) \tag{3.6.24}$$

where

$$\aleph = \frac{\kappa_{ij}}{2}\int_{\Omega}E_iE_j\mathrm{d}\Omega \tag{3.6.25}$$

In view of Eqs.(3.6.21)~(3.6.23), we arrive at the modified energy balance

$$\frac{\mathrm{d}}{\mathrm{d}t}\left(K + W + \mathcal{R} + \aleph + \rho_k\int_{\Omega}TE_k\mathrm{d}\Omega\right) + \chi_{\theta}$$

$$= \int_{\Gamma}\left(T_i\dot{u}_i + \frac{k_{ij}}{T_0}TT_{,j}n_i - \dot{D}_m\phi n_m\right)\mathrm{d}\Gamma \tag{3.6.26}$$

The energy functional (3.6.26) makes possible the proof of the uniqueness of the solutions.

Consider two distinct solutions (u_i',ϕ',T') and (u_i'',ϕ'',T'') which satisfy Eq.(3.6.1) and the appropriate boundary and initial conditions. Let

$$u_i^* = u_i' - u_i'', \qquad \phi^* = \phi' - \phi'', \qquad T^* = T' - T'' \tag{3.6.27}$$

Since the problem is linear, the difference variables in Eq.(3.6.27) are also solutions. Therefore, Eq. (3.6.26) holds for the solution (u_i^*, ϕ^*, T^*).

In view of the homogeneity of the equations and the boundary conditions, the right-hand side of Eq.(3.6.26) vanishes. Hence,

$$\frac{d}{dt}\left(K^* + W^* + \mathcal{R}^* + \aleph^* + \rho_k \int_\Omega T^* E_k^* d\Omega\right) = -\chi_\theta^* \leqslant 0 \tag{3.6.28}$$

where we have made use of the fact that the integrand of the energy-dissipation function χ_θ is a positive-definite quadratic form. The integral in the left-hand side of Eq.(3.6.28) vanishes at the outset, since the variables (u_i^*, ϕ^*, T^*) satisfy the homogeneous initial conditions. On the other hand, the inequality in Eq.(3.6.28) proves that its left-hand side is either negative or zero. The latter possibility occurs if the integrand is the sum of squares.

Consequently, we assume that

$$K^* = W^* = 0, \qquad \mathcal{R}^* + \aleph^* + \rho_k \int_\Omega T^* E_k^* d\Omega \geqslant 0 \tag{3.6.29}$$

These results imply that

$$u_i^* = \varepsilon_{ij}^* = T^* = E_k^* = 0 \tag{3.6.30}$$

Assuming that κ_{ij} is a known positive-definite symmetric tensor, ρ_k is a vector, and $\alpha > 0$. Consider the function

$$A(T, E_k) = \alpha T^2 + 2\rho_k T E_k + \kappa_{ij} E_i E_j \tag{3.6.31}$$

A is nonnegative $(A \geqslant 0)$ for every real pair (T, E_k), provided that

$$|\rho_i|^2 \leqslant \alpha \lambda_m \tag{3.6.32}$$

where λ_m is the smallest positive eigenvalue of the tensor κ_{ij}. Eq.(3.6.30) implies the uniqueness of the solutions of the thermo-piezoelectricity equations, i.e.,

$$u_i' = u_i'', \qquad \phi' = \phi'', \qquad T' = T'' \tag{3.6.33}$$

3.7   Solutions by Fourier transform method

The boundary-initial value problems described in the previous section can be solved by means of the Fourier transform approach. Hereafter, for simplicity, we assume that all variables do not vary with time. In this case the problem defined in the previous section is known as the boundary value problem. It should be noted that the Fourier transform approach to thermo-piezoelectric problems usually involves two basic steps: (1) solve a heat transfer problem first to obtain the steady-state T field; (2) calculate the electroelastic field caused by the T field, then add an isothermal solution to satisfy the corresponding electrical and mechanical boundary conditions, and finally, solve the modified problem for electroelastic fields. In this section, we first derive the Fourier transform formulation for temperature fields and then extend it to the case of thermo-electro-elasticity.

3.7.1 Fourier transform method and induced general solution

The Fourier transform pair used in this section is defined by [10,11]

$$\hat{f}(\xi) = \frac{1}{2\pi} \int_{-\infty}^{\infty} f(x)e^{i\xi x}dx, \qquad f(x) = \frac{1}{2\pi} \int_{-\infty}^{\infty} \hat{f}(\xi)e^{-i\xi x}d\xi \qquad (3.7.1)$$

where $i = \sqrt{-1}$. Applying the transform (3.7.1) to Eq.(3.6.2) with respect to x_1, leads to

$$\xi^2 k_{11}\hat{T} + 2i\xi k_{12}\frac{\partial \hat{T}}{\partial x_2} - k_{22}\frac{\partial^2 \hat{T}}{\partial x_2^2} = 0 \qquad (3.7.2)$$

Eq.(3.7.2) admits a solution of the form

$$\hat{T} = a_0(\xi)e^{-i\tau x_2} \qquad (3.7.3)$$

provided that τ satisfies the following eigenvalue equation:

$$k_{11}\xi^2 + 2k_{12}\xi\tau + k_{22}\tau^2 = 0 \qquad (3.7.4)$$

The roots of Eq.(3.7.4) are

$$p_1^* = -\frac{k_{12}}{k_{22}} + i\frac{k}{k_{22}}, \qquad p_2^* = \bar{p}_1^*, \qquad k = (k_{11}k_{22} - k_{12}^2)^{1/2} \qquad (3.7.5)$$

where $p_i^* = \tau_i/\xi$ and the overbar denotes the complex conjugate. For a given real ξ, τ_1 and τ_2 can be defined such that for $\mathrm{Im}(\tau_1) > 0$, the results are

$$\tau_1 = \begin{cases} p_1^*\xi, & \xi > 0 \\ \bar{p}_1^*\xi, & \xi < 0 \end{cases} \qquad (3.7.6)$$

As a consequence, the general solution of Eq. (3.7.2) can be written as

$$\hat{T} = \sqrt{2\pi}(F_0 f_0 + G_0 g_0) \qquad (3.7.7)$$

where

$$F_0 = e^{-i\tau x_2}, \qquad G_0 = e^{-i\bar{\tau} x_2} \qquad (3.7.8)$$

Recall that ξ and τ are contained in the eigenvalue $p = \tau/\xi$, while f_0 and g_0 are two functions of ξ to be determined from the boundary conditions of a given problem. The transformed heat flux $\hat{h}_i = -k_{ij}\hat{T}_{,j}$ is given by

$$\hat{h} = i\xi\sqrt{2\pi}\left[(\boldsymbol{k}_I + p_1^*\boldsymbol{k}_{II})F_0 f_0 + (\boldsymbol{k}_I + \bar{p}_1^*\boldsymbol{k}_{II})G_0 g_0\right]H(\xi) +$$

$$i\xi\sqrt{2\pi}\left[(\boldsymbol{k}_I + \bar{p}_1^*\boldsymbol{k}_{II})F_0 f_0 + (\boldsymbol{k}_I + p_1^*\boldsymbol{k}_{II})G_0 g_0\right]H(-\xi) \qquad (3.7.9)$$

where $\boldsymbol{h} = [h_1 \quad h_2]^{\mathrm{T}}$, $\boldsymbol{k}_I = [k_{11} \quad k_{12}]^{\mathrm{T}}$, $\boldsymbol{k}_{II} = [k_{21} \quad k_{22}]^{\mathrm{T}}$ and $H(\xi)$ is the Heaviside step function. Eq.(3.7.7) and Eq.(3.7.9) represent the general solution for the temperature and heat flux fields in the Fourier transform space. Taking the inverse Fourier transform, the results are

$$T(x_1, x_2) = \int_{-\infty}^{\infty} (F_0 f_0 + G_0 g_0) e^{-i\xi x_1} \mathrm{d}\xi \qquad (3.7.10)$$

$$h(x_1, x_2) = i\int_0^{\infty} \xi \left[(\boldsymbol{k}_I + p_1^*\boldsymbol{k}_{II})F_0 f_0 + (\boldsymbol{k}_I + \bar{p}_1^*\boldsymbol{k}_{II})G_0 g_0\right] e^{-i\xi x_1} \mathrm{d}\xi +$$

$$i\int_{-\infty}^{0} \xi \left[(\boldsymbol{k}_I + \bar{p}_1^*\boldsymbol{k}_{II})F_0 f_0 + (\boldsymbol{k}_I + p_1^*\boldsymbol{k}_{II})G_0 g_0\right] e^{-i\xi x_1} \mathrm{d}\xi \qquad (3.7.11)$$

Similarly, applying Eq.(3.7.1) to Eq.(3.6.2) with respect to x_1, we have

$$\xi^2 \boldsymbol{Q}\hat{\boldsymbol{U}} + i\xi(\boldsymbol{R} + \boldsymbol{R}^{\mathrm{T}})\frac{\partial \hat{\boldsymbol{U}}}{\partial x_2} - \boldsymbol{T}\frac{\partial^2 \hat{\boldsymbol{U}}}{\partial x_2^2} = i\xi\boldsymbol{\lambda}_1 T - \boldsymbol{\lambda}_2 \frac{\partial \hat{T}}{\partial x_2} \qquad (3.7.12)$$

where $\boldsymbol{\lambda}_i = [\lambda_{i1} \quad \lambda_{i2} \quad \lambda_{i3} \quad \rho_i]$. The solution to Eq.(3.7.12) can be assumed to consist of a particular part $\hat{\boldsymbol{U}}_p$ and a homogeneous part $\hat{\boldsymbol{U}}_h$ as

$$\hat{\boldsymbol{U}} = \boldsymbol{U}_h + \boldsymbol{U}_p \qquad (3.7.13)$$

since it is a linear problem.

Making use of the solution (3.7.7), the particular solution $\hat{\boldsymbol{U}}_p$, which satisfies Eq.(3.7.12) can be assumed in the form

$$\hat{\boldsymbol{U}}_p = -\frac{\sqrt{2\pi}}{i\xi}(\boldsymbol{A}_0 F_0 f_0 + \bar{\boldsymbol{A}}_0 G_0 g_0)H(\xi) - \frac{\sqrt{2\pi}}{i\xi}(\bar{\boldsymbol{A}}_0 F_0 f_0 + \boldsymbol{A}_0 G_0 g_0)H(-\xi)$$

$$(3.7.14)$$

where

$$\boldsymbol{A}_0 = \boldsymbol{D}^{-1}(p_1^*)(\boldsymbol{\lambda}_1 + p_1^*\boldsymbol{\lambda}_2), \qquad \boldsymbol{D}(x) = \boldsymbol{Q} + x(\boldsymbol{R} + \boldsymbol{R}^{\mathrm{T}}) + x^2\boldsymbol{T} \qquad (3.7.15)$$

The homogeneous part $\hat{\boldsymbol{U}}_h$ can be obtained by considering an arbitrary eigenfunction of the form

$$\hat{\boldsymbol{U}}_h = \boldsymbol{A}e^{-i\eta x_2} \qquad (3.7.16)$$

Substituting Eq.(3.7.16) into the left-hand side of Eq.(3.7.12), it is found that

$$\left[\boldsymbol{Q}\xi^2 + \xi\eta(\boldsymbol{R} + \boldsymbol{R}^{\mathrm{T}}) + \eta^2\boldsymbol{T}\right]\boldsymbol{A} = \boldsymbol{0} \qquad (3.7.17)$$

which is exactly the same as Eq.(3.3.7) if we put $p = \eta/\xi$. The eigenvalue p can be determined by considering the characteristic determinant of Eq.(3.7.17)

$$\left\| \mathbf{Q} + (\mathbf{R} + \mathbf{R}^{\mathrm{T}})p + p^2 \mathbf{T} \right\| = 0 \tag{3.7.18}$$

As was noted in Section 3.3, there are eight eigenvalues p from Eq.(3.7.18), which consists of four pairs of complex conjugates. Let

$$\eta_M = \begin{cases} p_M \xi, & \xi > 0 \\ \bar{p}_M \xi, & \xi < 0 \end{cases} \tag{3.7.19}$$

where $M=1, 2, 3, 4$. It is obvious that $\mathrm{Im}(\eta_M) > 0$ for all ξ. Such a definition is expedient for development of the subsequent derivation. Hence, a general solution of Eq.(3.7.12) can be obtained by simple summation of the two parts of the solution

$$\hat{U} = \sqrt{2\pi}(\mathbf{A}\mathbf{F}\mathbf{f} + \bar{\mathbf{A}}\mathbf{G}\mathbf{g})H(\xi) + \sqrt{2\pi}(\bar{\mathbf{A}}\mathbf{F}\mathbf{f} + \mathbf{A}\mathbf{G}\mathbf{g})H(-\xi) - $$
$$\frac{\sqrt{2\pi}}{i\xi}(\mathbf{A}_0 F_0 f_0 + \bar{\mathbf{A}}_0 G_0 g_0)H(\xi) - \frac{\sqrt{2\pi}}{i\xi}(\bar{\mathbf{A}}_0 F_0 f_0 + \mathbf{A}_0 G_0 g_0)H(-\xi) \tag{3.7.20}$$

where

$$\mathbf{F}(\xi, x_2) = \left\langle F_\alpha(\xi, x_2) \right\rangle = \left\langle \mathrm{e}^{-i\eta_\alpha x_2} \right\rangle \tag{3.7.21}$$

$$\mathbf{G}(\xi, x_2) = \left\langle G_\alpha(\xi, x_2) \right\rangle = \left\langle \mathrm{e}^{-i\bar{\eta}_\alpha x_2} \right\rangle \tag{3.7.22}$$

The transformed stress and electric displacements follow from the constitutive relation (3.6.6)

$$\hat{\Pi}_1 = i\xi\sqrt{2\pi}(\mathbf{B}\mathbf{P}\mathbf{F}\mathbf{f} + \bar{\mathbf{B}}\bar{\mathbf{P}}\mathbf{G}\mathbf{g})H(\xi) + i\xi\sqrt{2\pi}(\bar{\mathbf{B}}\bar{\mathbf{P}}\mathbf{F}\mathbf{f} + \mathbf{B}\mathbf{P}\mathbf{G}\mathbf{g})H(-\xi) - $$
$$\sqrt{2\pi}(\mathbf{B}_0 p_1^* F_0 f_0 + \bar{\mathbf{B}}_0 \bar{p}_1^* G_0 g_0)H(\xi) - \sqrt{2\pi}(\bar{\mathbf{B}}_0 \bar{p}_1^* F_0 f_0 + \mathbf{B}_0 p_1^* G_0 g_0)H(-\xi) \tag{3.7.23}$$

$$\hat{\Pi}_2 = -i\xi\sqrt{2\pi}(\mathbf{B}\mathbf{F}\mathbf{f} + \bar{\mathbf{B}}\mathbf{G}\mathbf{g})H(\xi) - i\xi\sqrt{2\pi}(\bar{\mathbf{B}}\mathbf{F}\mathbf{f} + \mathbf{B}\mathbf{G}\mathbf{g})H(-\xi) + $$
$$\sqrt{2\pi}(\mathbf{B}_0 F_0 f_0 + \bar{\mathbf{B}}_0 G_0 g_0)H(\xi) - \sqrt{2\pi}(\bar{\mathbf{B}}_0 F_0 f_0 + \mathbf{B}_0 G_0 g_0)H(-\xi) \tag{3.7.24}$$

where

$$\mathbf{B}_0 = \mathbf{R}^{\mathrm{T}} \mathbf{A}_0 + \mathbf{T}\mathbf{A}_0 p_1^* \tag{3.7.25}$$

The traction vector on a surface with normal $\mathbf{n} = [n_1 \quad n_2 \quad 0]$ can be found from Eq.(3.7.23) and Eq.(3.7.24) as

$$\hat{t} = \hat{\Pi}_1 n_1 + \hat{\Pi}_2 n_2 = i\xi\sqrt{2\pi}\left[\mathbf{B}(n_1\mathbf{P} - n_2\mathbf{I})\mathbf{F}\mathbf{f} + \bar{\mathbf{B}}(n_1\bar{\mathbf{P}} - n_2\mathbf{I})\mathbf{G}\mathbf{g} \right]H(\xi) + $$
$$i\xi\sqrt{2\pi}\left[\bar{\mathbf{B}}(n_1\bar{\mathbf{P}} - n_2\mathbf{I})\mathbf{F}\mathbf{f} + \mathbf{B}(n_1\mathbf{P} - n_2\mathbf{I})\mathbf{G}\mathbf{g} \right]H(-\xi) - $$
$$\sqrt{2\pi}\left[\mathbf{B}_0(n_1 p_1^* - n_2)F_0 f_0 + \bar{\mathbf{B}}_0(n_1 \bar{p}_1^* - n_2)G_0 g_0 \right]H(\xi) - $$
$$\sqrt{2\pi}\left[\bar{\mathbf{B}}_0(n_1 \bar{p}_1^* - n_2)F_0 f_0 + \mathbf{B}_0(n_1 p_1^* - n_2)G_0 g_0 \right]H(-\xi) \tag{3.7.26}$$

Eqs.(3.7.20), (3.7.23) and (3.7.24) represent the solution for the elastic and electric fields in the Fourier transform space. The general solution of electro-elastic fields in real space is obtained by applying the inverse Fourier transform to Eqs. (3.7.20), (3.7.23), (3.7.24) and (3.7.26). The results are

$$U(x_1,x_2) = \int_0^\infty \left[AFf + \bar{A}Gg - \frac{1}{i\xi}(A_0 F_0 f_0 + \bar{A}_0 G_0 g_0) \right] e^{-i\xi x_1} d\xi +$$

$$\int_{-\infty}^0 \left[\bar{A}Ff + AGg - \frac{1}{i\xi}(\bar{A}_0 F_0 f_0 + A_0 G_0 g_0) \right] e^{-i\xi x_1} d\xi \qquad (3.7.27)$$

$$\Pi_1(x_1,x_2) = \int_0^\infty \left[i(BPFf + \bar{B}\bar{P}Gg)\xi - B_0 p_1^* F_0 f_0 - \bar{B}_0 \bar{p}_1^* G_0 g_0 \right] e^{-i\xi x_1} d\xi +$$

$$\int_{-\infty}^0 \left[i(\bar{B}\bar{P}Ff + BPGg)\xi - \bar{B}_0 \bar{p}_1^* F_0 f_0 - B_0 p_1^* G_0 g_0 \right] e^{-i\xi x_1} d\xi$$

$$(3.7.28)$$

$$\Pi_2(x_1,x_2) = -\int_0^\infty \left[i(BFf + \bar{B}Gg)\xi - B_0 F_0 f_0 - \bar{B}_0 p G_0 g_0 \right] e^{-i\xi x_1} d\xi -$$

$$\int_{-\infty}^0 \left[i(\bar{B}Ff + BGg)\xi - \bar{B}_0 F_0 f_0 - B_0 G_0 g_0 \right] e^{-i\xi x_1} d\xi$$

$$(3.7.29)$$

$$t(x_1,x_2) = \int_0^\infty \left\{ i\xi \left[B(n_1 P - n_2 I)Ff + \bar{B}(n_1 \bar{P} - n_2 I)Gg \right] - \right.$$

$$\left. B_0(n_1 p_1^* - n_2)F_0 f_0 - \bar{B}_0(n_1 \bar{p}_1^* - n_2)G_0 g_0 \right\} e^{-i\xi x_1} d\xi +$$

$$\int_{-\infty}^0 \left\{ i\xi \left[(\bar{B}(n_1 \bar{P} - n_2 I)Ff + B(n_1 P - n_2 I)Gg \right] - \right.$$

$$\left. \bar{B}_0(n_1 \bar{p}_1^* - n_2)F_0 f_0 - B_0(n_1 p_1^* - n_2)G_0 g_0 \right\} e^{-i\xi x_1} d\xi \qquad (3.7.30)$$

For a given boundary value problem, the eight functions f_0, f, g_0 and g are determined from the appropriate boundary conditions. As an illustration, the general solutions (3.7.27)~(3.7.30) are now used for analyzing crack-tip singularities.

3.7.2   Crack-tip singularity

The singular behaviour at crack-tip can be found by considering a semi-infinite crack along the negative x_1-axis with the origin at the crack-tip under consideration. Assuming that the crack faces are traction-free and charge-free and thermally insulated, the boundary conditions at the crack faces are

$$\Pi_2(x_1,\ 0^+)=\Pi_2(x_1,\ 0^-)=0,\ \ h_2(x_1,\ 0^+)=h_2(x_1,\ 0^-)=0 \qquad (3.7.31)$$

The continuity conditions along $x_2=0$ and $x_1>0$ require that

$$T(x_1,\ 0^+)=T(x_1,\ 0^-),\quad h_2(x_1,\ 0^+)=h_2(x_1,\ 0^-)$$
$$U(x_1,\ 0^+)=U(x_1,\ 0^-),\quad \Pi_2(x_1,\ 0^+)=\Pi_2(x_1,\ 0^-) \qquad (3.7.32)$$

Since the solution must be bounded as $|x_2|\to\infty$, f_i and g_i (i=0~4) should be taken as

$$f_i(\xi)=0,\quad \text{when } x_2>0 \qquad (3.7.33)$$
$$g_i(\xi)=0,\quad \text{when } x_2<0 \qquad (3.7.34)$$

Further, satisfaction of the continuity conditions along x_2=0 requires that

$$f_i(-\xi)=\overline{f}_i(\xi),\quad g_i(-\xi)=\overline{g}_i(\xi),\quad i\text{=0~4} \qquad (3.7.35)$$

It follows then that

$$T_{(j)}(x_1,\ 0)=\frac{\partial Z_{0(j)}}{\partial x_1}+\frac{\partial \overline{Z}_{0(j)}}{\partial x_1} \qquad (3.7.36)$$

$$U_{(j)}(x_1,0)=A_{(j)}Z_{(j)}+\overline{A}_{(j)}\overline{Z}_{(j)}+A_{0(j)}Z_{0(j)}+\overline{A}_{0(j)}\overline{Z}_{0(j)} \qquad (3.7.37)$$

$$h_{2(j)}(x_1,\ 0)=-ik_{(j)}\frac{\partial^2 Z_{0(j)}}{\partial x_1^2}+ik_{(i)}\frac{\partial^2 \overline{Z}_{0(j)}}{\partial x_1^2} \qquad (3.7.38)$$

$$\Pi_{2(j)}(x_1,\ 0)=B_{(j)}\frac{\partial Z_{(j)}}{\partial x_1}+\overline{B}_{(j)}\frac{\partial \overline{Z}_{(j)}}{\partial x_1}+B_{0(j)}\frac{\partial Z_{0(j)}}{\partial x_1}+\overline{B}_{0(j)}\frac{\partial \overline{Z}_{0(j)}}{\partial x_1} \qquad (3.7.39)$$

where the subscript (j) is used to distinguish the lower and upper half-plane, j=1 corresponds to the domain $x_2>0$, and j=2 means that the point is located in the lower half-plane. The quantities $k_{(i)}$, $Z_{0(i)}$ and $Z_{(i)}$ are defined by

$$k_{(j)}=\sqrt{k_{11}^{(j)}k_{22}^{(j)}-(k_{12}^{(j)})^2} \qquad (3.7.40)$$

$$Z_{0(1)}=\int_0^\infty \frac{1}{i\xi}g_0(-\xi)e^{i\xi x_1}d\xi \qquad (3.7.41)$$

$$Z_{0(2)}=-\int_0^\infty \frac{1}{i\xi}f_0(\xi)e^{-i\xi x_1}d\xi \qquad (3.7.42)$$

$$Z_{(1)}=\int_0^\infty g(-\xi)e^{i\xi x_1}d\xi \qquad (3.7.43)$$

$$Z_{(2)}=\int_0^\infty f(\xi)e^{-i\xi x_1}d\xi \qquad (3.7.44)$$

The asymptotic form of the field variables can be obtained by setting $x_1\to0$ ($r\to0$) along $x_2=0$. Hence, $Z_{0(i)}$ and $Z_{(i)}$ become

$$Z_{0(i)}\approx (x_1)^{\delta+1}h_{0(i)},\quad i\text{=1,2} \qquad (3.7.45)$$

$$\boldsymbol{Z}_{(i)} \approx (x_1)^{\delta+1}\boldsymbol{h}_{(i)}, \quad i=1,2 \tag{3.7.46}$$

Substitution of Eq.(3.7.45) and Eq.(3.7.46) into Eqs. (3.7.36)~(3.7.39) yields

$$T_{(j)}(x_1,\,0) = (\delta+1)(q_{0(j)} + \overline{q}_{0(j)})|x_1|^{\delta}, \quad x_1 > 0 \tag{3.7.47}$$

$$h_{(j)}(x_1,\,0) = -ik_{(j)}\delta(\delta+1)(q_{0(j)} - \overline{q}_{0(j)})|x_1|^{\delta-1}, \quad x_1 > 0 \tag{3.7.48}$$

$$h_{(j)}(x_1,\,0) = -ik_{(j)}\delta(\delta+1)(e^{\delta(j)}q_{0(j)} - e^{-\delta(j)}\overline{q}_{0(j)})|x_1|^{\delta-1}, \quad x_1 < 0 \tag{3.7.49}$$

$$U_{(j)}(x_1,\,0) = 2\,\mathrm{Re}(A_{(j)}q_{(j)} + A_{0(j)}q_{0(i)})|x_1|^{\delta+1}, \quad x_1 > 0 \tag{3.7.50}$$

$$\Pi_{2(j)}(x_1,\,0) = 2\,\mathrm{Re}[(\delta+1)(B_{(j)}h_{(j)} + B_{0(j)}q_{0(j)})]|x_1|^{\delta}, \quad x_1 > 0 \tag{3.7.51}$$

$$\Pi_{2(j)}(x_1,\,0) = 2\,\mathrm{Re}[(\delta+1)e^{\delta(j)}(B_{(j)}q_{(j)} + B_{0(j)}q_{0(j)})]|x_1|^{\delta}, \quad x_1 < 0 \tag{3.7.52}$$

where $\delta(j) = i\pi\delta(-1)^{j+1}$. Summation convention does not apply to the repeated indices in Eqs.(3.7.47)~(3.7.52). Substitution of Eqs.(3.7.47)~(3.7.52) into Eq.(3.7.31) and Eq.(3.7.32) leads to a system of 20 homogeneous equations. The results can be written in matrix form as

$$\begin{bmatrix} \boldsymbol{K}_k & 0 \\ \boldsymbol{K}_m & \boldsymbol{K}_c \end{bmatrix}\begin{bmatrix} \boldsymbol{a}_k \\ \boldsymbol{a}_c \end{bmatrix} = \begin{bmatrix} 0 \\ 0 \end{bmatrix} \tag{3.7.53}$$

where

$$\boldsymbol{a}_k = \begin{bmatrix} q_{0(1)} \\ \overline{q}_{0(1)} \\ q_{0(2)} \\ \overline{q}_{0(2)} \end{bmatrix}, \quad \boldsymbol{a}_c = \begin{bmatrix} \boldsymbol{B}_{(1)}\boldsymbol{q}_{(1)} \\ \overline{\boldsymbol{B}}_{(1)}\overline{\boldsymbol{q}}_{(1)} \\ \boldsymbol{B}_{(2)}\boldsymbol{q}_{(2)} \\ \overline{\boldsymbol{B}}_{(2)}\overline{\boldsymbol{q}}_{(2)} \end{bmatrix} \tag{3.7.54}$$

$$\boldsymbol{K}_k = \begin{bmatrix} 1 & 1 & -1 & -1 \\ ik_{(1)} & -ik_{(1)} & -ik_{(2)} & ik_{(2)} \\ e^{2i\pi\delta}k_{(1)} & -k_{(1)} & 0 & 0 \\ 0 & 0 & k_{(2)} & -e^{2i\pi\delta}k_{(2)} \end{bmatrix}_{4\times4} \tag{3.7.55}$$

$$\boldsymbol{K}_c = \begin{bmatrix} \boldsymbol{A}_{(1)}\boldsymbol{B}_{(1)}^{-1} & \overline{\boldsymbol{A}}_{(1)}\overline{\boldsymbol{B}}_{(1)}^{-1} & -\boldsymbol{A}_{(2)}\boldsymbol{B}_{(2)}^{-1} & -\overline{\boldsymbol{A}}_{(2)}\overline{\boldsymbol{B}}_{(2)}^{-1} \\ \boldsymbol{I} & \boldsymbol{I} & -\boldsymbol{I} & -\boldsymbol{I} \\ e^{2i\pi\delta}\boldsymbol{I} & \boldsymbol{I} & 0 & 0 \\ 0 & 0 & \boldsymbol{I} & e^{2i\pi\delta}\boldsymbol{I} \end{bmatrix}_{16\times16} \tag{3.7.56}$$

$$\boldsymbol{K}_m = \begin{bmatrix} \boldsymbol{A}_{0(1)} & \overline{\boldsymbol{A}}_{0(1)} & -\boldsymbol{A}_{0(2)} & -\overline{\boldsymbol{A}}_{0(2)} \\ \boldsymbol{B}_{0(1)} & \overline{\boldsymbol{B}}_{0(1)} & -\boldsymbol{B}_{0(2)} & -\overline{\boldsymbol{B}}_{0(2)} \\ e^{2i\pi\delta}\boldsymbol{B}_{0(1)} & \overline{\boldsymbol{B}}_{0(1)} & 0 & 0 \\ 0 & 0 & \boldsymbol{B}_{0(2)} & e^{2i\pi\delta}\overline{\boldsymbol{B}}_{0(2)} \end{bmatrix}_{16\times4} \tag{3.7.57}$$

The order of singularity in the temperature and traction fields is determined by setting the determinant of the 20×20 matrix in Eq.(3.7.53) to zero. This is equivalent to

$$\left\| \boldsymbol{K}_k(\delta) \right\| = 0 \tag{3.7.58}$$

or

$$\left\| \boldsymbol{K}_c(\delta) \right\| = 0 \tag{3.7.59}$$

Note that the roots δ in Eq.(3.7.58) and Eq.(3.7.59) are uncoupled; they are given as δ_k and δ_c, respectively. For a semi-infinite crack in a homogeneous solid, Eq.(3.7.58) yields

$$\delta_k = -\frac{1}{2} \tag{3.7.60}$$

The roots of Eq.(3.7.59) are given in [42] as

$$\delta_c = -\frac{1}{2} \tag{3.7.61}$$

They are in multiples of 4. These result in five $r^{-1/2}$ singularities when $\delta_k = \delta_c$, which may convert into logarithmic singularities. For the present case, the stress and electric displacement singularities at the crack-tip may be one of the following cases: $r^{-1/2}$, $r^{-1/2} \ln^j r$ (j=0~3). For interface cracks, the root of Eq.(3.7.58) is still $-1/2$, and the roots of Eq.(3.7.59) will be [53]

$$\delta = -\frac{1}{2} \pm i\alpha, \quad -\frac{1}{2} \pm \kappa \tag{3.7.62}$$

where α and κ are real numbers depending on the constitutive constants. For certain special bimaterials, α may be zero [53]. In such a situation, three $r^{-1/2}$ singularities may prevail, they may also be converted into logarithmic singularities. The above analysis shows that the order of singularity for the temperature field is always of the inverse square root type for a crack in a homogeneous solid and lying at a bimaterial interface. The traction singularities at crack-tips, however, may vary with different materials.

3.7.3   Griffith crack in homogeneous piezoelectricity

As an application of the formulation developed in Subsection 3.7.1, consider a crack of length $2a$ with its tips located at $x_1 = -a$ and $x_1 = a$ in an infinite thermo-piezoelectric material subjected to uniform loading T^∞ and Π_2^∞ at infinity. The surface of the crack is traction-free and charge-free and is kept at zero temperature. The crack-tip behaviour can be found by considering the following conditions:

$$U_{(1)}(x_1,\, 0) = U_{(2)}(x_1,\, 0), \quad |x_1| > a \tag{3.7.63}$$

$$T_{(1)}(x_1,\, 0) = T_{(2)}(x_1,\, 0), \quad |x_1| > a \tag{3.7.64}$$

$$\Pi_{2(1)}(x_1,\, 0) = \Pi_{2(2)}(x_1,\, 0), \quad |x_1| > a \tag{3.7.65}$$

$$T_{(1)}(x_1,\, 0) = T_{(2)}(x_1,\, 0) = -T^{\infty}, \quad |x_1| < a \tag{3.7.66}$$

$$\Pi_{2(1)}(x_1,\, 0) = \Pi_{2(2)}(x_1,\, 0) = -\Pi_2^{\infty}, \quad |x_1| < a \tag{3.7.67}$$

$$T(x_1,\, x_2) \to 0, \quad \Pi_{iJ}(x_1,\, x_2) \to 0, \quad \text{when } (x_1^2 + x_2^2) \to \infty \tag{3.7.68}$$

1) Temperature field

Eqs.(3.7.63)~(3.7.68) can be applied to yield the governing dual integral equations with the aid of an additional continuity condition related to temperature field to supplement Eq.(3.7.63). This can be accomplished by introducing an auxiliary function, say d, in such a way that

$$d_{(1)}(x_1,\, 0) = d_{(2)}(x_1,\, 0), \quad |x_1| > a \tag{3.7.69}$$

$$d_{(j)}(z_t) = Q_{(j)}(z_t) + \overline{Q}_{(j)}(z_t), \quad j=1,2 \tag{3.7.70}$$

$$T_{(j)}(z_t) = A^* Q'_{(j)}(z_t) + \overline{A}^* \overline{Q}'_{(j)}(z_t) \tag{3.7.71}$$

where A^* is a complex constant to be determined, $z_t = x_1 + p_1^* x_2$, and

$$Q_{(1)}(z_t) = \int_0^{\infty} g_0^*(-\xi) e^{i\xi z_t} \, \mathrm{d}\xi \tag{3.7.72}$$

$$Q_{(2)}(z_t) = \int_0^{\infty} f_0^*(\xi) e^{-i\xi z_t} \, \mathrm{d}\xi \tag{3.7.73}$$

$$f_0^*(-\xi) = \overline{f}_0^*(\xi), \quad g_0^*(-\xi) = \overline{g}_0^*(\xi) \tag{3.7.74}$$

Substituting Eqs.(3.7.72)~(3.7.74) into Eq.(3.7.70) and Eq.(3.7.71) yields

$$d_{(1)}(z_t) = \int_0^{\infty} \left[g_0^*(\xi) e^{-i\xi \overline{z}_t} + \overline{g}_0^*(\xi) e^{i\xi z_t} \right] \mathrm{d}\xi, \quad x_2 > 0 \tag{3.7.75}$$

$$d_{(2)}(z_t) = \int_0^{\infty} \left[f_0^*(\xi) e^{-i\xi z_t} + \overline{f}_0^*(\xi) e^{i\xi \overline{z}_t} \right] \mathrm{d}\xi, \quad x_2 < 0 \tag{3.7.76}$$

$$T_{(1)}(z_t) = i \int_0^{\infty} \left[A^* \overline{g}_0^*(\xi) e^{i\xi z_t} - \overline{A}^* g_0^*(\xi) e^{-i\xi \overline{z}_t} \right] \xi \mathrm{d}\xi, \quad x_2 > 0 \tag{3.7.77}$$

$$T_{(2)}(z_t) = i \int_0^{\infty} \left[A^* f_0^*(\xi) e^{-i\xi z_t} - \overline{A}^* \overline{f}_0^*(\xi) e^{i\xi \overline{z}_t} \right] \xi \mathrm{d}\xi, \quad x_2 < 0 \tag{3.7.78}$$

Comparing Eq.(3.7.77) and Eq.(3.7.78) with Eq.(3.7.10) renders

$$f_0(\xi) = -i\xi A^* f_0^*(\xi) \tag{3.7.79}$$

$$g_0(\xi) = -i\xi \overline{A}^* g_0^*(\xi) \tag{3.7.80}$$

It follows from Eq.(3.7.64) that

$$f_0^*(\xi) = \overline{A}^* g_0^*(\xi) / A^* \tag{3.7.81}$$

Inserting Eqs.(3.7.75), (3.7.76) and (3.7.81) into Eq.(3.7.69), and Eq.(3.7.77) and Eq.(3.7.78) into Eq.(3.7.66), it is found that

$$\int_0^\infty g_0^*(\xi)\left(1 - \frac{\overline{A}^*}{A^*}\right) e^{-i\xi x_1} \,\mathrm{d}\xi + \int_0^\infty \overline{g}_0^*(\xi)\left(1 - \frac{A^*}{\overline{A}^*}\right) e^{i\xi x_1} \,\mathrm{d}\xi = 0, \quad |x_1| > a \tag{3.7.82}$$

$$-\int_0^\infty i\xi \overline{A}^* g_0^*(\xi) e^{-i\xi x_1} \,\mathrm{d}\xi + \int_0^\infty i\xi A^* \overline{g}_0^*(\xi) e^{i\xi x_1} \,\mathrm{d}\xi = -T^\infty, \quad |x_1| < a \tag{3.7.83}$$

It can be seen that Eq.(3.7.82) will be trivial if $1 - A^* / \overline{A}^* = 0$. Therefore, the constant A^* should be chosen so that $1 - A^* / \overline{A}^* \neq 0$, e.g., $A^* = i = \sqrt{-1}$. Now, denote the real and imaginary parts of $g_0^*(\xi)$ as

$$g_0^*(\xi) = q_s(\xi) + i q_a(\xi) \tag{3.7.84}$$

where $q_s(\xi)$ and $q_a(\xi)$ are two real functions of ξ. Since the temperature T^∞ is symmetric about $x_1 = 0$, it can be shown that the antisymmetric part on the left-hand side of Eq.(3.7.82) and Eq.(3.7.83) may be taken as zero, i.e., $q_a = 0$, and putting Eq.(3.7.84) into Eq.(3.7.82) and Eq.(3.7.83) leads to

$$\int_0^\infty q_s(\xi)\cos(x_1\xi)\,\mathrm{d}\xi = 0, \quad x_1 > 0 \tag{3.7.85}$$

$$\int_0^\infty \xi q_s(\xi)\cos(x_1\xi)\,\mathrm{d}\xi = \frac{1}{2}T^\infty, \quad 0 < x_1 < a \tag{3.7.86}$$

The pair of Eqs. (3.7.85) and (3.7.86) are the standard dual integral equations. A solution of the equations is given by [54]

$$q_s(\xi) = \frac{a}{2\xi} J_1(a\xi) T^\infty \tag{3.7.87}$$

where $J_1(a\xi)$ is the Bessel function of the first kind with order one. Substituting Eq.(3.7.84) and Eq.(3.7.87) into Eq.(3.7.79) and Eq.(3.7.80) gives

$$g_0^*(\xi) = -f_0^*(\xi) = \frac{a}{2\xi} J_1(a\xi) T^\infty \tag{3.7.88}$$

$$f_0(\xi) = g_0(\xi) = \frac{a}{2} J_1(a\xi) T^\infty \mathrm{sign}(\xi) \tag{3.7.89}$$

where $\mathrm{sign}(\xi)$ is defined as

$$\mathrm{sign}(\xi) = \begin{cases} 1, & \xi > 0 \\ -1, & \xi < 0 \\ 0, & \xi = 0 \end{cases} \tag{3.7.90}$$

Eq.(3.7.88) can be put into Eq.(3.7.77) and Eq.(3.7.78), yielding

$$T_{(1)} = \mathrm{Re}\left[\frac{iz_t}{(a^2 - z_t^2)^{1/2}} - 1\right]T^\infty, \quad x_2 > 0 \tag{3.7.91}$$

$$T_{(2)} = \mathrm{Re}\left[\frac{i\bar{z}_t}{(a^2 - \bar{z}_t^2)^{1/2}} - 1\right]T^\infty, \quad x_2 < 0 \tag{3.7.92}$$

Eq.(3.7.91) and Eq.(3.7.92) represent the temperature field of the boundary value problem stated by Eqs.(3.7.64), (3.7.66) and (3.7.68).

2) Elastic and electric fields

To simplify the derivation, we introduce the following notation:

$$\boldsymbol{g}^*(\xi) = \begin{cases} \boldsymbol{g}(\xi) - \bar{\boldsymbol{B}}^{-1}\bar{\boldsymbol{B}}_0 \boldsymbol{g}_0(\xi)/i\xi, & \xi > 0 \\ \boldsymbol{g}(\xi) - \boldsymbol{B}^{-1}\boldsymbol{B}_0 \boldsymbol{g}_0(\xi)/i\xi, & \xi < 0 \end{cases} \tag{3.7.93}$$

$$\boldsymbol{f}^*(\xi) = \begin{cases} \boldsymbol{f}(\xi) - \boldsymbol{B}^{-1}\boldsymbol{B}_0 \boldsymbol{f}_0(\xi)/i\xi, & \xi > 0 \\ \boldsymbol{f}(\xi) - \bar{\boldsymbol{B}}^{-1}\bar{\boldsymbol{B}}_0 \boldsymbol{f}_0(\xi)/i\xi, & \xi < 0 \end{cases} \tag{3.7.94}$$

Eq.(3.7.29) may thus be written as

$$\boldsymbol{\varPi}_{2(1)}(x_1,\, 0) = -i\int_0^\infty \left[\bar{\boldsymbol{B}}\boldsymbol{g}^*(\xi)\mathrm{e}^{-i\xi x_1} - \boldsymbol{B}\bar{\boldsymbol{g}}^*(\xi)\mathrm{e}^{i\xi x_1}\right]\xi\mathrm{d}\xi \tag{3.7.95}$$

$$\boldsymbol{\varPi}_{2(2)}(x_1,\, 0) = -i\int_0^\infty \left[\boldsymbol{B}\boldsymbol{f}^*(\xi)\mathrm{e}^{-i\xi x_1} - \bar{\boldsymbol{B}}\bar{\boldsymbol{f}}^*(\xi)\mathrm{e}^{i\xi x_1}\right]\xi\mathrm{d}\xi \tag{3.7.96}$$

As a consequence, Eq.(3.7.65) gives

$$\bar{\boldsymbol{B}}\boldsymbol{g}^*(\xi) = \boldsymbol{B}\boldsymbol{f}^*(\xi) \tag{3.7.97}$$

The crack opening displacement and electric potential along $x_2 = 0$ are obtained by substituting Eqs.(3.7.93), (3.7.94) and (3.7.97) into Eq.(3.7.27) to give the result

$$\Delta\boldsymbol{U}(x_1) = 2\,\mathrm{Re}\left[\int_0^\infty (i\boldsymbol{C}\boldsymbol{B}\boldsymbol{f}^* - \boldsymbol{b}f_0\xi^{-1})\,\mathrm{e}^{-i\xi x_1}\mathrm{d}\xi\right] \tag{3.7.98}$$

where

$$\Delta\boldsymbol{U}(x_1) = \boldsymbol{U}_{(1)}(x_1) - \boldsymbol{U}_{(2)}(x_1), \quad x_2 = 0 \tag{3.7.99}$$

$$\boldsymbol{C} = i(\boldsymbol{A}\boldsymbol{B}^{-1} - \bar{\boldsymbol{A}}\bar{\boldsymbol{B}}^{-1}) \tag{3.7.100}$$

$$\boldsymbol{b} = i\left[(\boldsymbol{A}_0 - \boldsymbol{A}\boldsymbol{B}^{-1}\boldsymbol{B}_0) - (\bar{\boldsymbol{A}}_0 - \bar{\boldsymbol{A}}\bar{\boldsymbol{B}}^{-1}\bar{\boldsymbol{B}}_0)\right] \tag{3.7.101}$$

The substitution

$$\boldsymbol{a}^* = \boldsymbol{f}^* - \frac{f_0}{i\xi}\boldsymbol{B}^{-1}\boldsymbol{C}^{-1}\boldsymbol{b} \tag{3.7.102}$$

leads to a pair of dual integral equations by combining Eqs.(3.7.63), (3.7.67), (3.7.99) and (3.7.102):

$$\int_0^\infty (\boldsymbol{B}\boldsymbol{a}^*\mathrm{e}^{-i\xi x_1} - \bar{\boldsymbol{B}}\bar{\boldsymbol{a}}^*\mathrm{e}^{i\xi x_1})\,\mathrm{d}\xi = 0, \quad |x_1| > a \tag{3.7.103}$$

$$-\int_0^\infty i\xi(\boldsymbol{Ba}^*\mathrm{e}^{-i\xi x_1} - \boldsymbol{\bar{B}\bar{a}}^*\mathrm{e}^{i\xi x_1})\,\mathrm{d}\xi = 2\boldsymbol{C}^{-1}\boldsymbol{b}\int_0^\infty f_0(\xi)\cos(x_1\xi)\mathrm{d}\xi - \boldsymbol{\Pi}_2^\infty,\quad |x_1|<a$$

$$(3.7.104)$$

For the sake of convenience, define

$$\boldsymbol{q}_a = \mathrm{Re}(\boldsymbol{Ba}^*),\quad \boldsymbol{q}_s = \mathrm{Im}(\boldsymbol{Ba}^*)\tag{3.7.105}$$

or

$$\boldsymbol{q} = \boldsymbol{Ba}^* = \boldsymbol{q}_a + i\boldsymbol{q}_s\tag{3.7.106}$$

where $\boldsymbol{q}_a$ and $\boldsymbol{q}_s$ are two real functions of ξ. Eq.(3.7.103) and Eq.(3.7.104) can thus be rewritten as

$$\int_0^\infty \left[\boldsymbol{q}_s(\xi)\cos(x_1\xi) - \boldsymbol{q}_a(\xi)\sin(x_1\xi)\right]\mathrm{d}\xi = 0,\quad |x_1|>a\tag{3.7.107}$$

$$\int_0^\infty \xi\left[\boldsymbol{q}_s(\xi)\cos(x_1\xi) - \boldsymbol{q}_a(\xi)\sin(x_1\xi)\right]\mathrm{d}\xi = \boldsymbol{C}^{-1}\boldsymbol{b}\int_0^\infty f_0(\xi)\cos(x_1\xi)\mathrm{d}\xi - \boldsymbol{\Pi}_2^\infty,\ |x_1|<a$$

$$(3.7.108)$$

The above pair of dual integral equations determines the functions $\boldsymbol{q}_a$ and $\boldsymbol{q}_s$. They are

$$\boldsymbol{q}_a(\xi) = 0\tag{3.7.109}$$

$$\boldsymbol{q} = i\boldsymbol{q}_s = \frac{ia}{2\xi}(\boldsymbol{C}^{-1}\boldsymbol{b}T^\infty - \boldsymbol{\Pi}_2^\infty)J_1(a\xi)\tag{3.7.110}$$

It can be seen from Eqs.(3.7.93), (3.7.94), (3.7.102), (3.7.109) and (3.7.110) that

$$f(\xi) = \frac{a\boldsymbol{B}^{-1}}{2i\xi}(\boldsymbol{\Pi}_2^\infty + \boldsymbol{B}_0 T^\infty)J_1(a\xi)\tag{3.7.111}$$

$$g(\xi) = \frac{a\boldsymbol{\bar{B}}^{-1}}{2i\xi}(\boldsymbol{\Pi}_2^\infty + \boldsymbol{\bar{B}}_0 T^\infty)J_1(a\xi)\tag{3.7.112}$$

Substituting Eq.(3.7.111) and Eq.(3.7.112) into Eqs.(3.7.27)~(3.7.29) and (3.7.98) results in

$$\boldsymbol{U}_{(1)} = \mathrm{Re}\left[\boldsymbol{\bar{A}}F(\bar{z})\boldsymbol{\bar{B}}^{-1}(\boldsymbol{\Pi}_2^\infty + \boldsymbol{\bar{B}}_0 T^\infty) - \boldsymbol{\bar{A}}_0 F(\bar{z}_t)T^\infty\right],\quad x_2>0\tag{3.7.113}$$

$$\boldsymbol{U}_{(2)} = \mathrm{Re}\left[\boldsymbol{A}F(z)\boldsymbol{B}^{-1}(\boldsymbol{\Pi}_2^\infty + \boldsymbol{B}_0 T^\infty) - \boldsymbol{A}_0 F(z_t)T^\infty\right],\quad x_2<0\tag{3.7.114}$$

$$\Delta\boldsymbol{U}(x_1,\,0) = (\boldsymbol{C}\boldsymbol{\Pi}_2^\infty - \boldsymbol{b}T^\infty)(a^2 - x_1^2)^{1/2},\quad |x_1|<a\tag{3.7.115}$$

$$\boldsymbol{\Pi}_{1(1)} = -\mathrm{Re}\left[\boldsymbol{\bar{B}\bar{P}}F^*(\bar{z})\boldsymbol{\bar{B}}^{-1}(\boldsymbol{\Pi}_2^\infty + \boldsymbol{B}_0 T^\infty) + \boldsymbol{\bar{B}}_0\boldsymbol{\bar{p}}_1^* F^*(\bar{z}_t)T^\infty\right],\quad x_2>0$$

$$(3.7.116)$$

$$\boldsymbol{\Pi}_{1(2)} = -\mathrm{Re}\left[\boldsymbol{B}\boldsymbol{P}F^*(z)\boldsymbol{B}^{-1}(\boldsymbol{\Pi}_2^\infty + \boldsymbol{B}_0 T^\infty) + \boldsymbol{B}_0\boldsymbol{p}_1^* F^*(z_t)T^\infty\right],\quad x_2<0$$

$$(3.7.117)$$

$$\boldsymbol{\Pi}_{2(1)} = \mathrm{Re}\left[\bar{\boldsymbol{B}}\boldsymbol{F}^{*}(\bar{z})\bar{\boldsymbol{B}}^{-1}(\boldsymbol{\Pi}_{2}^{\infty} + \boldsymbol{B}_{0}T^{\infty}) - \bar{\boldsymbol{B}}_{0}\boldsymbol{F}^{*}(\bar{z}_{t})T^{\infty}\right], \quad x_{2} > 0 \qquad (3.7.118)$$

$$\boldsymbol{\Pi}_{2(2)} = \mathrm{Re}\left[\boldsymbol{B}\boldsymbol{F}^{*}(z)\boldsymbol{B}^{-1}(\boldsymbol{\Pi}_{2}^{\infty} + \boldsymbol{B}_{0}T^{\infty}) - \boldsymbol{B}_{0}\boldsymbol{F}^{*}(z_{t})T^{\infty}\right], \quad x_{2} < 0 \qquad (3.7.119)$$

where

$$\boldsymbol{F}(z) = \mathrm{diag}\left[F(z_{1}) \quad F(z_{2}) \quad F(z_{3}) \quad F(z_{4})\right] \qquad (3.7.120)$$

$$\boldsymbol{F}^{*}(z) = \mathrm{diag}\left[F^{*}(z_{1}) \quad F^{*}(z_{2}) \quad F^{*}(z_{3}) \quad F^{*}(z_{4})\right] \qquad (3.7.121)$$

$$F(z) = (z^{2} - a^{2})^{1/2} - z \qquad (3.7.122)$$

$$F^{*}(z) = \frac{z}{(z^{2} - a^{2})^{1/2}} - 1 \qquad (3.7.123)$$

3) Crack-tip fields

A polar coordinate system (r, θ) centered at the crack tip with $(x_{1}, x_{2}) = (a, 0)$ and $\theta = 0$ along the crack line is taken. Let the variable z be given as

$$z = a + r(\cos\theta + p\sin\theta) \qquad (3.7.124)$$

The stress and electric displacement field near the crack-tip can be obtained by taking the asymptotic limit of Eqs.(3.7.113)~(3.7.119). Hence, $\boldsymbol{\Pi}_{2(j)}$ in Eq.(3.7.118) and Eq.(3.7.119) becomes

$$\boldsymbol{\Pi}_{1(2)} \approx \sqrt{\frac{a}{2r}}\,\mathrm{Re}\left\{\bar{\boldsymbol{B}}\boldsymbol{\beta}(\theta,\bar{p})\bar{\boldsymbol{B}}^{-1}\boldsymbol{\Pi}_{2}^{\infty} + \left[\bar{\boldsymbol{B}}\boldsymbol{\beta}(\theta,\bar{p})\bar{\boldsymbol{B}}^{-1} - \boldsymbol{I}\boldsymbol{\beta}^{*}(\theta,\bar{p}_{1}^{*})\right]\bar{\boldsymbol{B}}_{0}T^{\infty}\right\}, \quad x_{2} > 0$$

$$(3.7.125)$$

$$\boldsymbol{\Pi}_{2(2)} \approx \sqrt{\frac{a}{2r}}\,\mathrm{Re}\left\{\boldsymbol{B}\boldsymbol{\beta}(\theta,p)\boldsymbol{B}^{-1}\boldsymbol{\Pi}_{2}^{\infty} + \left[\boldsymbol{B}\boldsymbol{\beta}(\theta,p)\boldsymbol{B}^{-1} - \boldsymbol{I}\boldsymbol{\beta}^{*}(\theta,p_{1}^{*})\right]\boldsymbol{B}_{0}T^{\infty}\right\}, \quad x_{2} < 0$$

$$(3.7.126)$$

where

$$\boldsymbol{\beta}(\theta,p) = \mathrm{diag}\left[\beta^{*}(\theta,p_{1}) \quad \beta^{*}(\theta,p_{2}) \quad \beta^{*}(\theta,p_{3}) \quad \beta^{*}(\theta,p_{4})\right] \qquad (3.7.127)$$

$$\beta^{*}(\theta,p) = (\cos\theta + p\sin\theta)^{-1/2} \qquad (3.7.128)$$

3.8 Penny-shaped cracks

In the previous sections of this chapter, formulations were derived in terms of a rectangular coordinate system. The formulation is, however, inefficient for axisymmetric electroelastic problems. In this section, theoretical models are presented in terms of a cylindrical coordinate system and used to analyze (a) responses of elastic stress and electric displacement in a long piezoelectric cylinder with a concentric penny-shaped crack and (b) the effect of elastic coating on the fracture behaviour of piezoelectric fiber with a penny-shaped crack. The discussion follows the development in [55].

3.8.1 Problem statement and basic equation

Consider a piezoelectric cylinder of radius b containing a centered penny-shaped crack of radius a under axisymmetric electromechanical loads (Fig.3.6). For convenience, a cylindrical coordinate system (r, θ, z) originating at the center of the crack is used with the z-axis along the axis of symmetry of the cylinder. The cylinder is assumed to be a transversely isotropic piezoelectric material with the poling direction parallel to the z-axis. It is subjected to the far-field of a normal stress, $\sigma_z = \bar{\sigma}(r)$ and a normal electric displacement, $D_z = \bar{D}(r)$.

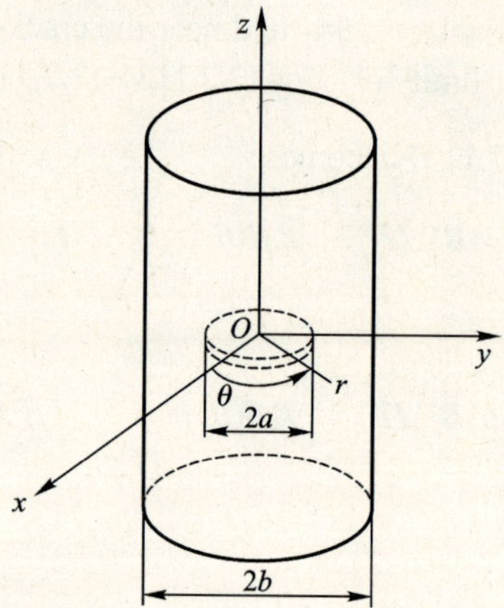

Fig.3.6   Penny-shaped crack in a piezoelectric cylinder.

The constitutive equations for a piezoelectric material which is transversely isotropic and poled along the z-axis can be written as [56]

$$\sigma_{rr} = c_{11}\frac{\partial u_r}{\partial r} + c_{12}\frac{u_r}{r} + c_{13}\frac{\partial u_z}{\partial z} + e_{31}\frac{\partial \phi}{\partial z} \tag{3.8.1}$$

$$\sigma_{\theta\theta} = c_{12}\frac{\partial u_r}{\partial r} + c_{11}\frac{u_r}{r} + c_{13}\frac{\partial u_z}{\partial z} + e_{31}\frac{\partial \phi}{\partial z} \tag{3.8.2}$$

$$\sigma_{zz} = c_{13}\frac{\partial u_r}{\partial r} + c_{13}\frac{u_r}{r} + c_{33}\frac{\partial u_z}{\partial z} + e_{33}\frac{\partial \phi}{\partial z} \tag{3.8.3}$$

$$\sigma_{rz} = c_{44}\left(\frac{\partial u_z}{\partial r} + \frac{\partial u_r}{\partial z}\right) + e_{15}\frac{\partial \phi}{\partial r} \qquad (3.8.4)$$

$$D_r = e_{15}\left(\frac{\partial u_z}{\partial r} + \frac{\partial u_r}{\partial z}\right) - \kappa_{11}\frac{\partial \phi}{\partial r} \qquad (3.8.5)$$

$$D_z = e_{31}\left(\frac{\partial u_r}{\partial r} + \frac{u_r}{r}\right) + e_{33}\frac{\partial u_z}{\partial z} - \kappa_{33}\frac{\partial \phi}{\partial z} \qquad (3.8.6)$$

in which u_r, u_z denote the displacements in the r-directions and z-directions respectively.

In the derivation of the analytical solution, the following potential functions are introduced [57]:

$$u_r = \sum_1^3 \frac{\partial \Phi_i}{\partial r}, \quad u_z = \sum_{i=1}^3 k_{1i}\frac{\partial \Phi_i}{\partial z}, \quad \phi = -\sum_{i=1}^3 k_{2i}\frac{\partial \Phi_i}{\partial z} \qquad (3.8.7)$$

where $\Phi_i(r,z)(i=1,2,3)$ are the potential functions to be determined, k_{1i} and k_{2i} ($i=1,2,3$) are unknown constants.

Substituting Eq. (3.8.7) into the constitutive equations (3.8.1)~(3.8.6), the field equations and gradient equations, we have the following governing equations:

$$c_{11}\sum_{i=1}^3\left(\frac{\partial \Phi_i}{\partial r^2} + \frac{1}{r}\frac{\partial \Phi_i}{\partial r}\right) + \sum_{i=1}^3\left\{[c_{44} + k_{1i}(c_{13} + c_{44}) + k_{2i}(e_{31} + e_{15})]\frac{\partial \Phi_i}{\partial z^2}\right\} = 0$$

$$(3.8.8)$$

$$\sum_{i=1}^3\left[(c_{44}k_{1i} + c_{13} + c_{44} + e_{15}k_{2i})\left(\frac{\partial \Phi_i}{\partial r^2} + \frac{1}{r}\frac{\partial \Phi_i}{\partial r}\right) + (c_{33}k_{1i} + e_{33}k_{2i})\frac{\partial \Phi_i}{\partial z^2}\right] = 0$$

$$(3.8.9)$$

$$\sum_{i=1}^3\left[(e_{15}k_{1i} + e_{31} + e_{15} - \kappa_{11}k_{2i})\left(\frac{\partial \Phi_i}{\partial r^2} + \frac{1}{r}\frac{\partial \Phi_i}{\partial r}\right) + (e_{33}k_{1i} - \kappa_{33}k_{2i})\frac{\partial \Phi_i}{\partial z^2}\right] = 0$$

$$(3.8.10)$$

Following the procedure presented in [57], the solution to Eqs.(3.8.8)~(3.8.10) can be assumed in the form:

$$\Phi_i(r,z) = \int_0^\infty \frac{1}{\xi}\left[A_i(\xi)I_0(\frac{\xi r}{s_i})\cos(\xi z) + B_i(\xi)\exp(-\xi s_i z)J_0(\xi r)\right]d\xi \qquad (3.8.11)$$

where $A_i(\xi)$, $B_i(\xi)$ ($i=1,2,3$) are the unknown functions to be determined, $J_n()$ is the Bessel functions of the first kind of order n, and $I_n()$ is the modified Bessel function of the first kind and the second kind of n order. In addition, s_i, k_{1i} and k_{2i} ($i=1,2,3$) are defined by

$$s_i = \frac{1}{\sqrt{n_i}}, \qquad i = 1,2,3 \tag{3.8.12}$$

$$n_i = \frac{c_{44} + (c_{13} + c_{44})k_{1i} - (e_{31} + e_{51})k_{2i}}{c_{11}}$$

$$= \frac{c_{33}k_{1i} - e_{33}k_{2i}}{c_{44}k_{1i} + c_{13} + c_{44} - e_{15}k_{2i}} = \frac{e_{33}k_{1i} + \kappa_{33}k_{2i}}{e_{15}k_{1i} + e_{15} + e_{31} + \kappa_{11}k_{2i}} \tag{3.8.13}$$

which n_i ($i = 1,2,3$) are determined by the following equation:

$$An_i^3 + Bn_i^2 + Cn_i + D = 0 \tag{3.8.14}$$

where

$$A = c_{44}\kappa_{11} + e_{15}^2 \tag{3.8.15}$$

$$B = (\kappa_{11}c_{13}^2 - c_{11}c_{33}\kappa_{11} + 2c_{13}c_{44}\kappa_{11} - c_{11}c_{44}\kappa_{33} +$$
$$2c_{13}e_{15}^2 + 2c_{13}e_{15}e_{31} - c_{44}e_{31}^2 - 2c_{11}e_{15}e_{33})/c_{11} \tag{3.8.16}$$

$$C = (c_{33}c_{44}\kappa_{11} - c_{13}^2\kappa_{33} + c_{11}c_{33}\kappa_{33} - 2c_{13}c_{44}\kappa_{33} + c_{33}e_{15}^2 + 2c_{33}e_{15}e_{31} +$$
$$c_{33}e_{31}^2 - 2c_{13}e_{15}e_{33} - 2c_{13}e_{31}e_{33} - 2c_{44}e_{31}e_{33} - 2c_{44}e_{31}e_{33} + c_{11}e_{33}^2)/c_{11}$$
$$\tag{3.8.17}$$

$$D = -c_{44}(c_{33}\kappa_{33} + e_{33}^2)/c_{11} \tag{3.8.18}$$

Using Eq.(3.8.11), the following expressions for electric and elastic fields in the cracked piezoelectric fibre composite can be obtained

$$u_z(r,z) = -\sum_{i=1}^{3} k_{1i} \int_0^\infty A_i(\xi)I_0\left(\frac{\xi r}{s_i}\right)\sin(\xi z)\mathrm{d}\xi -$$
$$\sum_{i=1}^{3} k_{1i}s_i \int_0^\infty B_i(\xi)J_0(\xi r)\mathrm{e}^{-\xi s_i z}\mathrm{d}\xi + \overline{a}(r)z \tag{3.8.19}$$

$$u_r(r,z) = \sum_{i=1}^{3} \frac{1}{s_i} \int_0^\infty A_i(\xi)I_1\left(\frac{\xi r}{s_i}\right)\cos(\xi z)\mathrm{d}\xi - \sum_{i=1}^{3} \int_0^\infty B_i(\xi)J_1(\xi r)\mathrm{e}^{-\xi s_i z}\mathrm{d}\xi \tag{3.8.20}$$

$$\phi(r,z) = \sum_{i=1}^{3} k_{2i} \int_0^\infty A_i(\xi)I_1\left(\frac{\xi r}{s_i}\right)\sin(\xi z)\mathrm{d}\xi +$$
$$\sum_{i=1}^{3} k_{2i}s_i \int_0^\infty B_i(\xi)J_0(\xi r)\mathrm{e}^{-\xi s_i z}\mathrm{d}\xi - \overline{b}(r)z \tag{3.8.21}$$

$$\sigma_{zz} = -\sum_{i=1}^{3} \frac{F_{1i}}{s_i^2} \int_0^\infty \xi A_i(\xi)I_0\left(\frac{\xi r}{s_i}\right)\cos(\xi z)\mathrm{d}\xi +$$
$$\sum_{i=1}^{3} F_{1i} \int_0^\infty \xi B_i(\xi)J_0(\xi r)\mathrm{e}^{-\xi s_i z}\mathrm{d}\xi + \overline{c}(r) \tag{3.8.22}$$

$$\sigma_{rr} = -\sum_{i=1}^{3} \frac{F_{5i}}{s_i^2} \int_0^\infty \xi A_i(\xi) I_0\left(\frac{\xi r}{s_i}\right) \cos(\xi z) \mathrm{d}\xi +$$

$$\frac{c_{11}-c_{12}}{2} \sum_{i=1}^{3} \frac{1}{s_i^2} \int_0^\infty \xi A_i(\xi) I_2\left(\frac{\xi r}{s_i}\right) \cos(\xi z) \mathrm{d}\xi +$$

$$\sum_{i=1}^{3} F_{5i} \int_0^\infty \xi B_i(\xi) J_0(\xi r) \mathrm{e}^{-\xi s_i z} \mathrm{d}\xi + \frac{c_{11}-c_{12}}{2} \sum_{i=1}^{3} \int_0^\infty \xi B_i(\xi) J_2(\xi r) \mathrm{e}^{-\xi s_i z} \mathrm{d}\xi$$

$$(3.8.23)$$

$$\sigma_{zr} = -\sum_{i=1}^{3} \frac{F_{3i}}{s_i^2} \int_0^\infty \xi A_i(\xi) I_1\left(\frac{\xi r}{s_i}\right) \sin(\xi z) \mathrm{d}\xi + \sum_{i=1}^{3} F_{3i} \int_0^\infty \xi B_i(\xi) J_1(\xi r) \mathrm{e}^{-\xi s_i z} \mathrm{d}\xi$$

$$(3.8.24)$$

$$D_z = -\sum_{i=1}^{3} \frac{F_{2i}}{s_i^2} \int_0^\infty \xi A_i(\xi) I_0\left(\frac{\xi r}{s_i}\right) \cos(\xi z) \mathrm{d}\xi +$$

$$\sum_{i=1}^{3} F_{2i} \int_0^\infty \xi B_i(\xi) J_0(\xi r) \mathrm{e}^{-\xi s_i z} \mathrm{d}\xi + d(r) \qquad (3.8.25)$$

$$D_r = -\sum_{i=1}^{3} \frac{F_{4i}}{s_i^2} \int_0^\infty \xi A_i(\xi) I_1\left(\frac{\xi r}{s_i}\right) \sin(\xi z) \mathrm{d}\xi + \sum_{i=1}^{3} F_{4i} \int_0^\infty \xi B_i(\xi) J_1(\xi r) \mathrm{e}^{-\xi s_i z} \mathrm{d}\xi$$

$$(3.8.26)$$

in which

$$F_{1i} = (c_{33}k_{1i} - e_{33}k_{2i})s_i^2 - c_{13}, \quad F_{2i} = (e_{33}k_{1i} + \kappa_{33}k_{2i})s_i^2 - e_{31}$$

$$F_{3i} = \left[c_{44}(1+k_{1i}) - e_{15}k_{2i}\right]s_i, \quad F_{4i} = \left[e_{15}(1+k_{1i}) + \kappa_{11}k_{2i}\right]s_i \qquad (3.8.27)$$

$$F_{5i} = (c_{13}k_{1i} - e_{31}k_{2i})s_i^2 - \frac{c_{11}+c_{12}}{2}$$

$$\bar{a}(r) = \frac{\kappa_{33}\bar{\sigma}(r) + \bar{e}_{33}\bar{D}(r)}{c_{33}\kappa_{33} + e_{33}^2}, \quad \bar{b}(r) = \frac{c_{33}\bar{D}(r) - e_{33}\bar{\sigma}(r)}{c_{33}\kappa_{33} + e_{33}^2} \qquad (3.8.28)$$

$$\bar{c}(r) = \bar{\sigma}(r), \quad \bar{d}(r) = \bar{D}(r)$$

3.8.2   Reduction of crack problem to the solution of a Fredholm integral equation

1. A piezoelectric cylinder with a penny-shaped crack

To illustrate applications of the formulations presented in Section 3.8.1, we consider a cracked cylinder shown in Fig. 3.6 subjected to the following two cases of boundary conditions:

(1) In the first case we assume that the piezoelectric cylindrical surface is free from shear and is supported in such a way that the radial component of the displacement vector vanishes on the surface. The problem of determining the distribution of stress and electric displacement in the vicinity of the crack is equivalent to that of finding the distribution of stress and electric displacement in the semi-infinite cylinder $z \geqslant 0, 0 \leqslant r \leqslant a$ when its plane boundary $z = 0$ is subjected to the conditions

$$\sigma_{zz}(r,0) = 0, \quad 0 \leqslant r < a \tag{3.8.29}$$

$$u_z(r,0) = 0, \quad a < r < b \tag{3.8.30}$$

$$\phi(r,0) = 0, \quad a < r < b \tag{3.8.31}$$

$$\sigma_{rz}(r,0) = 0, \quad 0 \leqslant r < b \tag{3.8.32}$$

$$D_z(r,0^+) = D_z(r,0^-), \quad 0 \leqslant r < a \tag{3.8.33}$$

$$E_r(r,0^+) = E_r(r,0^+), \quad 0 \leqslant r < a \tag{3.8.34}$$

and its curved boundary $r = b$ is subjected to the conditions

$$u_r(b,z) = 0, \quad \sigma_{rz}(b,z) = 0, \quad D_r(b,z) = 0, \quad z \geqslant 0 \tag{3.8.35}$$

From the boundary conditions (3.8.31)~(3.8.35), and making use of the Fourier inversion theorem as well as the Hankel inversion theorem, we obtain

$$B_1(\xi) = M_1 B_1(\xi), \quad B_2(\xi) = M_2 B_1(\xi), \quad B_3(\xi) = M_3 B_3(\xi) \tag{3.8.36}$$

$$A_1(\xi) = \frac{1}{\Delta(\xi)} \sum_{i=1}^{3} N_{1i}(\xi) f_{1i}(\xi), \quad A_2(\xi) = \frac{1}{\Delta(\xi)} \sum_{i=1}^{3} N_{2i}(\xi) f_{1i}(\xi),$$

$$A_3(\xi) = \frac{1}{\Delta(\xi)} \sum_{i=1}^{3} N_{3i}(\xi) f_{3i}(\xi) \tag{3.8.37}$$

where

$$M_1 = 1, \quad M_2 = \frac{F_{31}k_{23}s_3 - F_{33}k_{21}s_1}{F_{33}k_{22}s_2 - F_{32}k_{23}s_3}, \quad M_3 = \frac{F_{32}k_{21}s_1 - F_{31}k_{22}s_2}{F_{33}k_{22}s_2 - F_{32}k_{23}s_3} \tag{3.8.38}$$

$$f_{1i}(\xi) = \frac{2}{\pi} \int_0^\infty \frac{\eta B_1(\eta) J_1(\eta b)}{\eta^2 s_i^2 + \xi^2} \, \mathrm{d}\eta \tag{3.8.39}$$

$$\Delta(\xi) = \left[h_{12}(\xi)h_{33}(\xi) - h_{32}(\xi)h_{13}(\xi)\right]h_{21}(\xi) + \left[h_{31}(\xi)h_{13}(\xi) - h_{11}(\xi)h_{33}(\xi)\right] \times$$
$$h_{22}(\xi) + \left[h_{11}(\xi)h_{32}(\xi) - h_{31}(\xi)h_{12}(\xi)\right]h_{23}(\xi) \tag{3.8.40}$$

$$N_{1i}(\xi) = \left[(h_{13}(\xi)h_{22}(\xi) - h_{12}(\xi)h_{23}(\xi)\right]g_{3i} + \left[(h_{12}(\xi)h_{33}(\xi) - h_{13}(\xi)h_{32}(\xi)\right]g_{2i} +$$
$$\left[(h_{23}(\xi)h_{32}(\xi) - h_{22}(\xi)h_{33}(\xi)\right]g_{1i} \tag{3.8.41}$$

$$N_{2i}(\xi) = \left[(h_{11}(\xi)h_{23}(\xi) - h_{21}(\xi)h_{13}(\xi)\right]g_{3i} + \left[(h_{13}(\xi)h_{31}(\xi) - h_{11}(\xi)h_{33}(\xi)\right]g_{2i} + \\ \left[(h_{21}(\xi)h_{33}(\xi) - h_{31}(\xi)h_{23}(\xi)\right]g_{1i} \tag{3.8.42}$$

$$N_{3i}(\xi) = \left[(h_{12}(\xi)h_{21}(\xi) - h_{11}(\xi)h_{22}(\xi)\right]g_{3i} + \left[(h_{11}(\xi)h_{32}(\xi) - h_{31}(\xi)h_{12}(\xi)\right]g_{2i} + \\ \left[(h_{22}(\xi)h_{31}(\xi) - h_{21}(\xi)h_{32}(\xi)\right]g_{1i} \tag{3.8.43}$$

with

$$h_{1i}(\xi) = \frac{F_{4i}}{s_i^2}I_1\left(\frac{\xi b}{s_i}\right), \qquad g_{1i} = F_{4i}M_i \tag{3.8.44}$$

$$h_{2i}(\xi) = \frac{F_{3i}}{s_i^2}I_1\left(\frac{\xi b}{s_i}\right), \qquad g_{2i} = F_{3i}M_i s_i \tag{3.8.45}$$

$$h_{3i}(\xi) = \frac{1}{s_i}I_1\left(\frac{\xi b}{s_i}\right), \qquad g_{3i} = M_i s_i \tag{3.8.46}$$

From Eqs.(3.8.29)~(3.8.30), we can obtain a system of dual integral equations

$$-\int_0^\infty \xi\left[\frac{F_{11}}{s_1^2}I_0\left(\frac{\xi r}{s_1}\right)A_1(\xi) + \frac{F_{12}}{s_2^2}I_0\left(\frac{\xi r}{s_2}\right)A_2(\xi) + \frac{F_{13}}{s_3^2}I_0\left(\frac{\xi r}{s_3}\right)A_3(\xi)\right]d\xi +$$

$$\int_0^\infty \xi(M_1F_{11} + M_2F_{12} + M_3F_{13})B_1(\xi)J_0(\xi r)d\xi = -\bar{c}(r), \quad 0 \leqslant r < a \tag{3.8.47}$$

$$\int_0^\infty (M_1k_{11}s_1 + M_2k_{12}s_2 + M_3k_{13}s_3)B_1(\xi)J_0(\xi r)d\xi = 0, \quad a < r < b \tag{3.8.48}$$

Eq.(3.8.47) and Eq.(3.8.48) can be solved using the function $\psi(\alpha)$ defined by

$$B_1(\xi) = \int_0^a \psi(\alpha)\sin(\xi\alpha)d\alpha \tag{3.8.49}$$

where $\psi(0) = 0$.

Using the following solutions of integrals:

$$\int_0^\infty \sin(sz)e^{-uz}dz = \frac{s}{s^2 + u^2}, \qquad \int_0^\infty \cos(sz)e^{-uz}dz = \frac{u}{s^2 + u^2} \tag{3.8.50}$$

$$\int_0^t \frac{rI_0(\xi r)}{\sqrt{t^2 - r^2}}dr = \frac{\sinh(\xi t)}{\xi}, \qquad \int_0^\infty \frac{J_0(ru)\sin(ut)}{s^2 + u^2}du = \frac{\sinh(st)K_0(rs)}{s}, \quad t < r$$

$$\tag{3.8.51}$$

$$\int_0^\infty \frac{uJ_1(ru)\sin(ut)}{s^2 + u^2}du = \sinh(st)K_1(rs), \quad t < r \tag{3.8.52}$$

$$\int_0^\infty \frac{u^2 J_0(ru)\sin(ut)}{s^2 + u^2}du = -s \cdot \sinh(st)K_0(rs), \quad t < r \tag{3.8.53}$$

$$\int_0^\infty \frac{u^2 J_2(ru)\sin(ut)}{s^2+u^2}\,\mathrm{d}u = s\cdot\sinh(st)K_2(rs), \quad t<r \tag{3.8.54}$$

As well as the solution of integral equation

$$\int_0^\infty \frac{f(t)}{(x^2-t^2)^\alpha}\,\mathrm{d}t = g(x), \quad 0<\alpha<1, \quad a<x<b \tag{3.8.55}$$

which is given by

$$f(t)=\frac{2\sin\pi\alpha}{\pi}\frac{\mathrm{d}}{\mathrm{d}t}\int_0^t \frac{ug(u)}{(t^2-u^2)^{1-\alpha}}\,\mathrm{d}u, \quad a<t<b \tag{3.8.56}$$

We can obtain a Fredholm integral equation of the second kind in the following form:

$$\psi(\alpha)+\int_0^\alpha \psi(\beta)L(\alpha,\beta)\mathrm{d}\beta = \frac{2}{\pi m_0}\int_0^\alpha \frac{r\bar{c}(r)}{\sqrt{\alpha^2-r^2}}\,\mathrm{d}r \tag{3.8.57}$$

where

$$L(\alpha,\beta)=\frac{4}{\pi^2 m_0}\sum_{j=1}^3 \frac{F_{1j}}{s_j}\int_0^\infty \frac{1}{\Delta(\xi)}\sinh\left(\frac{\xi\alpha}{s_j}\right)\sum_{i=1}^3 \frac{1}{s_i^2}N_{ji}(\xi)\sinh\left(\frac{\xi\beta}{s_i}\right)K_1\left(\frac{\xi b}{s_i}\right)\mathrm{d}\xi$$

$$\tag{3.8.58}$$

(2) In the second case we assume that the piezoelectric cylindrical surface is traction-free. The conditions (3.8.29)~(3.8.34) remain the same, but the boundary condition (3.8.35) is replaced by

$$\sigma_{rr}(b,z)=0, \quad \sigma_{rz}(b,z)=0, \quad D_r(b,z)=0, \quad z\geqslant 0 \tag{3.8.59}$$

Using a procedure similar to that in the case (1), we have

$$A_1(\xi)=\frac{1}{\Delta(\xi)}\sum_{i=1}^3 \left[N_{1i}(\xi)f_{1i}(\xi)+P_{1i}(\xi)f_{2i}(\xi)+W_{1i}(\xi)f_{3i}(\xi)\right] \tag{3.8.60}$$

$$A_2(\xi)=\frac{1}{\Delta(\xi)}\sum_{i=1}^3 \left[N_{2i}(\xi)f_{1i}(\xi)+P_{2i}(\xi)f_{2i}(\xi)+W_{2i}(\xi)f_{3i}(\xi)\right] \tag{3.8.61}$$

$$A_3(\xi)=\frac{1}{\Delta(\xi)}\sum_{i=1}^3 \left[N_{3i}(\xi)f_{1i}(\xi)+P_{3i}(\xi)f_{2i}(\xi)+W_{3i}(\xi)f_{3i}(\xi)\right] \tag{3.8.62}$$

in which

$$f_{2i}(\xi)=\frac{2}{\pi}\int_0^\infty \frac{\eta^2 B_1(\eta)J_0(\eta b)}{\eta^2 s_i^2+\xi^2}\,\mathrm{d}\eta, \quad f_{3i}(\xi)=\frac{2}{\pi}\int_0^\infty \frac{\eta^2 B_1(\eta)J_2(\eta b)}{\eta^2 s_i^2+\xi^2}\,\mathrm{d}\eta \tag{3.8.63}$$

$$N_{1i}(\xi)=\left[h_{52}(\xi)-h_{42}(\xi)\right]\left[h_{33}(\xi)g_{2i}-h_{23}(\xi)g_{3i}\right]+$$

$$\left[h_{53}(\xi)-h_{43}(\xi)\right]\left[h_{22}(\xi)g_{3i}-h_{32}(\xi)g_{2i}\right] \tag{3.8.64}$$

$$P_{1i}(\xi)=\frac{1}{\xi}\left[h_{23}(\xi)h_{32}(\xi)-h_{22}(\xi)h_{33}(\xi)\right]g_{5i} \tag{3.8.65}$$

$$W_{1i}(\xi) = \frac{1}{\xi}\left[h_{23}(\xi)h_{32}(\xi) - h_{22}(\xi)h_{33}(\xi)\right]g_{4i} \qquad (3.8.66)$$

$$N_{2i}(\xi) = \left[h_{53}(\xi) - h_{43}(\xi)\right]\left[h_{31}(\xi)g_{2i} - h_{21}(\xi)g_{3i}\right] +$$
$$\left[h_{51}(\xi) - h_{41}(\xi)\right]\left[h_{23}(\xi)g_{3i} - h_{33}(\xi)g_{2i}\right] \qquad (3.8.67)$$

$$P_{2i}(\xi) = \frac{1}{\xi}\left[h_{21}(\xi)h_{33}(\xi) - h_{23}(\xi)h_{31}(\xi)\right]g_{5i} \qquad (3.8.68)$$

$$W_{2i}(\xi) = \frac{1}{\xi}\left[h_{21}(\xi)h_{33}(\xi) - h_{23}(\xi)h_{31}(\xi)\right]g_{4i} \qquad (3.8.69)$$

$$N_{3i}(\xi) = \left[h_{51}(\xi) - h_{41}(\xi)\right]\left[h_{32}(\xi)g_{2i} - h_{22}(\xi)g_{3i}\right] +$$
$$\left[h_{52}(\xi) - h_{42}(\xi)\right]\left[h_{21}(\xi)g_{3i} - h_{31}(\xi)g_{2i}\right] \qquad (3.8.70)$$

$$P_{3i}(\xi) = \frac{1}{\xi}\left[h_{22}(\xi)h_{31}(\xi) - h_{21}(\xi)h_{32}(\xi)\right]g_{5i} \qquad (3.8.71)$$

$$W_{3i}(\xi) = \frac{1}{\xi}\left[h_{22}(\xi)h_{31}(\xi) - h_{21}(\xi)h_{32}(\xi)\right]g_{4i} \qquad (3.8.72)$$

$$\Delta(\xi) = \left\{\left[-h_{53}(\xi) + h_{43}(\xi)\right]h_{32}(\xi) + \left[h_{52}(\xi) - h_{42}(\xi)\right]h_{33}(\xi)\right\}h_{21}(\xi) +$$
$$\left\{\left[h_{53}(\xi) - h_{43}(\xi)\right]h_{31}(\xi) + \left[-h_{51}(\xi) + h_{41}(\xi)\right]h_{33}(\xi)\right\}h_{22}(\xi) +$$
$$\left\{\left[h_{51}(\xi) - h_{41}(\xi)\right]h_{32}(\xi) + \left[-h_{52}(\xi) + h_{42}(\xi)\right]h_{31}(\xi)\right\}h_{23}(\xi) \qquad (3.8.73)$$

with

$$h_{1i}(\xi) = \frac{F_{4i}}{s_i^2}I_1\left(\frac{\xi b}{s_i}\right), \qquad g_{1i} = F_{4i}M_i \qquad (3.8.74)$$

$$h_{2i}(\xi) = \frac{F_{3i}}{s_i^2}I_1\left(\frac{\xi b}{s_i}\right), \qquad g_{2i} = F_{3i}M_i s_i \qquad (3.8.75)$$

$$h_{3i}(\xi) = \frac{1}{s_i}I_1\left(\frac{\xi b}{s_i}\right), \qquad g_{3i} = M_i s_i \qquad (3.8.76)$$

$$h_{4i}(\xi) = \frac{c_{11} - c_{12}}{2}\frac{1}{s_i^2}I_2\left(\frac{\xi b}{s_i}\right), \qquad g_{4i} = \frac{c_{11} - c_{12}}{2}M_i s_i \qquad (3.8.77)$$

$$h_{5i}(\xi) = \frac{F_{5i}}{s_i^2}I_0\left(\frac{\xi b}{s_i}\right), \qquad g_{5i} = F_{5i}M_i s_i \qquad (3.8.78)$$

In the above equations, $B_i(\eta)$ and $f_{1i}(\xi)$ are the same to those in the case (1).

Therefore, the corresponding Fredholm integral equation for the case (2) can be obtained and has exactly the same form as Eq.(3.8.57) except that the kernel $L(\alpha, \beta)$ is replaced by

$$L(\alpha,\beta) = \frac{4}{\pi^2 m_0} \sum_{j=1}^{3} \frac{F_{1j}}{s_j} \int_0^\infty \frac{1}{\Delta(\xi)} \sinh\left(\frac{\xi\alpha}{s_j}\right) \sum_{i=1}^{3} \frac{1}{s_i^2} \sinh\left(\frac{\xi\beta}{s_i}\right) \times$$

$$\left[N_{ji}(\xi)K_1\left(\frac{\xi b}{s_i}\right) - \frac{\xi}{s_i}P_{ji}(\xi)K_0\left(\frac{\xi b}{s_i}\right) + \frac{\xi}{s_i}W_{ji}(\xi)K_2\left(\frac{\xi b}{s_i}\right) \right] d\xi \quad (3.8.79)$$

The generalized stress intensity factor can thus be expressed in terms of $\psi(\xi)$ as [57]

$$K^\sigma = K_I = \lim_{r\to a^+} \sqrt{2\pi(r-a)}\,\sigma_z(r,0) = \sqrt{\frac{\pi}{a}}m_0\psi(a) \quad (3.8.80)$$

$$K^D = \lim_{r\to a^+} \sqrt{2\pi(r-a)}\,D_z(r,0) = \sqrt{\frac{\pi}{a}}m_1\psi(a) \quad (3.8.81)$$

$$K^\varepsilon = \lim_{r\to a^+} \sqrt{2\pi(r-a)}\,\varepsilon_z(r,0) = \sqrt{\frac{\pi}{a}}m_2\psi(a) \quad (3.8.82)$$

$$K^E = \lim_{r\to a^+} \sqrt{2\pi(r-a)}\,E_z(r,0) = \sqrt{\frac{\pi}{a}}m_3\psi(a) \quad (3.8.83)$$

in which

$$m_0 = -(M_1 F_{11} + M_2 F_{12} + M_3 F_{13}) \quad (3.8.84)$$

$$m_1 = -(F_{21}M_1 + F_{22}M_2 + F_{23}M_3) \quad (3.8.85)$$

$$m_2 = -(k_{11}s_1^2 M_1 + k_{12}s_1^2 M_2 + k_{13}s_1^3 M_3) \quad (3.8.86)$$

$$m_3 = -(k_{21}s_1^2 M_1 + k_{22}s_2^2 M_2 + k_{23}s_3^2 M_3) \quad (3.8.87)$$

and K^σ, K^D, K^ε and K^E are the stress intensity factor, electric displacement intensity factor, strain intensity factor and electric field intensity factor, respectively.

2. Coated piezoelectric fibre with a penny-shaped crack

To prevent piezoelectric cylinders such as that considered above from mechanical failure and to increase the bonding strength of the interface between the fibre and matrix during service, these cylinders are often coated with an elastic layer [55,58]. It is therefore desirable to understand effect of the coating layer on the fracture behavior of piezoelectric fibre composites. To this end, consider a piezoelectric fibre with a finite elastic coating and containing a centered penny-shaped crack of radius a under axisymmetric electromechanical loading (Fig. 3.7). The fibre is assumed to be a transversely isotropic piezoelectric material with the poling direction parallel to the z-axis, and the elastic coating is also a transversely isotropic material. They are subjected to the far-field of a normal strain, $\varepsilon_z = \overline{\varepsilon}(r)$ and a normal electric loading, $E_z = \overline{E}(r)$.

The boundary conditions of this problem are [55]

$$\sigma_{zz}(r,0) = 0, \quad 0 \leqslant r < a \tag{3.8.88}$$

$$u_z(r,0) = 0, \quad a < r < b \tag{3.8.89}$$

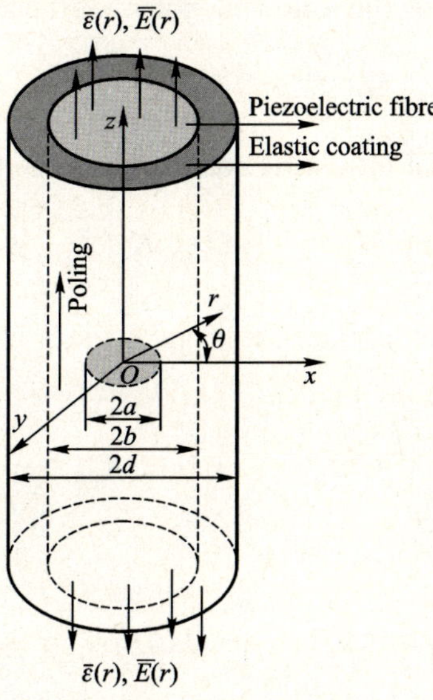

Fig.3.7 Piezoelectric fibre with a finite elastic coating and containing a penny-shaped crack under mechanical and electrical loading

$$\phi(r,0) = 0, \quad a < r < b \tag{3.8.90}$$

$$\sigma_{rz}(r,0) = 0, \quad 0 \leqslant r < b \tag{3.8.91}$$

$$D_z(r,0^+) = D_z(r,0^-), \quad 0 \leqslant r < a \tag{3.8.92}$$

$$E_r(r,0^+) = E_r(r,0^+), \quad 0 \leqslant r < a \tag{3.8.93}$$

$$D_r(b,z) = 0, \quad 0 < z < \infty \tag{3.8.94}$$

The continuity and loading conditions of this problem are defined by:

(1) The continuity conditions for elastic displacements and tractions at the interface between the fibre and elastic coating are

$$u_z(b,z) = u_z^c(b,z), \quad 0 < z < \infty, \quad u_r(b,z) = u_r^c(b,z), \quad 0 < z < \infty \tag{3.8.95}$$

$$\sigma_{rr}(b,z) = \sigma_{rr}^c(b,z), \quad 0 < z < \infty, \quad \sigma_{rz}(b,z) = \sigma_{rz}^c(b,z), \quad 0 < z < \infty \tag{3.8.96}$$

(2) Loading conditions at infinity are

$$\varepsilon_z(r,\infty)=\bar{\varepsilon}(r)\,,\quad E_z(r,\infty)=\bar{E}(r)\,,\quad \varepsilon_z^c(r,\infty)=\bar{\varepsilon}(r) \tag{3.8.97}$$

(3) Loading conditions on the outer surface of the coating are

$$u_r^c(d,z)=0\,,\quad \sigma_{rz}^c(d,z)=0\,,\quad 0<z<\infty \tag{3.8.98}$$

The expressions for electric and elastic fields in the cracked fibre are given by Eqs.(3.8.19)~(3.8.26). The elastic displacement field in the coating layer can be obtained by considering the following potential functions:

$$u_z^c=\sum_{i=1}^{2}k_i^c\frac{\partial\Phi_i^c}{\partial z}\,,\quad u_r^c=\sum_{i=1}^{2}\frac{\partial\Phi_i^c}{\partial r} \tag{3.8.99}$$

The potential functions for the elastic coating layer can then be written as:

$$\Phi_i^c(r,z)=\int_0^\infty\frac{1}{\xi}\left[C_i(\xi)I_0\left(\frac{\xi r}{s_i^c}\right)+D_i(\xi)K_0\left(\frac{\xi r}{s_i^c}\right)\right]\cos(\xi z)\mathrm{d}\xi \tag{3.8.100}$$

With Eq (3.8.100), elastic displacements and stresses in the elastic coating can be given in the form

$$u_z^c(r,z)=-\sum_{i=1}^{2}k_i^c\int_0^\infty\left[C_i(\xi)I_0\left(\frac{\xi r}{s_i^c}\right)+D_i(\xi)K_0\left(\frac{\xi r}{s_i^c}\right)\right]\sin(\xi z)\mathrm{d}\xi+\bar{a}(r)z$$

$$\tag{3.8.101}$$

$$u_r^c(r,z)=\sum_{i=1}^{2}\frac{1}{s_i^c}\int_0^\infty\left[C_i(\xi)I_1\left(\frac{\xi r}{s_i^c}\right)-D_i(\xi)K_1\left(\frac{\xi r}{s_i^c}\right)\right]\cos(\xi z)\mathrm{d}\xi \tag{3.8.102}$$

$$\sigma_{rr}^c(r,z)=-\sum_{i=1}^{2}\frac{F_{3i}^c}{s_i^{c2}}\int_0^\infty\xi\left[C_i(\xi)I_0\left(\frac{\xi r}{s_i^c}\right)+D_i(\xi)K_0\left(\frac{\xi r}{s_i^c}\right)\right]\cos(\xi z)\mathrm{d}\xi+$$

$$\frac{c_{11}^c-c_{12}^c}{2}\sum_{i=1}^{2}\frac{1}{s_i^{c2}}\int_0^\infty\xi\left[C_i(\xi)I_2\left(\frac{\xi r}{s_i^c}\right)+D_i(\xi)K_2\left(\frac{\xi r}{s_i^c}\right)\right]\cos(\xi z)\mathrm{d}\xi$$

$$\tag{3.8.103}$$

$$\sigma_{rz}^c(r,z)=-\sum_{i=1}^{2}\frac{F_{2i}^c}{s_i^{c2}}\int_0^\infty\xi\left[C_i(\xi)I_1\left(\frac{\xi r}{s_i^c}\right)-D_i(\xi)K_1\left(\frac{\xi r}{s_i^c}\right)\right]\sin(\xi z)\mathrm{d}\xi \tag{3.8.104}$$

in which

$$F_{1i}^c=c_{33}^c k_i^c s_i^{c2}-c_{13}^c\,,\quad F_{2i}^c=c_{44}^c(1+k_i^c)s_i^c\,,\quad F_{3i}^c=c_{13}^c k_i^c s_i^{c2}-\frac{c_{11}^c+c_{12}^c}{2}$$

$$\tag{3.8.105}$$

where the superscript "c" refers to the coating.

Using the boundary conditions (3.8.88)~(3.8.96) and (3.8.98), the Fourier inversion theorem and the Hankel inversion theorem, we obtain

$$A_1(\xi) = \frac{1}{\Delta(\xi)} \sum_{i=1}^{3} \left[N_{1i}(\xi) f_{1i}(\xi) + P_{1i}(\xi) f_{2i}(\xi) + W_{1i}(\xi) f_{3i}(\xi) + Y_{1i}(\xi) f_{4i}(\xi) \right]$$

$$(3.8.106)$$

$$A_2(\xi) = \frac{1}{\Delta(\xi)} \sum_{i=1}^{3} \left[N_{2i}(\xi) f_{1i}(\xi) + P_{2i}(\xi) f_{2i}(\xi) + W_{2i}(\xi) f_{3i}(\xi) + Y_{2i}(\xi) f_{4i}(\xi) \right]$$

$$(3.8.107)$$

$$A_3(\xi) = \frac{1}{\Delta(\xi)} \sum_{i=1}^{3} \left[N_{3i}(\xi) f_{1i}(\xi) + P_{3i}(\xi) f_{2i}(\xi) + W_{3i}(\xi) f_{3i}(\xi) + Y_{3i}(\xi) f_{4i}(\xi) \right]$$

$$(3.8.108)$$

$$C_1(\xi) = \sum_{i=1}^{3} \left[M_{3i} A_i(\xi) + M_{4i}(\xi) f_{2i}(\xi) \right]$$

$$(3.8.109)$$

$$C_2(\xi) = \sum_{i=1}^{3} \left[M_{5i} A_i(\xi) + M_{6i}(\xi) f_{2i}(\xi) \right]$$

$$D_1(\xi) = \sum_{i=1}^{2} M_{1i} C_i(\xi), \quad D_2(\xi) = \sum_{i=1}^{2} M_{2i} C_i(\xi) \qquad (3.8.110)$$

where

$$N_{1i}(\xi) = \left[H_{13}(\xi) H_{32}(\xi) - H_{12}(\xi) H_{33}(\xi) \right] g_{1i}(\xi) +$$
$$\left[h_{13}(\xi) H_{32}(\xi) - H_{12}(\xi) H_{33}(\xi) \right] h_{5i}(\xi) \qquad (3.8.111)$$

$$N_{2i}(\xi) = \left[H_{11}(\xi) H_{33}(\xi) - H_{31}(\xi) H_{13}(\xi) \right] g_{1i}(\xi) +$$
$$\left[h_{11}(\xi) H_{33}(\xi) - H_{13}(\xi) H_{31}(\xi) \right] h_{5i}(\xi) \qquad (3.8.112)$$

$$N_{3i}(\xi) = \left[H_{12}(\xi) H_{31}(\xi) - H_{11}(\xi) H_{32}(\xi) \right] g_{1i}(\xi) +$$
$$\left[h_{12}(\xi) H_{31}(\xi) - H_{11}(\xi) H_{32}(\xi) \right] h_{5i}(\xi) \qquad (3.8.113)$$

$$P_{1i}(\xi) = \left[h_{13}(\xi) H_{32}(\xi) - h_{12}(\xi) H_{33}(\xi) \right] H_{2i}(\xi) +$$
$$\left[h_{12}(\xi) H_{13}(\xi) - h_{13}(\xi) H_{12}(\xi) \right] H_{4i}(\xi) \qquad (3.8.114)$$

$$P_{2i}(\xi) = \left[h_{11}(\xi) H_{33}(\xi) - h_{13}(\xi) H_{31}(\xi) \right] H_{2i}(\xi) +$$
$$\left[h_{13}(\xi) H_{11}(\xi) - h_{11}(\xi) H_{13}(\xi) \right] H_{4i}(\xi) \qquad (3.8.115)$$

$$P_{3i}(\xi) = \left[h_{12}(\xi) H_{31}(\xi) - h_{11}(\xi) H_{32}(\xi) \right] H_{2i}(\xi) +$$
$$\left[h_{11}(\xi) H_{12}(\xi) - h_{12}(\xi) H_{11}(\xi) \right] H_{4i}(\xi) \qquad (3.8.116)$$

$$W_{1i}(\xi) = \left[h_{13}(\xi) H_{12}(\xi) - h_{12}(\xi) H_{13}(\xi) \right] h_{8i}(\xi) \qquad (3.8.117)$$

$$W_{2i}(\xi) = \left[h_{11}(\xi) H_{13}(\xi) - h_{13}(\xi) H_{11}(\xi) \right] h_{8i}(\xi) \qquad (3.8.118)$$

$$W_{3i}(\xi) = \left[h_{12}(\xi) H_{11}(\xi) - h_{11}(\xi) H_{12}(\xi) \right] h_{8i}(\xi) \qquad (3.8.119)$$

$$Y_{1i}(\xi) = \left[h_{13}(\xi) H_{12}(\xi) - h_{12}(\xi) H_{13}(\xi) \right] h_{9i}(\xi) \qquad (3.8.120)$$

$$Y_{2i}(\xi) = \left[h_{11}(\xi) H_{13}(\xi) - h_{13}(\xi) H_{11}(\xi) \right] h_{9i}(\xi) \tag{3.8.121}$$

$$Y_{3i}(\xi) = \left[h_{12}(\xi) H_{11}(\xi) - h_{11}(\xi) H_{12}(\xi) \right] h_{9i}(\xi) \tag{3.8.122}$$

$$\Delta(\xi) = h_{11}(\xi) \left[H_{13}(\xi) H_{32}(\xi) - H_{12}(\xi) H_{33}(\xi) \right] +$$
$$h_{12}(\xi) \left[H_{11}(\xi) H_{33}(\xi) - H_{31}(\xi) H_{13}(\xi) \right] +$$
$$h_{13}(\xi) \left[H_{12}(\xi) H_{31}(\xi) - H_{11}(\xi) H_{32}(\xi) \right] \tag{3.8.123}$$

where $f_{ji}(\xi)$, $h_{ji}(\xi)$, $g_{ji}(\xi)$, and $H_{ji}(\xi)$ are defined by

$$f_{1i}(\xi) = \frac{2}{\pi} \int_0^\infty \frac{\eta B_1(\eta) J_1(\eta b)}{\eta^2 s_i^2 + \xi^2} \, \mathrm{d}\eta, \quad f_{2i}(\xi) = \frac{2}{\pi} \int_0^\infty \frac{B_1(\eta) J_0(\eta b)}{\eta^2 s_i^2 + \xi^2} \, \mathrm{d}\eta \tag{3.8.124}$$

$$f_{3i}(\xi) = \frac{2}{\pi} \int_0^\infty \frac{\eta^2 B_1(\eta) J_0(\eta b)}{\eta^2 s_i^2 + \xi^2} \, \mathrm{d}\eta, \quad f_{4i}(\xi) = \frac{2}{\pi} \int_0^\infty \frac{\eta^2 B_1(\eta) J_2(\eta b)}{\eta^2 s_i^2 + \xi^2} \, \mathrm{d}\eta \tag{3.8.125}$$

$$h_{1i}(\xi) = \frac{F_{4i}}{s_i^2} I_1 \left(\frac{\xi b}{s_i} \right), \quad g_{1i}(\xi) = F_{4i} M_i(\xi) \tag{3.8.126}$$

$$h_{2i}(\xi) = k_{1i} I_0 \left(\frac{\xi b}{s_i} \right), \quad g_{2i} = k_i^c I_0 \left(\frac{\xi b}{s_i^c} \right) \tag{3.8.127}$$

$$h_{3i} = k_{1i} M_i s_i, \quad g_{3i} = k_i^c K_0 \left(\frac{\xi b}{s_i^c} \right) \tag{3.8.128}$$

$$h_{4i} = \frac{1}{s_i} I_1 \left(\frac{\xi b}{s_i} \right), \quad g_{4i} = \frac{1}{s_i^c} \left(\frac{\xi b}{s_i^c} \right) \tag{3.8.129}$$

$$h_{5i} = M_i s_i, \quad g_{5i} = \frac{1}{s_i^c} K_1 \left(\frac{\xi b}{s_i^c} \right) \tag{3.8.130}$$

$$h_{6i} = \frac{F_{5i}}{s_i^2} I_0 \left(\frac{\xi b}{s_i} \right), \quad g_{6i} = \frac{F_{3i}^c}{(s_i^c)^2} I_0 \left(\frac{\xi b}{s_i^c} \right) \tag{3.8.131}$$

$$h_{7i} = \frac{c_{11} - c_{12}}{2} \frac{1}{s_i^2} I_2 \left(\frac{\xi b}{s_i} \right), \quad g_{7i} = \frac{c_{11}^c - c_{12}^c}{2} \frac{1}{(s_i^c)^2} I_2 \left(\frac{\xi b}{s_i^c} \right) \tag{3.8.132}$$

$$h_{8i} = F_{5i} M_i s_i, \quad g_{8i} = \frac{F_{3i}^c}{s_i^c} K_0 \left(\frac{\xi b}{s_i^c} \right) \tag{3.8.133}$$

$$h_{9i} = \frac{c_{11} - c_{12}}{2} M_i s_i, \quad g_{9i} = \frac{c_{11}^c - c_{12}^c}{2} \frac{1}{(s_i^c)^2} K_2 \left(\frac{\xi b}{s_i^c} \right) \tag{3.8.134}$$

$$h_{10i} = \frac{F_{3i}}{s_i^2} I_1 \left(\frac{\xi b}{s_i} \right), \quad g_{10i} = \frac{F_{2i}^c}{(s_i^c)^2} I_1 \left(\frac{\xi b}{s_i^c} \right) \tag{3.8.135}$$

$$h_{11i} = F_{3i} M_i, \quad g_{11i} = \frac{F_{2i}^c}{(s_i^c)^2} K_1\left(\frac{\xi b}{s_i^c}\right) \tag{3.8.136}$$

$$g_{12i} = \frac{1}{s_i^c} I_1\left(\frac{\xi d}{s_i^c}\right), \quad g_{13i} = \frac{1}{s_i^c} K_1\left(\frac{\xi d}{s_i^c}\right) \tag{3.8.137}$$

$$g_{14i} = \frac{F_{2i}^c}{(s_i^c)^2} I_1\left(\frac{\xi d}{s_i^c}\right), \quad g_{15i} = \frac{F_{2i}^c}{(s_i^c)^2} K_1\left(\frac{\xi d}{s_i^c}\right) \tag{3.8.138}$$

$$M_{1i}(\xi) = \frac{g_{132}(\xi)g_{14i}(\xi) - g_{152}(\xi)g_{12i}(\xi)}{g_{132}(\xi)g_{151}(\xi) - g_{152}(\xi)g_{131}(\xi)} \tag{3.8.139}$$

$$M_{2i}(\xi) = \frac{g_{131}(\xi)g_{14i}(\xi) - g_{151}(\xi)g_{12i}(\xi)}{g_{131}(\xi)g_{152}(\xi) - g_{151}(\xi)g_{132}(\xi)} \tag{3.8.140}$$

$$G_{1i}(\xi) = g_{2i}(\xi) + g_{31}(\xi)M_{1i}(\xi) + g_{32}(\xi)M_{2i}(\xi) \tag{3.8.141}$$

$$G_{2i}(\xi) = g_{10i}(\xi) - g_{111}(\xi)M_{1i}(\xi) - g_{112}(\xi)M_{2i}(\xi) \tag{3.8.142}$$

$$M_{3i}(\xi) = \frac{G_{22}(\xi)h_{2i}(\xi) - G_{12}(\xi)h_{10i}(\xi)}{G_{22}(\xi)G_{11}(\xi) - G_{12}(\xi)G_{21}} \tag{3.8.143}$$

$$M_{4i}(\xi) = \frac{G_{22}(\xi)h_{3i}(\xi)\xi + G_{12}(\xi)h_{11i}(\xi)}{G_{22}(\xi)G_{11}(\xi) - G_{12}(\xi)G_{21}} \tag{3.8.144}$$

$$M_{5i}(\xi) = \frac{G_{21}(\xi)h_{2i}(\xi) - G_{11}(\xi)h_{10i}(\xi)}{G_{21}(\xi)G_{12}(\xi) - G_{11}(\xi)G_{22}} \tag{3.8.145}$$

$$M_{6i}(\xi) = \frac{G_{21}(\xi)h_{3i}(\xi) + G_{11}(\xi)h_{11i}(\xi)}{G_{21}(\xi)G_{12}(\xi) - G_{11}(\xi)G_{22}} \tag{3.8.146}$$

$$t_i(\xi) = g_{4i}(\xi) - g_{51}(\xi)M_{1i}(\xi) - g_{52}(\xi)M_{2i}(\xi) \tag{3.8.147}$$

$$r_i(\xi) = \left[g_{7i}(\xi) - g_{6i}(\xi)\right]\xi + \left[g_{91}(\xi) - g_{81}(\xi)\right]\xi M_{1i}(\xi) +$$
$$\left[g_{92}(\xi) - g_{82}(\xi)\right]\xi M_{2i}(\xi) \tag{3.8.148}$$

$$H_{1i}(\xi) = t_i(\xi)M_{3i}(\xi) + t_2(\xi)M_{5i}(\xi) - h_{4i}(\xi) \tag{3.8.149}$$

$$H_{2i}(\xi) = t_1(\xi)M_{4i}(\xi) + t_2(\xi)M_{6i}(\xi) \tag{3.8.150}$$

$$H_{3i}(\xi) = r_1(\xi)M_{3i}(\xi) + r_2(\xi)M_{5i}(\xi) - \left[h_{7i}(\xi) - h_{6i}(\xi)\right]\xi \tag{3.8.151}$$

$$H_{4i}(\xi) = r_1(\xi)M_{4i}(\xi) + r_2(\xi)M_{6i}(\xi) \tag{3.8.152}$$

From Eqs.(3.8.88), (3.8.89) and (3.8.106)~(3.8.110), the following system of dual integral equations can be deduced:

$$-\int_0^\infty \xi \left[\frac{F_{11}}{s_1^2} I_0\left(\frac{\xi r}{s_1}\right) A_1(\xi) + \frac{F_{12}}{s_2^2} I_0\left(\frac{\xi r}{s_2}\right) A_2(\xi) + \frac{F_{13}}{s_3^2} I_0\left(\frac{\xi r}{s_3}\right) A_3(\xi)\right] \mathrm{d}\xi +$$

$$\int_0^\infty \xi (M_1 F_{11} + M_2 F_{12} + M_3 F_{13}) B_1(\xi) J_0(\xi r)\mathrm{d}\xi = -\overline{c}(r), \quad 0 \leqslant r < a$$

$$\tag{3.8.153}$$

$$\int_0^\infty (M_1 k_{11} s_1 + M_2 k_{12} s_2 + M_3 k_{13} s_3)\, B_1(\xi) J_0(\xi r)\, \mathrm{d}\xi = 0, \qquad a < r < b$$

$$(3.8.154)$$

where M_i are given in Eq.(3.8.38).

It is found that the above dual integral equations (3.8.153) and (3.8.154) also yield the Fredholm integral equation (3.8.57), except that the kernel $L(\alpha, \beta)$ is replaced by

$$L(\alpha, \beta) = \frac{4}{\pi^2 m_0} \sum_{j=1}^{3} \frac{F_{1j}}{s_j} \int_0^\infty \frac{1}{\Delta(\xi)} \sinh\left(\frac{\xi\alpha}{s_j}\right) \sum_{i=1}^{3} \frac{1}{s_i^2} \sinh\left(\frac{\xi\beta}{s_i}\right) \times$$

$$\left[N_{ji}(\xi) K_1\left(\frac{\xi b}{s_i}\right) + \frac{s_i}{\xi} P_{ji}(\xi) K_0\left(\frac{\xi b}{s_i}\right) - \right.$$

$$\left. \frac{\xi}{s_i} W_{ji}(\xi) K_0\left(\frac{\xi b}{s_i}\right) + \frac{\xi}{s_i} Y_{ji}(\xi) K_2\left(\frac{\xi b}{s_i}\right) \right] \mathrm{d}\xi \qquad (3.8.155)$$

3.8.3   Numerical assessment

To illustrate applications of the solutions presented above, consider the cracked piezoelectric fibre composite as shown in Fig. 3.7. Material properties used in the analysis are

(1) Piezoelectric fibre.
Elastic constants $(10^{10}\ \mathrm{N/m^2})$: $c_{11} = 16.8$, $c_{12} = 6.0$, $c_{33} = 16.3$, $c_{44} = 2.71$.
Piezoelectric constants $(\mathrm{C/m^2})$: $e_{15} = 4.6$, $e_{31} = -0.9$, $e_{33} = 7.1$.
Dielectric permittivities $(10^{-10}\ \mathrm{F/m})$: $d_{11} = 36$, $d_{33} = 34$.

(2) Elastic coating.
Elastic constants $(10^{10}\ \mathrm{N/m^2})$: $c_{11} = 0.83$, $c_{12} = 0.28$, $c_{13} = 0.03$, $c_{33} = 8.68$, $c_{44} = 0.42$.

It can be seen from Eqs.(3.8.80)~(3.8.83) that determination of the stress intensity factor requires the solution of the function $\psi(\xi)$. The Fredholm integral equation of the second kind (3.8.57) can be solved numerically using a Gaussian quadrature formula. In the calculation, $b = 40\ \mathrm{mm}$, $\bar{\varepsilon}(r) = 1.0 \times 10^{-5}$, and $\bar{E}(r) = 10^6$ V/m are assumed.

The variations in the normalized stress intensity factor with the ratio of crack radius to fibre radius a/b under different thicknesses and elastic constants of the coating are shown in Fig.3.8 and Fig.3.9. It can be seen from

Fig.3.8 that the stress intensity factor decreases with increase of the ratio a/b. It is also evident that the thickness of the elastic coating has an important effect on the stress intensity factor, and greater thickness will lead to a higher decay rate and a smaller value of the stress intensity factor. This means that a thicker coating layer can reduce crack propagation.

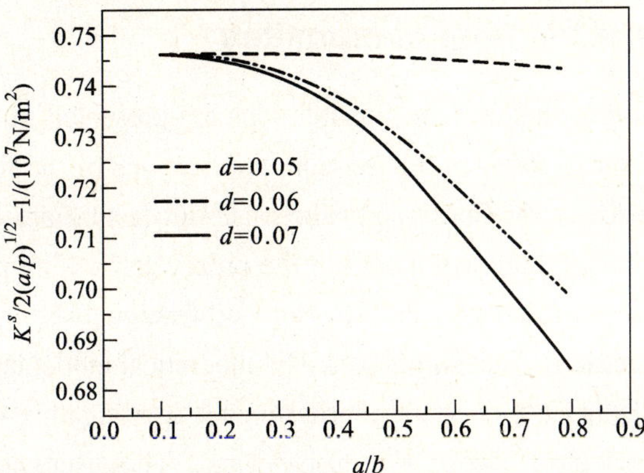

Fig.3.8   Variation of the stress intensity factor with the ratio a/b under different thicknesses of the coating

The variation in the stress intensity factor with the ratio a/b under different values of the elastic constant c_{33} of the coating layer is plotted in Fig.3.9. It can be seen from the figure that the stress intensity factor may increase or decrease with the ratio a/b depending on the value of c_{33} of the coating. When c_{33} of

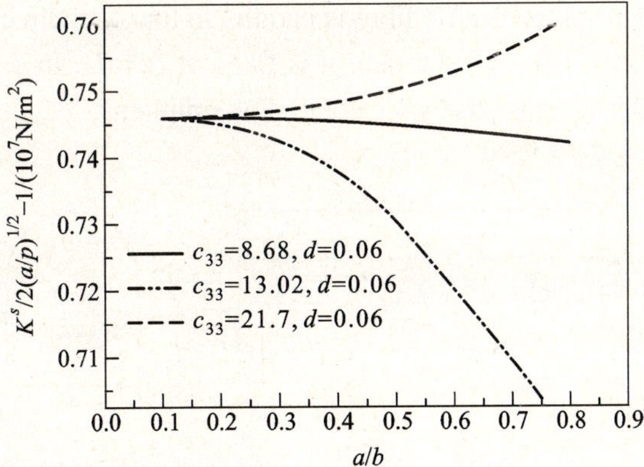

Fig.3.9   Variation of the stress intensity factor with the ratio a/b under different values of the elastic constant c_{33} of the coating.

the coating is greater than that of the piezoelectric fibre, the stress intensity factor will increase with an increase in the a/b. Obviously, the decay rate of the stress intensity depends strongly on the value of c_{33} when it is smaller than that of piezoelectric fibre.

3.9   Piezoelectric fibre composites

In the last section of this chapter, formulations are presented in terms of a cylindrical coordinate system. In this section, applications of the formulations to problems of piezoelectric fibre composite push-out testing are discussed. We start by presenting a theoretical model of the piezoelectric fibre push-out problem and use it to analyze elastic deformation and frictional sliding behaviour in a single piezoelectric fibre push-out test. The theoretical model is also used as a basis for establishing the debonding criterion for investigating the debonding process of piezoelectric fibre in the push-out test. The discussion follows the development in [59-61].

3.9.1   Theoretical model for piezoelectric fibre push-out

The physical problem to be studied is shown in Fig. 3.10, where a circular piezoelectric fibre polarized in the axial direction with radius a is embedded in the centre of a coaxial cylindrical matrix with radius b and total length L. The poling direction of the piezoelectric fibre is parallel to the axial direction. The load σ_a is applied at $z = 0$ and the matrix is fixed at $z = L$. In our analysis the matrix is considered transversely isotropic. For simplicity, a cylindrical coordinate system (r, θ, z) is used.

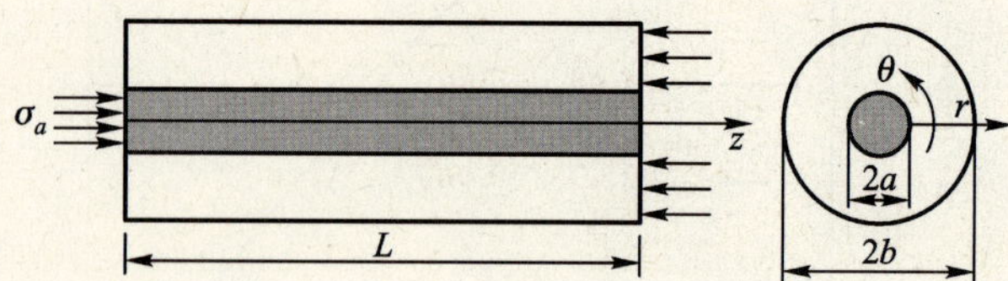

Fig.3.10   A piezoelectric fibre/matrix cylinder model in the fibre push-out test

It should be mentioned that the theoretical model developed here is based on the following two assumptions: (a) the axial stresses are independent of the radial distance in any cross-section of the composite; (b) the matrix shear strain is approximately given by $\gamma_m^{rz} = \dfrac{\partial u_m^z}{\partial r}$, where u_m^z is matrix displacement in the axial direction [62].

The general relation between strains and stresses is given by [59,61]

$$
\begin{bmatrix} \varepsilon_m^r \\ \varepsilon_m^\theta \\ \varepsilon_m^z \\ \gamma_m^{rz} \end{bmatrix} = \begin{bmatrix} f_{11} & f_{12} & f_{13} & 0 \\ f_{12} & f_{11} & f_{13} & 0 \\ f_{13} & f_{13} & f_{33} & 0 \\ 0 & 0 & 0 & f_{55} \end{bmatrix} \begin{bmatrix} \sigma_m^r \\ \sigma_m^\theta \\ \sigma_m^z \\ \tau_m^{rz} \end{bmatrix}
\tag{3.9.1}
$$

for the transversely isotropic matrix and

$$
\begin{bmatrix} \varepsilon_f^r \\ \varepsilon_f^\theta \\ \varepsilon_f^z \\ \gamma_f^{rz} \end{bmatrix} = \begin{bmatrix} f_{11}' & f_{12}' & f_{13}' & 0 \\ f_{12}' & f_{11}' & f_{13}' & 0 \\ f_{13}' & f_{13}' & f_{33}' & 0 \\ 0 & 0 & 0 & f_{55}' \end{bmatrix} \begin{bmatrix} \sigma_f^r \\ \sigma_f^\theta \\ \sigma_f^z \\ \tau_f^{rz} \end{bmatrix} + \begin{bmatrix} 0 & g_{31} \\ 0 & g_{31} \\ 0 & g_{33} \\ g_{15} & 0 \end{bmatrix} \begin{bmatrix} D^r \\ D^z \end{bmatrix}
\tag{3.9.2}
$$

$$
\begin{bmatrix} E^r \\ E^z \end{bmatrix} = - \begin{bmatrix} 0 & 0 & 0 & g_{15} \\ g_{31} & g_{31} & g_{33} & 0 \end{bmatrix} \begin{bmatrix} \sigma_f^r \\ \sigma_f^\theta \\ \sigma_f^z \\ \tau_f^{rz} \end{bmatrix} + \begin{bmatrix} \kappa_{11} & 0 \\ 0 & \kappa_{33} \end{bmatrix}^{-1} \begin{bmatrix} D^r \\ D^z \end{bmatrix}
\tag{3.9.3}
$$

for the piezoelectric fibre, where subscripts "f" and "m" refer to fibre and matrix, the superscripts stand for coordinate direction (r,θ,z), ε_j^i and σ_j^i ($i=r,\theta,z$, and $j=m,f$) are strain and stress components, respectively. In Eq.(3.9.2) and Eq.(3.9.3), D^i and E^i are components of the electric displacement ($NV^{-1}m^{-1}$) and electric field (Vm^{-1}), g_{ij} and κ_{ij} are piezoelectric coefficients (VmN^{-1}) and dielectric constants (NV^{-2}), and f_{ij} and f_{ij}' are components of elastic compliance [61].

The field equations of the piezoelectric fibre undergoing axially symmetric deformations about the z-axis can be expressed as

$$
\frac{\partial \sigma_j^z}{\partial z} + \frac{\partial \tau_j^{rz}}{\partial r} + \frac{\tau_j^{rz}}{r} = 0
\tag{3.9.4}
$$

$$
\frac{\partial \sigma_j^r}{\partial r} + \frac{\partial \tau_j^{rz}}{\partial z} + \frac{\sigma_j^r - \sigma_j^\theta}{r} = 0
\tag{3.9.5}
$$

$$
\frac{\partial D^r}{\partial r} + \frac{D^r}{r} + \frac{\partial D^z}{\partial z} = 0
\tag{3.9.6}
$$

The equilibriums between the axial stress and the interfacial stress can be expressed as

$$\sigma_a = \sigma_f^z + \frac{1}{\gamma}\sigma_m^z \quad \text{or} \quad \frac{d\sigma_m^z}{dz} = \frac{2\gamma}{a}\tau_i(z) \tag{3.9.7}$$

$$\frac{d\sigma_f^z}{dz} = -\frac{2}{a}\tau_i(z) \tag{3.9.8}$$

in which $\gamma = a^2/(b^2 - a^2)$ and $\tau_i(z)$ is the interfacial shear stress.

The electric field, E^i, is defined by

$$E^r = -\frac{\partial\phi}{\partial r}, \qquad E^z = -\frac{\partial\phi}{\partial z} \tag{3.9.9}$$

To simplify the derivation of the theoretical model and without loss of generality, the axial stresses σ_f^z and σ_m^z are assumed to be functions of z only, and the electric potential which is caused by elastic deformation of the fibre is also independent of r [61], i.e.,

$$\sigma_f^z = \sigma_f^z(z), \quad \sigma_m^z = \sigma_m^z(z), \quad \phi = \phi(z) \tag{3.9.10}$$

Using Eqs.(3.9.3), (3.9.6), (3.9.9), and (3.9.10), the electric displacements in the fibre can be expressed in terms of matrix stresses as

$$D^z = d_{15}\sigma_f^z, \qquad D^r = d_{15}\tau_f^{rz} \tag{3.9.11}$$

in which $d_{15} (= \kappa_{11}g_{15})$ is the piezoelectric coefficient.

It is now necessary to find the expression of fibre stress in terms of some derivable functions. In the push-out test, although the electro-mechanical coupling effect in Eq.(3.9.2) and Eq.(3.9.3) is considered, the following assumption is still acceptable [59,61]:

$$\sigma_f^r(z) = \sigma_f^\theta(z) = q_i(z) \tag{3.9.12}$$

where $q_i(z)$ is the interfacial radial stress arising from Poisson contraction between the fibre and the matrix.

Interfacial shear stress in the frictional sliding interface is governed by Coulomb's friction law [59]. That is,

$$\tau_i(z) = -\mu[q_0 - q_i(z)] \tag{3.9.13}$$

where q_0 is the residual fibre clamping (compressive) stress in the radial direction caused by matrix shrinkage and differential thermal contraction of the constituents upon cooling from the processing temperature, and μ is the friction coefficient.

The remaining task is to derive the differential equation for σ_f^z and radial stress $q_i(z)$ due to elastic deformation in composites with a perfectly bonded interface or in the frictional sliding process after the interface is completely debonded. The detailed derivations for these two processes are provided in the following two subsections.

3.9.2 Stress transfer in the bonded interface

Stress transfer is of fundamental importance in determining the mechanical properties of fibre-reforced composite materials [63]. At the first stage of the push-out process, elastic stress usually transfers from an elastic matrix to an elastic fibre through the bonded interface which predominates in an elastic matrix/elastic fibre composite, and it depends largely on the micromechanical characteristics of the fibre/matrix interface. In the interface in piezoelectric fibre reinforced composites, stress transfer is affected by the piezoelectric coefficient in addition to the micromechanical properties. To investigate the effect of the piezoelectric coefficient on the elastic stress transfer, we consider the inner and outer boundary conditions of the matrix

$$\sigma_m^r(a,z) = q_i(z), \quad \tau_m^{rz}(a,z) = \tau_i(z), \quad \sigma_m^r(b,z) = 0, \quad \tau_m^{rz}(b,z) = 0 \quad (3.9.14)$$

Then from Eqs.(3.9.4),(3.9.8) and (3.9.14), we obtain

$$\tau_m^{rz}(r,z) = \frac{\gamma(b^2 - r^2)}{ar}\tau_i(z) \quad (3.9.15)$$

Following the procedure given in [62], we have

$$\sigma_m^r(r,z) = \frac{\gamma}{4a}\frac{d\tau_i}{dz}\left\{2\eta_1 b^2\left[\ln(r/b) + \gamma(b^2/r^2 - 1)\ln(b/a)\right] + \right.$$
$$\left. \gamma q_i(z)(b^2/r^2 - 1) + \eta_2(b^2 - r^2)(1 - a^2/r^2)\right\} \quad (3.9.16)$$

$$\sigma_m^\theta(r,z) = -\gamma q_i(z)(1 + r^2/b^2) + \frac{\gamma}{4a}\frac{d\tau_i}{dz}\left\{\eta_2(b^2 + r^2)(1 + a^2/r^2) + 4b^2 + \right.$$
$$\left. 2\eta_1 b^2\left[\ln(r/b) - \gamma(b^2/r^2 + 1)\ln(b/a)\right] + 2\eta_1(b^2 - r^2)\right\} \quad (3.9.17)$$

Substituting Eq.(3.9.16) and Eq.(3.9.17) into Eq.(3.9.1) yields

$$\frac{u_m^r}{r} = \gamma q_i(z)\left[(f_{12} - f_{11})b^2/r^2 - f_{11} - f_{12}\right] + f_{13}\sigma_m^z + $$
$$\frac{\gamma}{4a}\frac{d\tau_i}{dz}\left\{2\eta_1 b^2(f_{11} + f_{12})\left[\ln(r/b) - \gamma\ln(b/a)\right] + \right.$$
$$2\eta_1 b^2\gamma\ln(b/a)(f_{12} - f_{11})b^2/r^2 + f_{11}\eta_2(b^2 + r^2)(1 + a^2/r^2) + $$
$$\left. f_{12}\eta_2(b^2 - r^2)(1 - a^2/r^2) + 4f_{11}b^2 + 2f_{11}\eta_1(b^2 - r^2)\right\} \quad (3.9.18)$$

$$\frac{\partial u_m^z}{\partial z} = -2f_{13}\gamma q_i(z) + f_{33}\sigma_m^z + 2f_{13}\frac{\gamma}{4a}\frac{\mathrm{d}\tau_i}{\mathrm{d}z}\left\{2\eta_1 b^2\left[\ln(r/b) - \gamma\ln(b/a)\right]+\right.$$

$$\left.\eta_1(b^2 - r^2) + 2b^2 + \eta_2(a^2 + b^2)\right\}$$

$$(3.9.19)$$

For a fully bonded interface, the continuity conditions of axial and radial deformation between fibre and matrix are given by

$$u_m^r(a,z) = u_f^r(a,z) \qquad\qquad (3.9.20)$$

$$u_m^z(a,z) = u_f^z(a,z) \qquad\qquad (3.9.21)$$

From Eqs.(3.9.19), (3.9.21) and (3.9.2), the radial stress of the fibre is obtained as

$$q_i(z) = \frac{1}{2(f_{13}' + \gamma f_{13})}\left\{\gamma f_{33}\sigma_a - (\gamma f_{33} + f_{33}' + g_{33}d_{15})\sigma_f^z +\right.$$

$$\left. 2f_{13}\frac{\gamma}{4a}\frac{\mathrm{d}\tau_i}{\mathrm{d}z}\left[2\eta_1 b^2(1+\gamma)\ln(a/b) + \eta_1(b^2 - a^2) + 2b^2 + \eta_2(a^2 + b^2)\right]\right\}$$

$$(3.9.22)$$

Then, combining Eqs.(3.9.2), (3.9.18), (3.9.20) and (3.9.22) yields the differential equation of σ_f^z as

$$\frac{\mathrm{d}^2\sigma_f^z(z)}{\mathrm{d}z^2} - A_1\sigma_f^z(z) = A_2\sigma_a \qquad\qquad (3.9.23)$$

where A_1 and A_2 are two constants

$$A_1 = -\frac{\gamma f_{13} + f_{13}' + g_{31}d_{15} - B_1(\gamma f_{33} + f_{33}' + g_{33}d_{15})}{\dfrac{\gamma}{8}(C_1 - 2B_1 f_{13}C_2)} \qquad (3.9.24)$$

$$A_2 = \frac{\gamma f_{13} - B_1\gamma f_{33}}{\dfrac{\gamma}{8}(C_1 - 2B_1 f_{13}C_2)} \qquad\qquad (3.9.25)$$

with

$$B_1 = \frac{f_{12}' + f_{11}' - \gamma(f_{12} - f_{11})b^2/a^2 + \gamma(f_{11} + f_{12})}{2(f_{13}' + \gamma f_{13})} \qquad (3.9.26)$$

$$C_1 = 2\eta_1 b^2(f_{11} + f_{12})(1+\gamma)\ln(a/b) + 2\eta_1 b^2\gamma\ln(b/a)(f_{12} - f_{11})b^2/a^2 +$$

$$2f_{11}\eta_2(a^2 + b^2) + 4f_{11}b^2 + 2f_{11}\eta_1(b^2 - a^2) \qquad (3.9.27)$$

$$C_2 = 2\eta_1 b^2(1+\gamma)\ln(a/b) + \eta_1(b^2 - a^2) + 2b^2 + \eta_2(a^2 + b^2) \qquad (3.9.28)$$

$$\eta_1 = \frac{f_{55}}{f_{13}} \qquad\qquad (3.9.29)$$

$$\eta_2 = \frac{1}{2}\frac{f_{55}}{f_{13}} - 1 \qquad\qquad (3.9.30)$$

Using the stress boundary conditions

$$\sigma_f^z(0) = \sigma_a, \quad \sigma_f^z(L) = 0 \tag{3.9.31}$$

the distribution of the axial stress in the piezoelectric fibre is given by

$$\sigma_f^z(z) = K_1 \sinh(\sqrt{A_1}\,z) + K_2 \cosh(\sqrt{A_1}\,z) - \frac{A_2}{A_1}\sigma_a \tag{3.9.32}$$

where K_1 and K_2 are defined by

$$K_1 = \frac{\dfrac{A_2}{A_1} - \left(1 + \dfrac{A_2}{A_1}\right)\cosh(\sqrt{A_1}\,L)}{\sinh(\sqrt{A_1}\,L)}\,\sigma_a \tag{3.9.33}$$

$$K_2 = \left(1 + \frac{A_2}{A_1}\right)\sigma_a \tag{3.9.34}$$

In addition, using Eqs.(3.9.4), (3.9.5) and (3.9.12), $q_i(z)$ can be expressed as

$$q_i(z) = N_1\sigma_a - N_2\sigma_f^z + N_3\frac{\mathrm{d}^2\sigma_f^z}{\mathrm{d}z^2} \tag{3.9.35}$$

where N_i (i=1,2,3) are given by

$$N_1 = \frac{\gamma f_{33}}{2(f_{13}' + \gamma f_{13})}, \quad N_2 = \frac{\gamma f_{33} + f_{33}' + g_{33}d_{15}}{2(f_{13}' + \gamma f_{13})}, \quad N_3 = -\frac{\gamma C_2 f_{13}}{4} \tag{3.9.36}$$

From Eqs.(3.9.3), (3.9.11), (3.9.32) and (3.9.35), the electric field E^z can be calculated by

$$E^z(z) = -2g_{31}q_i(z) + \left(\frac{d_{15}}{\kappa_{33}} - g_{33}\right)\sigma_f^z(z) \tag{3.9.37}$$

3.9.3 Frictional sliding

Once the interface debonds completely, the frictional sliding of the fibre out of the surrounding matrix will begin, which is the last stage of the push-out process. To better characterize this stage, theoretical analysis was conducted with the micromechanical model shown in Fig.3.11.

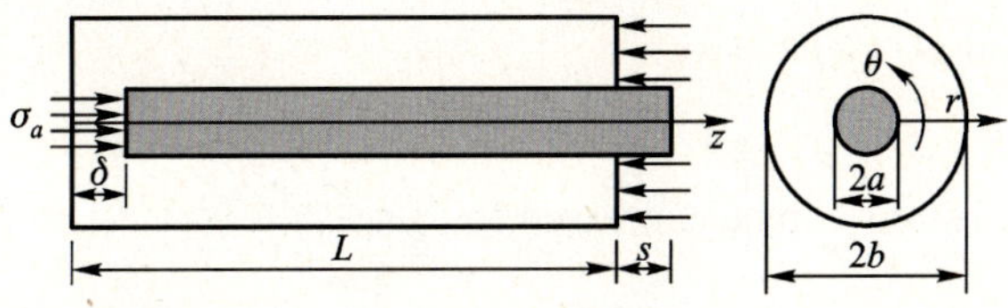

Fig.3.11　A fibre-matrix cylinder model for frictional sliding in the push-out test

For the sake of simplicity, we assume a small elastic deformation and a large displacement at the fibre-loaded end during sliding. Therefore, the elastic deformation of the fibre can be neglected, and the fibre axial displacement δ approximately equals the fibre sliding distance s, that is, $\delta \approx s$.

Similar to the mathematical operation for Eq.(3.9.23), we have from the continuity condition of radial displacement Eq.(3.9.20)

$$q_i(z) = \frac{\gamma f_{13}}{D_1}\sigma_a - \frac{\gamma f_{13} + f_{13}' + g_{31}d_{15}}{D_1}\sigma_f^z + \frac{\gamma}{4a}\frac{C_1}{D_1}\frac{d\tau_i}{dz} \tag{3.9.38}$$

Subsitituting Eq.(3.9.38) into Eq.(3.9.13) gives the governing equation for σ_f^z

$$\frac{d^2\sigma_f^z(z)}{dz^2} + E_1\frac{d\sigma_f^z(z)}{dz} + E_2\sigma_f^z(z) = E_3 \tag{3.9.39}$$

where E_i $(i=1,2,3)$ are given by

$$E_1 = -\frac{4aD_1}{\mu\gamma C_1}, \quad E_2 = \frac{8(\gamma f_{13} + f_{13}' + g_{31}d_{15})}{\gamma C_1}, \quad E_3 = \frac{8D_1}{\gamma C_1}\left(\frac{\gamma f_{13}}{D_1}\sigma_a - q_0\right)$$

$$\tag{3.9.40}$$

with

$$D_1 = f_{12}' + f_{11}' - \gamma(f_{12} - f_{11})b^2/a^2 + \gamma(f_{11} + f_{12}) \tag{3.9.41}$$

Using the stress boundary conditions

$$\sigma_f^z(s) = \sigma_a, \quad \sigma_f^z(L) = 0 \tag{3.9.42}$$

Eq (3.9 39) is solved and the solution is obtained

$$\sigma_f^z(z) = K_3 e^{\lambda_1 z} + K_4 e^{\lambda_2 z} + \frac{E_3}{E_2} \tag{3.9.43}$$

where s is the fibre sliding distance and K_3, K_4, λ_1, λ_2 are given by

$$K_3 = E_3 m_3 + n_3\sigma_a, \quad K_4 = E_4 m_4 + n_4\sigma_a \tag{3.9.44}$$

$$\lambda_1 = \frac{1}{2}E_1 + \frac{1}{2}\sqrt{E_1^2 - 4E_2}, \quad \lambda_2 = \frac{1}{2}E_1 - \frac{1}{2}\sqrt{E_1^2 - 4E_2} \tag{3.9.45}$$

In the process, the interfacial shear stress is governed by Coulomb's frictional law given in Eq.(3.9.13). Noting that fibre and matrix maintain contact in the radial direction, we have

$$u_f^r(a,z) = u_m^r(a,z) \tag{3.9.46}$$

Then, the radial stress can be expressed as

$$q_i(z) = M_1 - M_2\sigma_a + M_3\frac{d^2\sigma_f^z}{dz^2} \tag{3.9.47}$$

where M_i (i=1,2,3) are given by

$$M_1 = \frac{\gamma f_{13}}{D_1}, \quad M_2 = \frac{\gamma f_{13} + f_{13}' + g_{31}d_{15}}{D_1}, \quad M_3 = -\frac{\gamma C_1}{8D_1} \tag{3.9.48}$$

Using Eqs.(3.9.3), (3.9.11), (3.9.43) and (3.9.47), the electric field can be obtained as

$$E^z(z) = -2g_{31}q_i(z) + \left(\frac{d_{15}}{\kappa_{33}} - g_{33}\right)\sigma_f^z(z) \tag{3.9.49}$$

Noting that the sum of the radial stress of the fibre should be negative, and the fibre and matrix can contact each other during the fibre sliding process, the radial stress must satisfy the expression

$$q_0 - q_i(z) \leqslant 0 \tag{3.9.50}$$

According to the distribution of the fibre stress fields in the push-out test, the axial stress reaches its maximum value at the fibre-loaded end $z = s$ ($s \geqslant 0$, and s is defined in Fig. 3.7), while the interfacial shear stress reaches its minimum value at the same location. Then Eq.(3.9.50) yields

$$q_0 - q_i(s) = 0 \tag{3.9.51}$$

Therefore the relationship between the applied stress σ_a and the axial displacement δ at the fibre-loaded end can be given as

$$\sigma_a = \frac{q_0(1 - m_3\lambda_1^2 e^{\lambda_1 s} - m_4\lambda_2 e^{\lambda_2 s})}{\dfrac{f_{13}' + g_{31}d_{15}}{D_1} + \dfrac{\gamma f_{13}}{D_1}(m_3\lambda_1^2 e^{\lambda_1 s} + m_4\lambda_2^2 e^{\lambda_2 s}) + \dfrac{\gamma C_1}{8D_1}(n_3\lambda_1^2 e^{\lambda_1 s} + n_4\lambda_2^2 e^{\lambda_2 s})}$$

$$\tag{3.9.52}$$

where m_3, m_4, n_3, and n_4 are defined as

$$m_3 = \frac{e^{\lambda_2 L} - e^{\lambda_2 s}}{E_2(e^{\lambda_1 L + \lambda_2 s} - e^{\lambda_1 s + \lambda_2 L})}, \quad m_4 = \frac{e^{\lambda_1 s} - e^{\lambda_1 L}}{E_2(e^{\lambda_1 L + \lambda_2 s} - e^{\lambda_1 s + \lambda_2 L})} \tag{3.9.53}$$

$$n_3 = -\frac{e^{\lambda_2 L}}{e^{\lambda_1 L + \lambda_2 s} - e^{\lambda_1 s + \lambda_2 L}}, \quad n_4 = \frac{e^{\lambda_1 L}}{e^{\lambda_1 L + \lambda_2 s} - e^{\lambda_1 s + \lambda_2 L}} \tag{3.9.54}$$

3.9.4 Partially debonding model

Consider again the physical set-up shown in Fig.3.10, but now an interfacial debonding crack of length l is situated on the interface between piezoelectric fibre and matrix (see Fig.3.12).

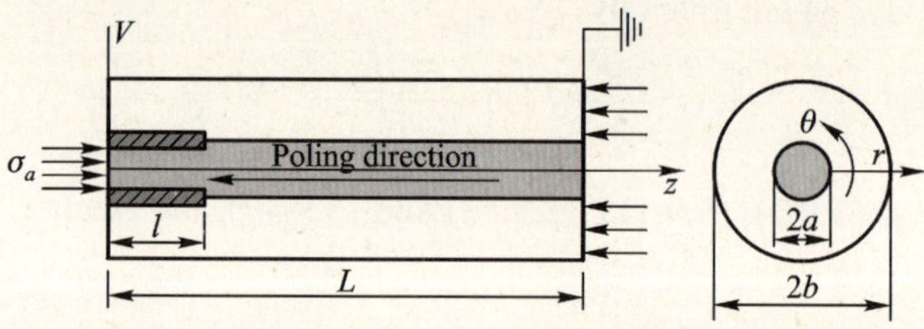

Fig.3.12   Piezoelectric fibre push-out model under electrical and mechanical loading

1. Mechanical loading

Using the mechanical boundary conditions (3.9.31) and a procedure similar to that in [64], the stress fields in the debonded region ($0 \leqslant z \leqslant l$) are obtained as

$$q_i(z) = -\frac{E_m(f_{13} + g_{31}d_{15})\sigma_f^z + \nu_m\sigma_m^z}{E_m(f_{11} + f_{12}) + 1 + 2\gamma + \nu_m} \tag{3.9.55}$$

$$\sigma_m^z(z) = \gamma\omega\left[\sigma^* + \sigma_a\right]\left[1 - \exp(-\lambda z)\right] \tag{3.9.56}$$

$$\sigma_f^z(z) = \sigma_a - \omega\left[\sigma^* + \sigma_a\right]\left[1 - \exp(-\lambda z)\right] \tag{3.9.57}$$

$$\tau_i(z) = \frac{a\lambda\omega}{2}(\sigma^* + \sigma_a)\exp(-\lambda z) \tag{3.9.58}$$

where

$$\kappa = -\frac{E_m(f_{13} + g_{31}d_{15}) - \gamma\nu_m}{E_m(f_{11} + f_{12}) + 1 + 2\gamma + \nu_m} \tag{3.9.59}$$

$$\omega = \frac{E_m(f_{13} + g_{31}d_{15})}{E_m(f_{13} + g_{31}d_{15}) - \gamma\nu_m} \tag{3.9.60}$$

$$\lambda = \frac{2\mu\kappa}{a} \tag{3.9.61}$$

$$\sigma^* = -\frac{q_0}{\omega\kappa} \tag{3.9.62}$$

The solutions of the stress fields in the bonded region ($l \leqslant z \leqslant L$) are given by

$$\sigma_f^z(z) = \frac{\left(\dfrac{A_2}{A_1}\sigma_a + \sigma_l\right)\sinh\sqrt{A_1}(L-z) - \dfrac{A_2}{A_1}\sigma_a\sinh\sqrt{A_1}(l-z)}{\sinh\sqrt{A_1}(L-l)} - \frac{A_2}{A_1}\sigma_a \tag{3.9.63}$$

$$\sigma_m^z(z) = \gamma\left(1 + \frac{A_2}{A_1}\right)\sigma_a - \gamma\,\frac{\left(\dfrac{A_2}{A_1}\sigma_a + \sigma_l\right)\sinh\sqrt{A_1}\,(L-z) - \dfrac{A_2}{A_1}\sigma_a \sinh\sqrt{A_1}\,(l-z)}{\sinh\sqrt{A_1}\,(L-l)}$$

$$(3.9.64)$$

$$\tau_i(z) = \frac{\lambda a\left[\left(\dfrac{A_2}{A_1}\sigma_a + \sigma_l\right)\cosh\sqrt{A_1}\,(L-z) - \dfrac{A_2}{A_1}\sigma_a \cosh\sqrt{A_1}\,(l-z)\right]}{2\sinh\sqrt{A_1}\,(L-l)} \qquad (3.9.65)$$

where

$$\sigma_f^z(l) = \sigma_l = \sigma_a - \omega\left[\sigma^* + \sigma_a\right]\left[1 - \exp(-\lambda l)\right] \qquad (3.9.66)$$

$$A_1 = \frac{2\left[\gamma + E_m(f_{33} + g_{33}d_{15}) - 2\kappa(\gamma\nu_m - E_m f_{13})\right]}{(1 + \nu_m)\left[2\gamma b^2 \ln(b/a) - a^2\right]} \qquad (3.9.67)$$

$$A_2 = -\frac{2\left[\gamma + 2(\gamma\nu_m - E_m f_{13})(\omega - 1)\kappa\right]}{(1 + \nu_m)\left[2\gamma b^2 \ln(b/a) - a^2\right]} \qquad (3.9.68)$$

The electrical field E_z in both the debonded and bonded regions is given as

$$E_z = -2g_{31}\kappa(\omega - 1)\sigma_a + \left(\frac{d_{15}}{\kappa_{33}} - g_{33} - 2g_{31}\kappa\right)\sigma_f^z \qquad (3.9.69)$$

2. Electrical and mechanical loading

The solutions presented above apply for problems with mechanical loading only. To obtain solutions due to electrical and mechanical loading, we rewrite the constitutive equations (3.9.2) and (3.9.3) in the following form [56,60]:

$$\begin{bmatrix} \varepsilon_f^r \\ \varepsilon_f^\theta \\ \varepsilon_f^z \\ 2\varepsilon_f^{rz} \end{bmatrix} = \begin{bmatrix} f_{11} & f_{12} & f_{13} & 0 \\ f_{12} & f_{11} & f_{13} & 0 \\ f_{13} & f_{13} & f_{33} & 0 \\ 0 & 0 & 0 & f_{55} \end{bmatrix}\begin{bmatrix} \sigma_f^r \\ \sigma_f^\theta \\ \sigma_f^z \\ \tau_f^{rz} \end{bmatrix} + \begin{bmatrix} 0 & d_{13} \\ 0 & d_{13} \\ 0 & d_{33} \\ d_{15} & 0 \end{bmatrix}\begin{bmatrix} E_r \\ E_z \end{bmatrix} \qquad (3.9.70)$$

$$\begin{bmatrix} D_r \\ D_z \end{bmatrix} = \begin{bmatrix} 0 & 0 & 0 & d_{15} \\ d_{13} & d_{13} & d_{33} & 0 \end{bmatrix}\begin{bmatrix} \sigma_f^r \\ \sigma_f^\theta \\ \sigma_f^z \\ \tau_f^{rz} \end{bmatrix} + \begin{bmatrix} \kappa_{11} & 0 \\ 0 & \kappa_{33} \end{bmatrix}\begin{bmatrix} E_r \\ E_z \end{bmatrix} \qquad (3.9.71)$$

Considering Eqs.(3.9.6), (3.9.9), (3.9.10), (3.9.12), (3.9.70) and (3.9.71), we can deduce that the electric potential can be obtained and written in the form

$$\phi(z) = \frac{1}{\kappa_{33}} \int_0^z \left[2d_{13}q_i(z) + (d_{33} - d_{15})\sigma_f^z(z) \right] dz + C_1 z + C_2 \qquad (3.9.72)$$

where

$$C_1 = -\frac{V}{L} - \frac{1}{e_{33}L} \int_0^L \left[2d_{13}q_i(z) + (d_{33} - d_{15})\sigma_f^z(z) \right] dz, \quad C_2 = V \qquad (3.9.73)$$

Electrical boundary conditions at the ends of the piezoelectric fibre are given as

$$\phi(0) = V, \quad \phi(L) = 0 \qquad (3.9.74)$$

where mechanical boundary conditions are given by Eq. (3.9.31).

The solutions for stress distribution in the constituents are obtained in the bonded region and the debonded region, and are exactly the same as those given in Eqs. (3.9.56)~(3.9.58) and (3.9.63)~(3.9.65), except that certain variables and   parameters are replaced as follows:

$$q_i(z) = -\frac{E_m \left[f_{13} - d_{13}(d_{33} - d_{15})/\kappa_{33} \right] \sigma_f^z(z) + \nu_m \sigma_m^z(z) - E_m d_{13} C_1}{E_m (f_{11} + f_{12} - 2d_{13}^2/\kappa_{33}) + 1 + 2\gamma + \nu_m} \qquad (3.9.75)$$

$$\kappa = -\frac{E_m \left[f_{13} - d_{13}(d_{33} - d_{15})/\kappa_{33} \right] - \gamma\nu_m}{E_m (f_{11} + f_{12} - 2d_{13}^2/\kappa_{33}) + 1 + 2\gamma + \nu_m} \qquad (3.9.76)$$

$$\omega = \frac{E_m \left[f_{13} - d_{13}(d_{33} - d_{15})/\kappa_{33} \right]}{E_m \left[f_{13} - d_{13}(d_{33} - d_{15})/\kappa_{33} \right] - \gamma\nu_m} \qquad (3.9.77)$$

$$q^* = \frac{E_m d_{13} C_1}{E_m (f_{11} + f_{12} - 2d_{13}^2/\kappa_{33}) + 1 + 2\gamma + \nu_m} \qquad (3.9.78)$$

$$\sigma^* = -\frac{q_0 - q^*}{\kappa\omega} \qquad (3.9.79)$$

$$\lambda = \frac{2\mu\kappa}{a} \qquad (3.9.80)$$

$$A_1 = \frac{2\left\{ \gamma + E_m \left[f_{33} - d_{33}(d_{33} - d_{15})/\kappa_{33} \right] - 2\kappa(\gamma\nu_m - E_m f_{13} + E_m d_{13}d_{33}/\kappa_{33}) \right\}}{(1+\nu_m)\left[2\gamma b^2 \ln(b/a) - a^2 \right]}$$

$$(3.9.81)$$

$$A_2 = -\frac{2\left\{ \begin{array}{l} \left[\gamma + 2(\gamma\nu_m - E_m f_{13} + E_m d_{13}d_{33}/\kappa_{33})(\omega - 1)\kappa \right] + \\ 2\left[\gamma\nu_m - E_m (f_{13} - d_{13}d_{33}/\kappa_{33}) \right] q^*/\sigma_a + E_m d_{33}C_1/\sigma_a \end{array} \right\}}{(1+\nu_m)\left[2\gamma b^2 \ln(b/a) - a^2 \right]} \qquad (3.9.82)$$

The electrical field E_z in both the debonded and bonded regions is given as

$$E_z = -\frac{1}{\kappa_{33}}\left[2d_{13}q_i(z) + (d_{33} - d_{15})\sigma_f^z(z)\right] - C_1 \qquad (3.9.83)$$

3.9.5   Interfacial debonding criterion

In piezoelectric fibre composites (PFCs), unlike non-piezoelectric fibre composites, there are electromechanical couplings caused by piezoelectric or inverse piezoelectric effects. Therefore, the existing debonding criterion based on non-piezoelectric fibre composites is not applicable to PFCs. To incorporate the piezoelectric effect in the debonding criterion we consider a cracked piezoelectric elastic body of volume V in which traction P, frictional stress t and surface electrical charge ω are applied. S_P, S_t and S_ω are the corresponding surfaces respectively, as shown in Fig.3.13. For the sake of simplicity, the matrix is assumed to be a piezoelectric material whose piezoelectric coefficients and dielectric constants equal zero. In our analysis, the debonding region is taken to be a crack (see Fig.3.13).

Based on the principle of energy balance, the variation of the energy in the piezoelectric system for crack growth dA along the friction surface under electromechanical loading is

$$d\Theta = -G_c dA - dW_f \qquad (3.9.84)$$

where G_c is the fracture energy, W_f is the work done by friction stress during crack growth

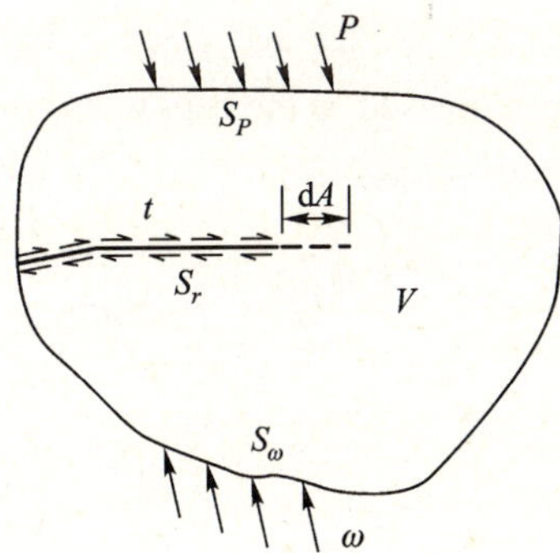

Fig.3.13   A piezoelectric elastic body with a frictional crack under electromechanical loading

$$W_f = \int_{S_T} (t + t_0) v \mathrm{d}S \qquad (3.9.85)$$

and Θ is the generalized mechanical and electrical energy stored inside the piezoelectric body

$$\Theta = \frac{1}{2}\int_{\Omega}(\varepsilon + \varepsilon_0):(\sigma + \sigma_0)\mathrm{d}\Omega -$$

$$\frac{1}{2}\int_{\Omega}(D + D_0):(E + E_0)\mathrm{d}\Omega - \int_{S_P} Pu\mathrm{d}S + \int_{S_\omega} \omega\phi\mathrm{d}S \qquad (3.9.86)$$

in which v is the relative slip of crack surfaces, and t_0 is the tangential component of pre-stress (or initial stress) on the crack surfaces. s_0, D_0 are self-equilibrium stress and generalized stress states, respectively, and $s + s_0$, $D + D_0$ balances the applied stress and generalized stress.

Using the basic theory of piezoelectricity [Eqs.(3.9.1)~(3.9.9)], one can easily prove the corresponding reciprocal principle of work and the principle of virtual work for piezoelectric material

$$\int_{\Gamma} t_i^1 u_i^2 \mathrm{d}\Gamma - \int_{\Gamma} \omega^1 \phi^2 \mathrm{d}\Gamma + \int_{\Omega} b_i^1 u_i^2 \mathrm{d}\Omega = \int_{\Gamma} t_i^2 u_i^1 \mathrm{d}\Gamma - \int_{\Gamma} \omega^2 \phi^1 \mathrm{d}\Gamma + \int_{\Omega} b_i^2 u_i^1 \mathrm{d}\Omega$$

$$(3.9.87)$$

$$\int_{\Gamma_t} \bar{t}_i \delta u_i \mathrm{d}\Gamma + \int_{\Omega} b_i \delta u_i \mathrm{d}\Omega - \int_{\Gamma_\omega} \bar{\omega}\delta\phi\mathrm{d}\Gamma + \int_{\Omega} b_e \delta\phi\mathrm{d}\Omega = \int_{\Omega}(\sigma_{ij}\delta\varepsilon_{ij} - D_i\delta E_i)\mathrm{d}\Omega$$

$$(3.9.88)$$

Using the two principles (3.9.87) and (3.9.88), it can be proved that the energy release rate against the incremental debonding length is equal to the interfacial fracture toughness $G_{ic.}$, that is [60]

$$2\pi a G_i = \frac{\partial U_t}{\partial l} \qquad (3.9.89)$$

in which U_t is the total elastic energy and electrical energy stored in the fibre and matrix, which can be expressed in the following form:

$$U_t = \int_0^l \int_0^a \left(\sigma_f^z \varepsilon_f^z - D^z E^z\right)\pi r \mathrm{d}r\mathrm{d}z + \int_l^L \int_0^a \left(\sigma_f^z \varepsilon_f^z - D^z E^z\right)\pi r \mathrm{d}r\mathrm{d}z +$$

$$\int_0^l \int_a^b \left[\frac{(\sigma_m^z)^2}{E_m} + \frac{2(1+\nu_m)}{E_m}(\tau_m^{rz})^2\right]\pi r \mathrm{d}r\mathrm{d}z +$$

$$\int_l^L \int_a^b \left[\frac{(\sigma_m^z)^2}{E_m} + \frac{2(1+\nu_m)}{E_m}(\tau_m^{rz})^2\right]\pi r \mathrm{d}r\mathrm{d}z \qquad (3.9.90)$$

Then the following energy criterion is introduced:

$$G_i \geqslant G_{ic} \qquad (3.9.91)$$

where G_{ic} is a critical interface debonding energy release rate.

In Equation (3.9.90), U_t is a complex function of the material properties of the constituents and geometric factors. Performing some mathematical manipulations on Eq. (3.9.90) over the debonded and bonded regions for the piezoelectric-epoxy composite system by utilizing a numerical quadrature approach and then substituting the result into Eq.(3.9.89), we can obtain G_i as a second-order function of the applied stress σ_a for a fibre/matrix system with given debonding length l.

3.9.6 Numerical examples

To illustrate applications of the formulations developed in this section, numerical assessment is presented for a hypothetical piezoelectric fibre/epoxy composite system. The material properties and geometrical characteristics of the piezoelectric fibre, matrix and interface are [61]

$$s_{11} = 0.019 \ (\text{GPa})^{-1}, \quad s_{33} = 0.015 \ (\text{GPa})^{-1}, \quad s_{12} = -0.0057 \ (\text{GPa})^{-1}$$

$$s_{13} = -0.0045 \ (\text{GPa})^{-1}, \quad s_{55} = 0.039 \ (\text{GPa})^{-1}, \quad d_{33} = 390 \times 10^{-12} \ \text{mV}^{-1}$$

$$d_{31} = -d_{15} = -190 \times 10^{-12} \ \text{mV}^{-1}, \quad g_{33} = 24 \times 10^{-3} \ \text{VmN}^{-1}$$

$$g_{31} = -11.6 \times 10^{-3} \ \text{VmN}^{-1}, \quad e_{33} = 16.25 \times 10^{-9} \ \text{NV}^{-2}$$

$$E_m = 3\text{GPa}, \nu_m = 0.4$$

The radii of fibre and matrix are: $a = 0.065 \ \text{mm}$, $b = 3 \ \text{mm}$, and $l = 0.6 \ \text{mm}$, $L = 2 \ \text{mm}$. The residual fibre clamping stress in the radial direction q_0 is assumed to be -0.01 GPa and $\mu=0.8$ [59].

Fig.3.14 shows the distribution of stresses and electric field as functions of dimensionless axial distance z/L for a partially debonded piezoelectric composite system subject to a constant external stress $\sigma_a = 1.5$ GPa in the fibre push-out test. In the calculation, the debonding length is assumed to be $l = 0.6\text{mm}$. For comparison and illustration of the effect of electromechanical coupling on stress transfer behaviour, the corresponding distribution of stresses for non-piezoelectric fibre composite (NPFC) is also plotted in Fig. 3.14. It is shown that the curves for PFCs and NPFCs have similar shapes. When subjected to applied stress of same value, the axial stress σ_f^z in PFC is smaller that in NPFC (Fig.3.14a). It can also be seen from Fig.3.14a and Fig. 3.14d that both axial and radial stresses in the fibre gradually decrease as z/L increases. Fig. 3.14c demonstrates that there is a greater radial stress in PFC and it decays more rapidly than that in NPFC, which leads to a larger interface shear stress in the debonded region of PFC in Fig.3.14b due to the Coulomb friction law [Eq.(3.9.13)]. This phenomenon can be attributed to the piezoelectric effect in piezoelectric fibre; greater applied stress is required in PFC to produce the same axial stress as in non-piezoelectric fibre composites. The difference in the stress fields between these two composite systems is controlled by piezoelectric coefficients, which were investigated in our previous work [59] for fully bonded composites. When the piezoelectric coefficients and dielectric constants are set to be zero, piezoelectric fibre degenerates to non-piezoelectric fibre. Fig. 3.14d shows the variation of electrical field as a function of axial distance z/L. The variation of E_z with z/L is very similar to that of the fibre axial stress.

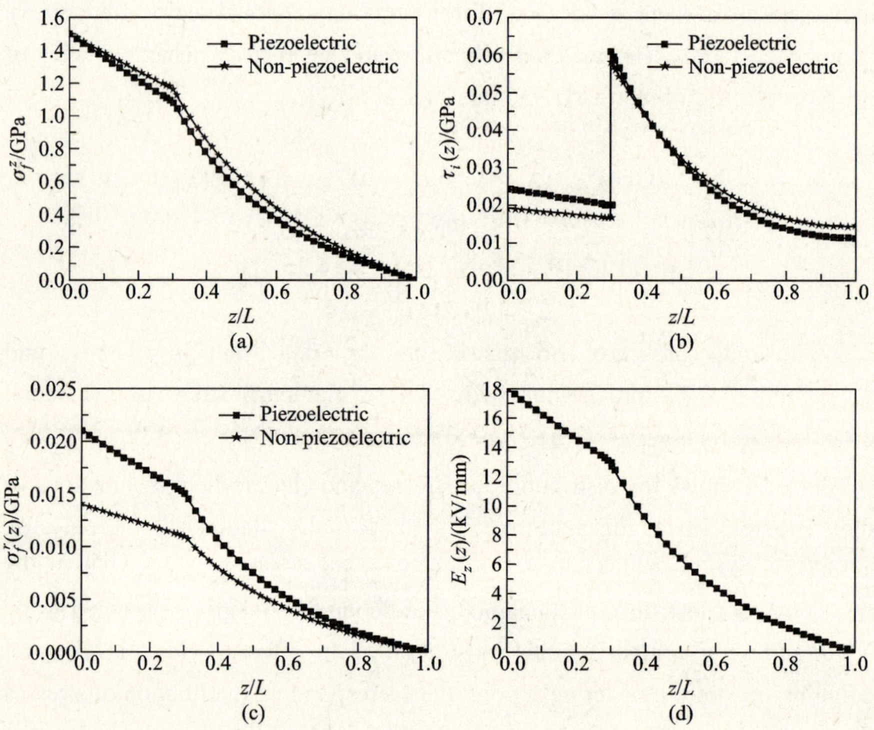

Fig.3.14   Plot of (a) fibre axial stress, (b) interface shear stress, (c) fibre radial stress, (d) electric field for the piezoelectric fibre push-out under mechanical loading

Fig.3.15 shows the distribution of stresses and electric field as functions of dimensionless axial distance z/L for a partially debonded piezoelectric composite system subject to electrical loading and a constant external stress

$\sigma_a = 1.5$ GPa in the fibre push-out test. To study the effect of positive and negative electric loading on stress transfer, electric potentials of 5000 V, 0 V and –5000 V are applied on the end of a piezoelectric fibre ($z = 0$). Fig.3.15a shows that the fibre axial stress under negative electric potential decays more rapidly than under positive electric potential. It can also be seen from Fig.3.15c that negative electric potential leads to a larger radial stress in piezoelectric fibre than does applied positive electric potential, causing greater interface frictional shear stress accordingly in the debonded region in Fig.3.15b. This is because when piezoelectric fibre is subjected to an electric potential applied parallel to the polarization direction, expansion occurs in the same direction and shrinkage occurs in the transverse direction [25]. For a positive applied electric potential, the hoop stress developed is compressive, while for a negative applied electric potential, the hoop stress developed is tensile. In Fig.3.15d, the distribution of electric field in piezoelectric fibre is plotted via a/L, and it depends strongly on the applied electric field.

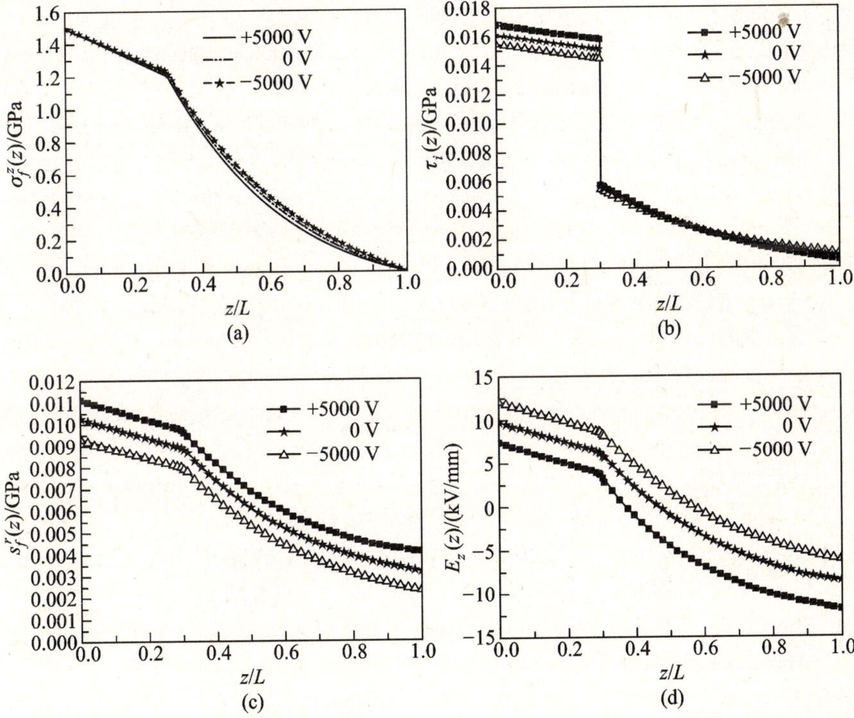

Fig.3.15 Plot of (a) fibre axial stress, (b) interface shear stress, (c) fibre radial stress, (d) electric field for the piezoelectric fibre push-out under electrical and mechanical loading

References

[1] Mindlin R D. Equations of high frequency vibrations of thermopiezoelectric crystal plates.Int. J. Solids Struct., 1974, 10:625-637.

[2] Nowacki W. Some general theorems of thermopiezoelectricity. J. Thermal Stresses, 1978, 1:171-182.

[3] Dunn M L. Micromechanics of coupled electroelastic composites: effective thermal expansion and pyroelectric coefficients. J. Appl. Phys., 1993, 73:5131-5140.

[4] Benveniste Y. Exact results in the micromechanics of fibrous piezoelectric composites exhibiting pyroelectricity. Proc. R. Soc. Lond., 1993, A441:59-81.

[5] Ashida F, Tauchert T R, Noda N. Potential function method for piezothermoelastic problems of solids of crystal class 6 mm in cylindrical coordinates. J. Thermal Stresses, 1994, 17:361-375.

[6] Altay G A, Dökmeci M C. Fundamental variational equations of discontinuous thermopiezoelectric fields. Int. J. Eng. Sci., 1996, 34:769-782.

[7] Noda N, Kimura S. Deformation of a piezothermoelectric composite plate considering the coupling effect. J. Thermal Stresses, 1998, 21:359-379.

[8] Ashida F, Tauchert T R. Finite difference scheme for inverse transient piezothermoe-lasticity problems, J. Thermal Stresses, 1998, 21:271-293.

[9] Shang F L, Wang Z K, Li Z H. Thermal stress analysis of a three dimensional crack in a thermopiezoelectric solid. Eng. Frac. Mech., 1996, 55:737-750.

[10] Yu Shouwen, Qin Qinghua. Damage analysis of thermopiezoelectric properties: part I —crack tip singularities. Theore. Appl. Frac. Mech., 1996, 25:263-277.

[11] Yu Shouwen, Qin Qinghua. Damage analysis of thermopiezoelectric properties: part II —effective crack model. Theore. Appl. Frac. Mech., 1996, 25:279-288.

[12] Qin Qinghua. Using GSC theory for effective thermal expansion and pyroelectric coefficient of cracked piezoelectric solids. Int. J. Frac., 1996, 82:R41-R46.

[13] Qin Qinghua. Yu Shouwen. Effective moduli of thermopiezoelectric material with microcavities. Int. J. Solids Struct., 1998, 35:5085-5095.

[14] Qin Qinghua. Mai Y W, Yu Shouwen, Effective moduli for thermopiezoelectric materials with microcracks. Int. J. Frac., 1998, 91:359-371.

[15] Qin Qinghua. Mai Y W. Crack growth prediction of an inclined crack in a half-plane thermopiezoelectric solid. Theore. Appl. Frac. Mech., 1997, 26:185-191.

[16] Qin Qinghua. Mai Y W. Multiple cracks in thermoelectroelastic bimaterials. Theore Appl. Frac. Mech., 1998, 29:141-150.

[17] Qin Qinghua. Thermoelectroelastic Green's function for a piezoelectric plate containing an elliptic hole. Mech. Mater., 1998, 30:21-29.

[18] Qin Qinghua, Mai Y W. Thermoelectroelastic Green's function and its application for

bimaterial of piezoelectric materials. Arch. Appl. Mech., 1998, 68: 433-444.

[19] Qin Qinghua, Mai Y W. A closed crack tip model for interface crack in thermo-piezoelectric materials. Int. J. Solids Struct., 1999, 36:2463-2479.

[20] Qin Qinghua. Green's function and its application for piezoelectric plate with various openings. Arch. Appl. Mech., 1999, 69:133-144.

[21] Qin Qinghua. Thermoelectroelastic analysis of cracks in piezoelectric half-plane by BEM. Compu. Mech., 1999, 23:353-360.

[22] Qin Qinghua. Green's function for thermopiezoelectric materials with holes of various shapes. Arch. Appl. Mech., 1999, 69:406-418.

[23] Qin Qinghua. Thermoelectroelastic Green's function for thermal load inside or on the boundary of an elliptic inclusion. Mech. Mater., 1999, 31:611-626.

[24] Qin Qinghua. Thermopiezoelectric interaction of macro-and micro-cracks in piezoelectric medium. Theore. Appl. Frac. Mech., 1999, 32:129-135.

[25] Tiersten H F. Linear piezoelectric plate vibrations. New York: Plenum Press, 1969.

[26] Parton V Z, Kudryavtsev B A. Electromagnetoelasticity, piezoelectrics and electrically conductive solids. New York: Gordon and Breach Science Publishers, 1988.

[27] Mason W P. Piezoelectric crystals and their application to ultrasonics. New York: Van Nostrand, 1950.

[28] Biot M A. Thermoelasticity and irreversible thermodynamics. J. Appl. Phys., 1956, 27: 240-253.

[29] Nye J F. Physical properties of crystals. Oxford: Oxford University Press, 1957.

[30] Lekhnitskii S G. Theory of elasticity of an anisotropic elastic body, San Francisco: Holden-Day, 1963.

[31] Stroh A N. Dislocations and cracks in anisotropic elasticity. Phil. Mag., 1958, 3:625-646.

[32] Suo Z. Singularities, interfaces and cracks in dissimilar anisotropic media. Proc. R. Soc. Lond., 1990, A427:331-358.

[33] Barnett D M, Lothe J. Dislocations and line charges in anisotropic piezoelectric insulators. Phys. Stat. Sol.(b), 1975, 67:105-111.

[34] Sosa H. Plane problems in piezoelectric media with defects, Int. J. Solids Struct., 1991, 28:491-505.

[35] Qin Qinghua. A new solution for thermopiezoelectric solid with an insulated elliptic hole. Acta Mechanica Sinica, 1998, 14:157-170.

[36] Qin Qinghua. Mai Y W. A new thermoelectroelastic solution for piezoelectric materials with various openings. Acta Mechanica, 1999, 138:97-111.

[37] Wang S S, Choi I. The interface crack between dissimilar anisotropic composite materials. J. Appl. Mech., 1983, 50:169-178.

[38] Lothe J, Barnett D M. Integral formalism for surface waves in piezoelectric crystals: existence considerations. J. Appl. Phys., 1976, 47:1799-1807.

[39] Chadwick P, Smith G D. Foundations of the theory of surface waves in anisotropic elastic materials. Adv. Appl. Mech., 1977, 17:303-376.

[40] Ting T C T. Some identities and structure of Ni in the Stroh formalism of anisotropic elasticity. Q. Appl. Mathe., 1988, 46:109-120.

[41] Ting T C T. Explicit solution and invariance of the singularities at an interface crack in anisotropic composites. Int. J. Solids Struct., 1986, 22:965-983.

[42] Qin Qinghua. Yu S W. Logarithmic singularity at crack tips in piezoelectric media. Chinese Science Bulletin, 1996, 41:563-566.

[43] Ting T C T. Effects of change of reference coordinates on the stress analysis of anisotropic elastic materials. Int. J. Solids Struct., 1982, 18:139-152.

[44] Dempsey J P, Sinclair G B. On the stress singularities in the plane elasticity of the composite wedge. J. Elasticity, 1979, 9:373-391.

[45] Jirousek J, Leon N. A powerful finite element for plate bending. Comp. Meth. Appl. Mech. Eng., 1977, 12: 77-96.

[46] Qin Qinghua. The Trefftz finite and boundary element method. Southampton: WIT Press, 2000.

[47] Qin Qinghua. Solving anti-plane problems of piezoelectric materials by the Trefftz finite element approach. Compu. Mech., 2003, 31:461-468.

[48] Qin Qinghua. Variational formulations for TFEM of piezoelectricity. Int. J. Solids Struc., 2003, 40:6335-6346.

[49] Trefftz E. Ein gegenstück zum ritzschen verfahren. In: Proceedings 2nd Int. Congress of Applied mechanics, Zurich, 1926, pp.131-137.

[50] Simpson H C. Spector S J. On the positive of the second variation of finite elasticity, Arch. Rational Mech. Anal., 1987, 98:1-30.

[51] Zielinski A P, Zienkiewicz O C. Generalized finite element analysis with T-complete boundary solution functions. Int. J. Numer. Meth. Eng., 1985, 21:509-528.

[52] Ding H J, Wang G Q, Chen W Q. A boundary integral formulation and 2D fundamental solutions for piezoelectric media. Comput. Meth. Appl. Mech. Eng., 1998, 158:65-80.

[53] Suo Z, Kuo C M, Barnett D M, et al. Fracture mechanics for piezoelectric ceramics. J. Mech. Phys. Solids, 1992, 40:739-765.

[54] Sneddon I N. Fourier transforms. McGraw-Hill Book Company, Inc., 1951.

[55] Qin Qinghua. Wang J S, Li X L. Effect of elastic coating on fracture behaviour of piezoelectric fibre with a penny-shaped crack. Composites Structures, 2006, 75:465-471.

[56] Qin Qinghua. Fracture mechanics of piezoelectric materials. Southampton: WIT Press, 2001.

[57] Yang J H, Lee K Y. Penny shaped crack in a piezoelectric cylinder surrounded by an elastic medium subjected to combined in-plane mechanical and electrical loads. Int. J. Solids Struct., 2003, 40:573-590.

[58] Xiao Z M, Bai J. Numerical simulation on a coated piezoelectric sensor interacting with a crack. Finite Elem. Anal. Des., 2002, 38: 691-706.

[59] Qin Qinghua, Wang J S, Kang Y L. A theoretical model for electroelastic analysis in piezoelectric fibre push-out test. Adv. Appl. Mech., 2006, 75:527-540.

[60] Wang J S, Qin Qinghua. Debonding criterion for the piezoelectric fibre push-out test. Phil. Mag. Letters, 2006, 86:123-136.

[61] Liu H Y, Qin Qinghua, Mai Y W. Theoretical model of piezoelectric fibre pull-out. Int. J. Solid Struct., 2003, 47:5511-5519.

[62] Zhang X, Liu H Y, Mai Y W, et al. On steady-state fibre pull-out I: the stress field with an elastic interfacial coating. Compo. Sci. Tech., 1999, 59:2179-2189.

[63] Jiang, X Y, Kong X G. Micro-mechanical characteristics of fibre/matrix interfaces in composite materials. Compo. Sci. Tech., 1999, 59:635-642.

[64] Zhou L M, Mai Y W, Ye L. Analyses of fibre push-out test based on the fracture mechanics approach. Compo. Eng., 1995, 5:1199-1219.

Chapter 4 Thermo-magneto-electro-elastic problems

4.1 Introduction

In the previous chapter we presented the linear theory of piezoelectricity and its application to various engineering problems. Extension of the theory and the methodology to thermo-magneto-electro-elastic problems is described in this chapter. First, we present a brief review of the developments in this field. As mentioned in [1], Suchtelen [2] appears to have been the first to report the magnetoelectric coupling effect in piezoelectric-piezomagnetic composites. He indicated that the magnetoelectric effect is a product property that results from the interaction between different properties of the two phases in composites. Later, Boomgaaed et al [3] further explored the magneto-electric effect of $BaTiO_3$-$CoFe_2O_4$ composites. To investigate fundamental theories and solution procedures, Lee [4], He [5], and Qing et al [6] constructed a variety of variational principles for magneto-electro-elastic materials. Alshits et al [7] studied the existence of surface waves in piezoelectric and piezomagnetic composites. Using a perturbation method, Lee et al [8] investigated stress effects on the electromagnetic resonance of circular dielectric disks. Li and Dunn [9] and Li [10] obtained formulas for predicting the average magneto-electro-elastic field and effective material properties of magneto-electro-elastic solids containing a multi-inclusion or inhomogeneity using the micromechanics approach. The investigation of general solution procedures should also be mentioned, it includes: eight sets of constitutive equations [1]; analytical solutions for simply-supported and multilayered magneto-electro-elastic plates [11, 12], for magneto-electro-elastic plates with polygonal inclusions [13], and for functionally graded and layered magneto-electro-elastic plates [14]; general solutions of three-dimensional magneto-electro-elastic solids based on the potential function approach [15, 16]; and hyperboloidal notch problems [17]. Recently, Liu et al [18] presented closed-form expressions of elastic, electric and magnetic fields for a moving dislocation in a magneto-electro-elastic solid and found that the magneto-electro-elastic field exhibits the singularity of r^{-1} near the dislocation core. Using the methods of Laplace and finite sine transformations, Ootao and Tanigawa [19] obtained the transient solution for a simply supported and multilayered magneto-electro-thermo-elastic strip due to unsteady and nonuniform heat supply in the width direction. Soh and Liu [20] presented an analytical expression for the interfacial debonding problem of a piezoelectric-piezomagnetic composite with a circular inclusion. In addition, it should be noted that the application of fracture mechanics to magneto-electro-elastic problems has been a fruitful subject, including but not limited to the work on interfacial cracks [21], plane cracks under out-of-plane deformation [22], crack-tip fields and energy release rate [23], collinear cracks [24], parallel cracks in a bimaterial solid [25], constant moving cracka under anti-plane deformation [26], microcrack-microcrack interaction [27], dynamic anti-plane crack problems [28], J-integral and dislocation model for crack problems [29], dynamic behaviour of two collinear cracks [30], and finite cracks in a piezoelectromagnetic strip [31]. Regarding Green's functions of magneto-electro-elastic problems, reports can be found on elliptic hole and rigid inclusion problems [32,33], three-dimensional problems [34,35], general inclusion problems [36], bimaterial problems [37], and an infinite magneto-electro-elastic solid with various defects [38, 39].

We begin this chapter with a discussion of the general theory of magneto-electro-elastic problems, followed by an introduction of the variational principle and potential approach. Then, Green's functions are presented for half-plane, bimaterial, and wedge problems. Finally, solutions for an antiplane shear crack in a magneto-electro-elastic layer are derived.

4.2   Basic field equations for magneto-electro-elastic solids

4.2.1   Basic equations of general anisotropy

In this subsection we review briefly basic equations of three-dimensional magneto-electro-elastic solids. For a linear magneto-electro-elastic solid of general anisotropy, the governing equations of the mechanical and electric fields are in the same form as those of Eqs.(3.2.3), (3.2.5), (3.2.12), and (3.2.13). For the sake of completeness, we list all these equations together with equations governing magnetic fields below [4, 40]:

Force equilibrium equation:

$$\sigma_{ij,j} + b_i = 0 \tag{4.2.1}$$

Strain-displacement relationship:

$$\varepsilon_{ij} = \frac{1}{2}(u_{i,j} + u_{j,i}) \tag{4.2.2}$$

Magnetoelectric Maxwell equation:

$$D_{i,i} + b_e = 0, \quad B_{i,i} + b_m = 0 \tag{4.2.3}$$

Magnetoelectric gradient equation:

$$E_i = -\phi_{,i}, \quad H_i = -\psi_{,i} \tag{4.2.4}$$

Eqs.(4.2.1), (4.2.2) are coupled to Eqs.(4.2.3), (4.2.4) with the following constitutive relations [4]:

$$\begin{aligned}
\sigma_{ij} &= c_{ijkl}\varepsilon_{kl} - e_{lij}E_l - \tilde{e}_{lij}H_l \\
D_i &= e_{ikl}\varepsilon_{kl} + \kappa_{il}E_l + \alpha_{il}H_l \\
B_i &= \tilde{e}_{ikl}\varepsilon_{kl} + \alpha_{il}E_l + \mu_{il}H_l
\end{aligned} \tag{4.2.5}$$

In the above equations, b_m is the body electric current; B_i, H_i, and ψ are the magnetic induction, magnetic field, and magnetic potential respectively, and $\tilde{e}_{lij}$, α_{ij}, and μ_{ij} are piezomagnetic constants, electromagnetic constants, and magnetic permeabilities, respectively. Let Γ be the boundary of the solution domain Ω of the magneto-electro-elastic solid considered. Then the boundary conditions can be given in the form [40]

$$u_i = \overline{u}_i, \qquad\qquad \text{on } \Gamma_u \tag{4.2.6}$$

$$t_i = \sigma_{ij}n_j = \overline{t}_i, \qquad\qquad \text{on } \Gamma_t \tag{4.2.7}$$

$$\phi = \overline{\phi}, \qquad\qquad \text{on } \Gamma_\phi \tag{4.2.8}$$

$$D_n = D_i n_i = -\overline{q}_s = \overline{D}_n, \qquad \text{on } \Gamma_D \tag{4.2.9}$$

$$\psi = \overline{\psi}, \qquad\qquad \text{on } \Gamma_\psi \tag{4.2.10}$$

$$B_n = B_i n_i = \overline{m}, \qquad\qquad \text{on } \Gamma_B \tag{4.2.11}$$

where a bar over a variable indicates that the variable is prescribed, and $\Gamma = \Gamma_u \cup \Gamma_t = \Gamma_\phi \cup \Gamma_D = \Gamma_\psi \cup \Gamma_B$. The introduction of the symbol $\overline{D}_n$ is for simplicity of the writing that follows. Obviously, the first four equations

(4.2.6)~(4.2.9) are the electric and mechanical boundary conditions which are the same as those presented in Section 3.5, and the remaining two are for the magnetic field. Eqs.(4.2.1)~(4.2.11) constitute the complete set of equations of a linear magneto-electro-elastic solid. This set of equations is coupled among magnetic, electric, and mechanical fields.

4.2.2   Eight forms of constitutive equations

In addition to the constitutive equation (4.2.5), Soh and Liu [1] presented the other seven equivalent constitutive representations commonly used in the stationary theory of linear magneto-electro-elastic solid to describe the coupled interaction among elastic, electric, and magnetic variables. Table 4.1 lists the eight forms of constitutive relations, the corresponding independent variables, and the generalized Gibbs energy functionals. In Table 4.1, c and s are elastic stiffness and compliance tensors, κ and β are permittivity and impermittivity tensors, μ and ν are permeability and reluctivity tensors, α, λ, η, and ζ are magnetoelectric constants, e, h, d, and g are piezoelectric constants, and $\tilde{e}$, $\tilde{h}$, $\tilde{d}$ and $\tilde{g}$ are piezomagnetic constants. It can be seen from Table 4.1 that each form of constitutive representation has its own distinct independent variables and corresponding thermodynamic potential. But they are dependent on each other and any one form can be deduced from another through the Legendre transform. For example, if we choose σ, E, and H as the independent variables, Θ_8 can be deduced through the Legendre transform from Θ_1 as follows: $\Theta_8 = \Theta_1 - \sigma\varepsilon$, which results in the following constitutive relations in terms of Θ_8:

$$\varepsilon = -\frac{\partial \Theta_8}{\partial \sigma}, \quad D = -\frac{\partial \Theta_8}{\partial E}, \quad B = -\frac{\partial \Theta_8}{\partial H} \qquad (4.2.12)$$

Table 4.1   Eight forms of constitutive models [1]

Independent variables	Constitutive relations	Thermodynamic potentials
ε, E, H	$\sigma = c\varepsilon - e^{\mathrm{T}}E - \tilde{e}^{\mathrm{T}}H$ $D = e\varepsilon + \kappa E + \alpha H$ $B = \tilde{e}\varepsilon + \alpha E + \mu H$	$\Theta_1 = \frac{1}{2}(c\varepsilon^2 - \kappa E^2 - \mu H^2) -$ $e\varepsilon E - \tilde{e}\varepsilon H - \alpha EH$
ε, D, H	$\sigma = c\varepsilon - h^{\mathrm{T}}D - \tilde{e}^{\mathrm{T}}H$ $E = -h\varepsilon + \beta D - \zeta H$ $B = \tilde{e}\varepsilon + \zeta D + \mu H$	$\Theta_2 = \Theta_1 + DE$

Table 4.1   (continued)

Independent variables	Constitutive relations	Thermodynamic potentials
ε, E, B	$\begin{cases} \sigma = c\varepsilon - e^{\mathrm{T}}E - \tilde{h}^{\mathrm{T}}B \\ D = e\varepsilon + \kappa E + \eta B \\ H = -\tilde{h}\varepsilon - \eta E + vB \end{cases}$	$\Theta_3 = \Theta_1 + BH$
ε, D, B	$\begin{cases} \sigma = c\varepsilon - h^{\mathrm{T}}D - \tilde{h}^{\mathrm{T}}B \\ E = -h\varepsilon + \beta D - \lambda B \\ H = \tilde{h}\varepsilon - \lambda D + vB \end{cases}$	$\Theta_4 = \Theta_1 + DE + BH$
σ, D, B	$\begin{cases} \varepsilon = s\sigma + g^{\mathrm{T}}D + \tilde{g}^{\mathrm{T}}B \\ E = -g\sigma + \beta D - \lambda B \\ H = \tilde{g}\sigma - \lambda D + vB \end{cases}$	$\Theta_5 = \Theta_1 + DE + BH - \sigma\varepsilon$
σ, E, B	$\begin{cases} \varepsilon = s\sigma + d^{\mathrm{T}}E + \tilde{g}^{\mathrm{T}}B \\ D = d\sigma + \kappa E + \eta B \\ H = -\tilde{g}\sigma - \eta E + vB \end{cases}$	$\Theta_6 = \Theta_1 + BH - \sigma\varepsilon$
σ, D, H	$\begin{cases} \varepsilon = s\sigma + g^{\mathrm{T}}D + \tilde{d}^{\mathrm{T}}H \\ E = -g\sigma + \beta D - \zeta H \\ B = \tilde{d}\sigma + \zeta D + \mu H \end{cases}$	$\Theta_7 = \Theta_1 + DE - \sigma\varepsilon$
σ, E, H	$\begin{cases} \varepsilon = s\sigma + d^{\mathrm{T}}E + \tilde{d}^{\mathrm{T}}H \\ D = d\sigma + \kappa E + \alpha H \\ B = \tilde{d}\sigma + \alpha E + \mu H \end{cases}$	$\Theta_8 = \Theta_1 - \sigma\varepsilon$

4.2.3 Transversely isotropic simplification

If the magneto-electro-elastic solid considered is transversely isotropic, the equations described in Subsection 4.2.1 can be further simplified [41]. In a fixed rectangular coordinate system (x,y,z), the constitutive equations (4.2.5), and governing equations (4.2.1) and (4.2.3) of a transversely isotropic magneto-electro-elastic solid with the isotropic plane perpendicular to the z axis can, respectively, be expressed in the following form [15, 16]:

$$\begin{aligned}
\sigma_{xx} &= c_{11}u_{1,x} + c_{12}u_{2,y} + c_{13}u_{3,z} + e_{31}\phi_{,z} + \tilde{e}_{31}\psi_{,z} \\
\sigma_{yy} &= c_{12}u_{1,x} + c_{11}u_{2,y} + c_{13}u_{3,z} + e_{31}\phi_{,z} + \tilde{e}_{31}\psi_{,z} \\
\sigma_{zz} &= c_{13}u_{1,x} + c_{13}u_{2,y} + c_{33}u_{3,z} + e_{33}\phi_{,z} + \tilde{e}_{33}\psi_{,z} \\
\sigma_{yz} &= c_{44}(u_{2,z} + u_{3,y}) + e_{15}\phi_{,y} + \tilde{e}_{15}\psi_{,y} \\
\sigma_{xz} &= c_{44}(u_{1,z} + u_{3,x}) + e_{15}\phi_{,x} + \tilde{e}_{15}\psi_{,x} \\
\sigma_{xy} &= c_{66}(u_{1,y} + u_{2,x})
\end{aligned} \tag{4.2.13}$$

$$D_x = e_{15}(u_{1,z} + u_{3,x}) - \kappa_{11}\phi_{,x} - \alpha_{11}\psi_{,x}$$

$$D_y = e_{15}(u_{2,z} + u_{3,y}) - \kappa_{11}\phi_{,y} - \alpha_{11}\psi_{,y} \qquad (4.2.14)$$

$$D_z = e_{31}(u_{1,x} + u_{2,y}) - e_{33}u_{3,z} - \kappa_{33}\phi_{,z} - \alpha_{33}\psi_{,z}$$

$$B_x = \tilde{e}_{15}(u_{1,z} + u_{3,x}) + \alpha_{11}\phi_{,x} - \mu_{11}\psi_{,x}$$

$$B_y = \tilde{e}_{15}(u_{2,z} + u_{3,y}) + \alpha_{11}\phi_{,y} - \mu_{11}\psi_{,y} \qquad (4.2.15)$$

$$B_z = \tilde{e}_{31}(u_{1,x} + u_{2,y}) + \tilde{e}_{33}u_{3,z} + \alpha_{33}\phi_{,z} - \mu_{33}\psi_{,z}$$

and

$$c_{11}u_{1,xx} + \frac{1}{2}(c_{11} - c_{12})u_{1,yy} + \frac{1}{2}(c_{11} + c_{12})u_{2,xy} + (c_{13} + c_{44})u_{3,xz} +$$

$$c_{44}u_{1,zz} + (e_{31} + e_{15})\phi_{,xz} - (\tilde{e}_{15} + \tilde{e}_{31})\psi_{,xz} + b_1 = 0 \qquad (4.2.16)$$

$$c_{11}u_{2,yy} + \frac{1}{2}(c_{11} - c_{12})u_{2,xx} + \frac{1}{2}(c_{11} + c_{12})u_{1,xy} + (c_{13} + c_{44})u_{3,yz} +$$

$$c_{44}u_{2,zz} + (e_{31} + e_{15})\phi_{,yz} - (\tilde{e}_{15} + \tilde{e}_{31})\psi_{,yz} + b_2 = 0 \qquad (4.2.17)$$

$$c_{44}(u_{3,xx} + u_{3,yy}) + (c_{44} + c_{13})(u_{1,xz} + u_{2,yz}) + c_{33}u_{3,33} +$$

$$e_{15}(\phi_{,xx} + \phi_{,yy}) + e_{33}\phi_{,zz} - \tilde{e}_{33}\psi_{,zz} - \tilde{e}_{15}(\psi_{,xx} + \psi_{,yy}) + b_3 = 0 \quad (4.2.18)$$

$$e_{15}(u_{3,xx} + u_{3,yy}) + (e_{15} + e_{31})(u_{1,xz} + u_{2,yz}) + e_{33}u_{3,zz} -$$

$$\kappa_{11}(\phi_{,xx} + \phi_{,yy}) - \kappa_{33}\phi_{,zz} - \alpha_{11}(\psi_{,xx} + \psi_{,yy}) - \alpha_{33}\psi_{,zz} + b_e = 0$$

$$\qquad (4.2.19)$$

$$\tilde{e}_{15}(u_{3,xx} + u_{3,yy}) + (\tilde{e}_{15} + \tilde{e}_{31})(u_{1,xz} + u_{2,yz}) + \tilde{e}_{33}u_{3,zz} +$$

$$\alpha_{11}(\phi_{,xx} + \phi_{,yy}) + \alpha_{33}\phi_{,zz} - \mu_{11}(\psi_{,xx} + \psi_{,yy}) - \mu_{33}\psi_{,zz} + b_m = 0 \quad (4.2.20)$$

Eqs.(4.2.13)~(4.2.20) are used as a basis in later sections.

4.2.4 Extension to include thermal effect

The equations presented in Section 4.2.1 can be straightforwardly extended to include thermal effects if the temperature field does not fully couple with the magneto-electro-elastic field, that is, if the magneto-electro-elastic field can be affected by the temperature field through constitutive relations but the temperature field is not affected by the magneto-electro-elastic field. Under such assumptions the governing equations of thermo-magneto-electro-elastic problems can be expressed as [42]

$$\sigma_{ij,j} + b_i = 0, \quad D_{i,i} + b_e = 0, \quad B_{i,i} + b_m = 0 \qquad (4.2.21)$$

$$\left. \begin{array}{l} \sigma_{ij} = c_{ijkl}\varepsilon_{kl} - e_{lij}E_l - \tilde{e}_{lij}H_l - \lambda_{ij}T \\ D_i = e_{ikl}\varepsilon_{kl} + \kappa_{il}E_l + \alpha_{il}H_l - \rho_i T \\ B_i = \tilde{e}_{ikl}\varepsilon_{kl} + \alpha_{il}E_l + \mu_{il}H_l - \nu_i T \end{array} \right\} \qquad (4.2.22)$$

$$\varepsilon_{ij} = \frac{1}{2}(u_{i,j} + u_{j,i}), \quad E_i = -\phi_{,i}, \quad H_i = -\psi_{,i} \tag{4.2.23}$$

$$h_{i,i} = 0, \quad h_i = -k_{ij}T_{,j} \tag{4.2.24}$$

where ν_i are pyromagnetic coefficients. The boundary conditions of the thermo-magneto-electro-elastic problem are still defined by Eqs.(4.2.6)~(4.2.11) together with the boundary conditions for the thermal field. To obtain the solution of the boundary value problem defined by Eqs.(4.2.6)~(4.2.11) and (4.2.21)~(4.2.24), we usually first solve the heat transfer problem to obtain the steady-state T field, and then calculate the magneto-electro-elastic field caused by the T field, add an isothermal solution to satisfy the corresponding magnetic, electrical and mechanical boundary conditions, and finally solve the modified problem for magneto-electro-elastic fields.

4.3 Variational formulation

Taking $\Theta_1(\boldsymbol{\varepsilon}, \boldsymbol{E}, \boldsymbol{H})$ (see Table 4.1) as an example, we now present a variational principle for a magneto-electro-elastic solid in the domain Ω bounded by Γ. The variational principle is based on the independent variables $(\boldsymbol{\varepsilon}, \boldsymbol{E}, \boldsymbol{H})$. First, the explicit expression of $\Theta_1(\boldsymbol{\varepsilon}, \boldsymbol{E}, \boldsymbol{H})$ in terms of ε_{ij}, E_i and H_i is presented

$$\Theta_1 = \frac{1}{2}c_{ijkl}\varepsilon_{ij}\varepsilon_{kl} - \frac{1}{2}\kappa_{ij}E_iE_j - \frac{1}{2}\mu_{ij}H_iH_j - e_{ijk}E_i\varepsilon_{jk} - \tilde{e}_{ijk}H_i\varepsilon_{jk} - \alpha_{ij}E_iH_j$$

$$\tag{4.3.1}$$

which results in the following constitutive relations:

$$\sigma_{ij} = \frac{\partial \Theta_1}{\partial \varepsilon_{ij}}, \quad D_i = -\frac{\partial \Theta_1}{\partial E_i}, \quad B_i = -\frac{\partial \Theta_1}{\partial H_i} \tag{4.3.2}$$

Then, based on the variational functional $\Theta_1(\boldsymbol{\varepsilon}, \boldsymbol{E}, \boldsymbol{H})$ and the basic equations (4.2.1)~(4.2.11), a variational functional can be constructed as

$$\int_\Omega (\Theta_1 - b_iu_i - b_e\phi - b_m\psi)\,\mathrm{d}\Omega - \int_{\Gamma_t} \bar{t}_iu_i\,\mathrm{d}\Gamma - \int_{\Gamma_D} \bar{D}_n\phi\,\mathrm{d}\Gamma - \int_{\Gamma_B} \bar{m}\psi\,\mathrm{d}\Gamma$$

$$\tag{4.3.3}$$

in which Eqs.(4.2.5), (4.2.6), (4.2.8), and (4.2.10) are assumed to be satisfied, a priori.

We now proceed to show that Eqs.(4.2.1), (4.2.3), (4.2.7), (4.2.9), and (4.2.11) can be derived from Eq.(4.3.3) for the independent variables

δu_i, $\delta\phi$, and $\delta\psi$. Taking the vanishing variation of Eq.(4.3.3), we have

$$\int_\Omega(\delta\Theta_1 - b_i\delta u_i - b_e\delta\phi - b_m\delta\psi)\mathrm{d}\Omega - \int_{\Gamma_t}\bar{t}_i\delta u_i\mathrm{d}\Gamma - \int_{\Gamma_D}\bar{D}_n\delta\phi\mathrm{d}\Gamma - \int_{\Gamma_B}\bar{m}\delta\psi\mathrm{d}\Gamma = 0$$

$$(4.3.4)$$

Noting that the variation of Θ_1 expressed by Eq.(4.3.1) can be further written as

$$\begin{aligned}
\delta\Theta_1 &= c_{ijkl}\varepsilon_{kl}\delta\varepsilon_{ij} - e_{ijk}(E_k\delta\varepsilon_{ij} + \varepsilon_{jk}\delta E_i) - \kappa_{ij}E_j\delta E_i - \mu_{ij}H_j\delta H_i - \\
&\quad \tilde{e}_{ijk}(H_k\delta\varepsilon_{ij} + \varepsilon_{jk}\delta H_i) - \alpha_{ij}(E_j\delta H_i + H_j\delta E_i) \\
&= \sigma_{ij}\delta\varepsilon_{ij} - D_i\delta E_i - B_i\delta H_i
\end{aligned} \qquad (4.3.5)$$

The substitution of Eq.(4.3.5) into Eq.(4.3.4) yields

$$\int_\Omega(\sigma_{ij}\delta\varepsilon_{ij} - D_i\delta E_i - B_i\delta H_i - b_i\delta u_i - b_e\delta\phi - b_m\delta\psi)\mathrm{d}\Omega - $$

$$\int_{\Gamma_t}\bar{t}_i\delta u_i\mathrm{d}\Gamma - \int_{\Gamma_D}\bar{D}_n\delta\phi\mathrm{d}\Gamma - \int_{\Gamma_B}\bar{m}\delta\psi\mathrm{d}\Gamma = 0 \qquad (4.3.6)$$

Making use of Eqs.(4.2.2) and (4.2.4), the variables can be expressed as

$$\delta\varepsilon_{ij} = \frac{1}{2}(\delta u_{j,i} + \delta u_{i,j}), \quad \delta E_i = -\delta\phi_{,i}, \quad \delta H_i = -\delta\psi_{,i} \qquad (4.3.7)$$

By substituting Eq.(4.3.7) into Eq.(4.3.6) and employing the chain rule of differentiation, integration by parts, the divergence theorem, and the boundary conditions (4.2.6), (4.2.8), (4.2.10), the first three terms of Eq.(1.106), with the substitution of Eq.(4.3.6), become

$$\int_\Omega\sigma_{ij}\delta\varepsilon_{ij}\mathrm{d}\Omega = \int_{\Gamma_t}\sigma_{ij}n_j\delta u_i\mathrm{d}\Gamma - \int_\Omega\sigma_{ij,j}\delta u_j\mathrm{d}\Gamma$$

$$= \int_{\Gamma_t}t_i\delta u_i\mathrm{d}\Gamma - \int_\Omega\sigma_{ij,j}\delta u_j\mathrm{d}\Gamma \qquad (4.3.8)$$

$$-\int_\Omega D_i\delta E_i\mathrm{d}\Omega = \int_{\Gamma_D}D_in_i\delta\phi\mathrm{d}\Gamma - \int_\Omega D_{i,i}\delta\phi\mathrm{d}\Gamma$$

$$= \int_{\Gamma_D}D_n\delta\phi\mathrm{d}\Gamma - \int_\Omega D_{i,i}\delta\phi\mathrm{d}\Gamma \qquad (4.3.9)$$

$$-\int_\Omega B_i\delta H_i\mathrm{d}\Omega = \int_{\Gamma_D}B_in_i\delta\psi\mathrm{d}\Gamma - \int_\Omega B_{i,i}\delta\psi\mathrm{d}\Gamma$$

$$= \int_{\Gamma_D}B_n\delta\psi\mathrm{d}\Gamma - \int_\Omega B_{i,i}\delta\psi\mathrm{d}\Gamma \qquad (4.3.10)$$

Then, with substitution of Eqs.(4.3.8)~(4.3.10) into Eq.(4.3.6), we have

$$\int_\Omega[(\sigma_{ij,j} + b_i)\delta u_i + (D_{i,i} + b_e)\delta\phi + (B_{i,i} + b_m)\delta\psi]\,\mathrm{d}\Omega - $$

$$\int_{\Gamma_t}(t_i - \bar{t})_i\delta u_i\mathrm{d}\Gamma - \int_{\Gamma_D}(D_n - \bar{D}_n)\delta\phi\mathrm{d}\Gamma - \int_{\Gamma_B}(B_n - \bar{m})\delta\psi\mathrm{d}\Gamma = 0$$

$$(4.3.11)$$

The above equation must be satisfied for the independent variables δu_i, $\delta\phi$, and $\delta\psi$. Hence, the volume integral in Eq.(4.3.11) leads to Eqs.(4.2.1) and (4.2.3), and the corresponding surface integrals yield boundary conditions (4.2.7), (4.2.9), and (4.2.11). Similarly, we can present variational principles for the other seven thermodynamic functionals ($\Theta_2 \sim \Theta_8$) in a straightforward way. It should be mentioned that the above variational principles yield half of the boundary conditions only. This is because only three variables are taken as independent variables. To obtain all field equations from the variational principle we need to construct the corresponding variational functional in such a way that all variables, i.e., u_i, ε_{ij}, σ_{ij}, ϕ, E_i, D_i, ψ, H_i, B_i are independent variables. Taking $\Theta_1(\boldsymbol{\varepsilon}, \boldsymbol{E}, \boldsymbol{H})$ (see Table 4.1) as an example again, Yao [40] presented a generalized variational functional as follows:

$$\int_\Omega [\sigma_{ij}\varepsilon_{ij} - \Theta_1 + u_i(\sigma_{ij,j} + b_i) - D_iE_i - B_iH_i + \phi(D_{i,i} + b_e) +$$

$$\psi(B_{i,i} + b_m)]\mathrm{d}\Omega - \int_{\Gamma_u} t_i\bar{u}_i\mathrm{d}\Gamma - \int_{\Gamma_\phi} D_n\bar{\phi}\mathrm{d}\Gamma - \int_{\Gamma_\psi} B_n\bar{\psi}\mathrm{d}\Gamma -$$

$$\int_{\Gamma_t} (t_i - \bar{t}_i)u_i\mathrm{d}\Gamma - \int_{\Gamma_D} (D_n - \bar{D}_n)\phi\mathrm{d}\Gamma - \int_{\Gamma_B} (B_n - \bar{m})\psi\mathrm{d}\Gamma \qquad (4.3.12)$$

It can be shown that the vanishing variation of the functional (4.3.12) leads to Eqs.(4.2.1)~(4.2.11) [40].

4.4   General solution for 3D transversely isotropic magneto-electro-elastic solids

Based on the developments in [15, 41], this subsection describes the potential function approach for solving three-dimensional magneto-electro-elastic problems. To this end, Wang and Shen [15] assumed that the general solution to Eqs.(4.2.16)~(4.2.20) is in the form

$$u_1 = \Phi_{,x} - \Psi_{,y}, \quad u_2 = \Phi_{,y} + \Psi_{,x}, \quad u_3 = k_1\Phi_{,z}, \quad \phi = k_2\Phi_{,z}, \quad \psi = k_3\Phi_{,z}$$

$$(4.4.1)$$

where Φ and Ψ are two potential functions to be solved for, and k_1, k_2, k_3 are three constants to be determined. Substituting Eq.(4.4.1) into Eqs.(4.2.16)~(4.2.20), we obtain

$$\Psi_{,xx} + \Psi_{,yy} + \frac{c_{44}}{c_{66}}\Psi_{,zz} = 0 \qquad (4.4.2)$$

$$c_{11}(\varPhi_{,xx}+\varPhi_{,yy})+[c_{44}+k_1(c_{13}+c_{44})+k_2(e_{15}+e_{31})-k_3(\tilde{e}_{15}+\tilde{e}_{31})]\varPhi_{,zz}=0$$

$$(c_{13}+c_{44}+k_1c_{44}+k_2e_{15}-k_3\tilde{e}_{15})(\varPhi_{,xx}+\varPhi_{,yy})+(k_1c_{33}+k_2e_{33}-k_3\tilde{e}_{33})\varPhi_{,zz}=0$$

$$(e_{15}+e_{31}+k_1e_{15}-k_2\kappa_{11}-k_3\alpha_{11})(\varPhi_{,xx}+\varPhi_{,yy})+(k_1e_{33}-k_2\kappa_{33}-k_3\alpha_{33})\varPhi_{,zz}=0$$

$$(\tilde{e}_{15}+\tilde{e}_{31}+k_1\tilde{e}_{15}+k_2\alpha_{11}-k_3\mu_{11})(\varPhi_{,xx}+\varPhi_{,yy})+(k_1\tilde{e}_{33}+k_2\alpha_{33}-k_3\mu_{33})\varPhi_{,zz}=0$$

$$(4.4.3)$$

which results in the following equation:

$$\frac{c_{44}+k_1(c_{13}+c_{44})+k_2(e_{15}+e_{31})-k_3(\tilde{e}_{15}+\tilde{e}_{31})}{c_{11}}$$

$$=\frac{k_1c_{33}+k_2e_{33}-k_3\tilde{e}_{33}}{c_{13}+c_{44}+k_1c_{44}+k_2e_{15}-k_3\tilde{e}_{15}}=\frac{k_1e_{33}-k_2\kappa_{33}-k_3\alpha_{33}}{e_{15}+e_{31}+k_1e_{15}-k_2\kappa_{11}-k_3\alpha_{11}}$$

$$=\frac{k_1\tilde{e}_{33}+k_2\alpha_{33}-k_3\mu_{33}}{\tilde{e}_{15}+\tilde{e}_{31}+k_1\tilde{e}_{15}+k_2\alpha_{11}-k_3\mu_{11}}=\lambda$$

$$(4.4.4)$$

Eliminating k_1, k_2 and k_3 from Eq.(4.4.4) yields the following equation for λ:

$$\boldsymbol{u}^{\mathrm{T}}(\boldsymbol{A}-\lambda\boldsymbol{B})^{-1}\boldsymbol{u}=c_{11}-c_{44}\lambda^{-1} \qquad (4.4.5)$$

where

$$\boldsymbol{u}=\begin{bmatrix}c_{11}+c_{44}\\ e_{15}+e_{31}\\ -\tilde{e}_{15}-\tilde{e}_{31}\end{bmatrix},\quad \boldsymbol{A}=\begin{bmatrix}c_{33} & e_{33} & -\tilde{e}_{33}\\ e_{33} & -\kappa_{33} & -\alpha_{33}\\ -\tilde{e}_{33} & -\alpha_{33} & \mu_{33}\end{bmatrix},\quad \boldsymbol{B}=\begin{bmatrix}c_{44} & e_{15} & -\tilde{e}_{15}\\ e_{15} & -\kappa_{11} & -\alpha_{11}\\ -\tilde{e}_{15} & -\alpha_{11} & \mu_{11}\end{bmatrix}$$

$$(4.4.6)$$

It is formidable and injudicious to directly expand the left-hand side of Eq.(4.4.5). In order to obtain the algebraic equation for λ in an elegant manner, Wang and Shen [15] considered the following eigenvalue problem:

$$\boldsymbol{A}\boldsymbol{\xi}=\delta\boldsymbol{B}\boldsymbol{\xi} \qquad (4.4.7)$$

Then the following orthogonal relationship with respect to $\mathbf{A}$ and $\mathbf{B}$ establishes:

$$\mathbf{Y}^{\mathrm{T}}\mathbf{B}\mathbf{Y}=\boldsymbol{\Lambda}_b,\quad \mathbf{Y}^{\mathrm{T}}\mathbf{A}\mathbf{Y}=\boldsymbol{\Lambda}_a=\boldsymbol{\Lambda}_0\boldsymbol{\Lambda}_b \qquad (4.4.8)$$

where

$$\boldsymbol{Y}=\begin{bmatrix}\xi_1 & \xi_2 & \xi_3\end{bmatrix},\quad \boldsymbol{\Lambda}_b=\mathrm{diag}\begin{bmatrix}b_1 & b_2 & b_3\end{bmatrix},\quad \boldsymbol{\Lambda}_0=\mathrm{diag}\begin{bmatrix}\delta_1 & \delta_2 & \delta_3\end{bmatrix} \qquad (4.4.9)$$

Making use of the relationships (4.4.8), Eq.(4.4.5) can be rewritten as

$$\tilde{\boldsymbol{u}}(\boldsymbol{\Lambda}_0-\lambda\boldsymbol{I})^{-1}\boldsymbol{\Lambda}_b^{-1}=c_{11}-c_{44}\lambda^{-1} \qquad (4.4.10)$$

where

$$\tilde{\boldsymbol{u}}=[\tilde{u}_1\ \tilde{u}_2\ \tilde{u}_3]^{\mathrm{T}}=\boldsymbol{Y}^{\mathrm{T}}\boldsymbol{u} \qquad (4.4.11)$$

while $k_i (i=1,\ 2,\ 3)$ can be expressed in terms of λ as

$$k = \begin{bmatrix} k_1 \\ k_2 \\ k_3 \end{bmatrix} = \lambda Y (\Lambda_0 - \lambda I)^{-1} \Lambda_b^{-1} \tilde{u} = \lambda Y \begin{bmatrix} \dfrac{\tilde{u}_1}{(\delta_1 - \lambda)b_1} \\[2mm] \dfrac{\tilde{u}_2}{(\delta_2 - \lambda)b_2} \\[2mm] \dfrac{\tilde{u}_3}{(\delta_3 - \lambda)b_3} \end{bmatrix} \tag{4.4.12}$$

Expanding Eq.(4.4.10) yields the following algebraic equation for λ

$$\frac{c_{44}}{\lambda} + \sum_{i=1}^{3} \frac{\tilde{u}_i^2}{(\delta_i - \lambda)b_i} = c_{11} \tag{4.4.13}$$

Eq.(4.4.13) is a quartic equation in λ. Denote the four roots of Eq.(4.4.13) as $\lambda_i (i = 1, 2, 3, 4)$ and let $\lambda_0 = c_{44}/c_{66}$. Then, there exist five potential functions $\Phi_i (i = 1 \sim 4)$ and $\Phi_0 = \Psi$, which satisfy

$$\Phi_{i,xx} + \Phi_{i,yy} + \lambda_i \Phi_{i,zz} = 0, \quad i = 0, 1, 2, 3, 4, 5 \tag{4.4.14}$$

Based on the above derivation, displacements, electric potential and magnetic potential in the fixed Cartesian coordinate system can be expressed in terms of the five potential functions $\Phi_i (i = 0 \sim 4)$ as

$$u_1 = \sum_{i=1}^{4} \Phi_{i,x} - \Phi_{0,y}, \quad u_2 = \sum_{i=1}^{4} \Phi_{i,y} + \Phi_{0,x}$$

$$u_3 = \sum_{i=1}^{4} k_{1i} \Phi_{i,z}, \quad \phi = \sum_{i=1}^{4} k_{2i} \Phi_{i,z}, \quad \psi_3 = \sum_{i=1}^{4} k_{3i} \Phi_{i,z} \tag{4.4.15}$$

where k_{1i}, k_{2i} and k_{3i} can be obtained from Eq.(4.4.4) as [41]

$$k_{1i} = \frac{\Sigma_{1i}}{\Sigma^i}, \quad k_{2i} = \frac{\Sigma_{2i}}{\Sigma^i}, \quad k_{3i} = \frac{\Sigma_{3i}}{\Sigma^i} \tag{4.4.16}$$

with

$$\Sigma^i = \begin{vmatrix} c_{13} + c_{44} & e_{15} + e_{31} & \tilde{e}_{15} + \tilde{e}_{31} \\ e_{33} - e_{15}\lambda_i & \kappa_{11}\lambda_i - \kappa_{33} & \alpha_{11}\lambda_i - \alpha_{33} \\ \tilde{e}_{33} - \tilde{e}_{15}\lambda_i & \alpha_{11}\lambda_i - \alpha_{33} & \mu_{11}\lambda_i - \mu_{33} \end{vmatrix} \tag{4.4.17}$$

$$\Sigma_{1i} = \begin{vmatrix} c_{11}\lambda_i - c_{44} & e_{15} + e_{31} & \tilde{e}_{15} + \tilde{e}_{31} \\ (e_{15} + e_{31})\lambda_i & \kappa_{11}\lambda_i - \kappa_{33} & \alpha_{11}\lambda_i - \alpha_{33} \\ (\tilde{e}_{15} + \tilde{e}_{31})\lambda_i & \alpha_{11}\lambda_i - \alpha_{33} & \mu_{11}\lambda_i - \mu_{33} \end{vmatrix} \tag{4.4.18}$$

$$\Sigma_{2i} = \begin{vmatrix} c_{13} + c_{44} & c_{13}\lambda_i - c_{44} & \tilde{e}_{15} + \tilde{e}_{31} \\ e_{33} - e_{15}\lambda_i & (e_{15} + e_{31})\lambda_i & \alpha_{11}\lambda_i - \alpha_{33} \\ \tilde{e}_{33} - \tilde{e}_{15}\lambda_i & (\tilde{e}_{15} + \tilde{e}_{31})\lambda_i & \mu_{11}\lambda_i - \mu_{33} \end{vmatrix} \tag{4.4.19}$$

$$\Sigma_{3i} = \begin{vmatrix} c_{13}+c_{44} & e_{15}+e_{31} & c_{13}\lambda_i - c_{44} \\ e_{33}-e_{15}\lambda_i & \kappa_{11}\lambda_i - \kappa_{33} & (e_{15}+e_{31})\lambda_i \\ \tilde{e}_{33}-\tilde{e}_{15}\lambda_i & \alpha_{11}\lambda_i - \alpha_{33} & (\tilde{e}_{15}+\tilde{e}_{31})\lambda_i \end{vmatrix} \tag{4.4.20}$$

Eq.(4.4.15) can also be written in matrix form as

$$U = \Lambda(\tilde{J}P + i\Phi_0), \quad W = KP_{,z} \tag{4.4.21}$$

where

$$U = u_1 + iu_2, \quad W = \begin{bmatrix} u_3 \\ \phi \\ \psi \end{bmatrix}, \quad P = \begin{bmatrix} \Phi_1 \\ \Phi_2 \\ \Phi_3 \\ \Phi_4 \end{bmatrix}, \quad \tilde{J} = \begin{bmatrix} 1 \\ 1 \\ 1 \\ 1 \end{bmatrix}^{\mathrm{T}} \tag{4.4.22}$$

$$K = \begin{bmatrix} k_1 & k_2 & k_3 & k_4 \end{bmatrix}, \quad \Lambda = \frac{\partial}{\partial x} + i\frac{\partial}{\partial y}$$

Making use of Eqs.(4.4.21) and (4.2.12)~(4.2.14), we can obtain stress, electric displacement, and magnetic induction in terms of Φ_i as

$$\sigma_{xx} + \sigma_{yy} = 2(c_{66}\tilde{J}H - c_{44}\tilde{J} - \tilde{I}_0^{\mathrm{T}}BK)P_{,zz}$$

$$\sigma_{xx} - \sigma_{yy} + 2i\sigma_{xy} = 2c_{66}\Lambda^2(\tilde{J}P + i\Phi_0)$$

$$\begin{bmatrix} \sigma_{zz} \\ D_z \\ -B_z \end{bmatrix} = B(\tilde{I}_0\tilde{J} + K)HP_{,zz}, \quad \begin{bmatrix} \sigma_{zx} + i\sigma_{zy} \\ D_x + iD_y \\ -B_x - iB_y \end{bmatrix} = \Lambda B\left[(\tilde{I}_0\tilde{J} + K)P_{,z} + i\tilde{I}_0\Phi_{0,z} \right]$$

$$\tag{4.4.23}$$

where

$$\tilde{I}_0 = \begin{bmatrix} 1 & 0 & 0 \end{bmatrix}^{\mathrm{T}}, \quad H = \mathrm{diag}\begin{bmatrix} \lambda_1 & \lambda_2 & \lambda_3 & \lambda_4 \end{bmatrix} \tag{4.4.24}$$

Eqs.(4.4.21) and (4.4.23) are the general expressions of magneto-electro-elastic fields in terms of the five potential functions $\Phi_i (i = 0, 1, 2, 3, 4)$.

Therefore, with the potential function method, the 3D transversely isotropic magneto-electro-elastic problem is reduced to one of finding five complex potential functions.

4.5　Green's function for half-plane and bimaterial problems

In this and next sections, Green's functions of defective magneto-electro-elastic solids are derived based on the Stroh formalism. The defects considered here include bimaterial interface, half plane boundary, and wedge boundary.

4.5.1 Preliminary formulations

To simplify subsequent writing, the shorthand notation described in Subsection 3.3.1 is used here. In the stationary case where no free electric charge, electric current, and body force are assumed to exist, the complete set of governing equations for coupled electromagnetoelastic problems are [38]

Governing equation:
$$\Pi_{iJ,i} = 0 \tag{4.5.1}$$

Constitutive relationship:
$$\Pi_{iJ} = E_{iJMn} U_{M,n} \tag{4.5.2}$$

where

$$\Pi_{iJ} = \begin{cases} \sigma_{ij}, & J \leqslant 3 \\ D_i, & J = 4, \\ B_i, & J = 5 \end{cases} \qquad U_M = \begin{cases} u_m, & M \leqslant 3 \\ \phi, & M = 4 \\ \psi, & M = 5 \end{cases} \tag{4.5.3}$$

$$E_{iJMn} = \begin{cases} c_{ijmn}, & J, M \leqslant 3 \\ e_{nij}, & J \leqslant 3, M = 4 \\ \tilde{e}_{nij}, & J \leqslant 3, M = 5 \\ e_{imn}, & J = 4, M \leqslant 3 \\ -\kappa_{in}, & J = 4, M = 4 \\ -\alpha_{in}, & J = 4, M = 5 \\ \tilde{e}_{imn}, & J = 5, M \leqslant 3 \\ -\alpha_{in}, & J - 5, M = 4 \\ -\mu_{in}, & J = 5, M = 5 \end{cases} \tag{4.5.4}$$

A general solution to Eq.(4.5.1) can be expressed as [38]
$$U = 2\operatorname{Re}[Af(z)q] \tag{4.5.5}$$

where

$$A = [A_1 \quad A_2 \quad A_3 \quad A_4 \quad A_5]$$
$$f(z) = \langle f(z_\alpha) \rangle = \operatorname{diag}[f(z_1) \quad f(z_2) \quad f(z_3) \quad f(z_4) \quad f(z_5)]$$
$$q = [q_1 \ q_2 \ q_3 \ q_4 \ q_5]^{\mathrm{T}}$$
$$z_i = x_1 + p_i x_2 \tag{4.5.6}$$

where f is an arbitrary function to be determined, q denotes unknown constants to be found by boundary conditions, and p_i and A_i are constants determined by

$$\left[Q + (R + R^{\mathrm{T}})p_i + Tp_i^2\right] A_i = 0 \tag{4.5.7}$$

with Q, R and T being 5×5 constant matrices defined by

$$(Q)_{IK} = E_{1IK1}, \quad (R)_{IK} = E_{1IK2}, \quad (T)_{IK} = E_{2IK2} \tag{4.5.8}$$

The stress-electric displacement-magnetic induction (SEDMI), Π, obtained from Eq.(4.5.2) can be written as

$$\Pi_{1J} = -\varphi_{J,2}, \quad \Pi_{2J} = \varphi_{J,1} \tag{4.5.9}$$

where φ is the SEDMI function given as

$$\varphi = 2\,\mathrm{Re}[Bf(z)q] \tag{4.5.10}$$

with

$$\begin{aligned}
B &= R^{\mathrm{T}}A + TAP = -(QA + RAP)P^{-1} \\
P &= \langle p_\alpha \rangle = \mathrm{diag}[p_1\ p_2\ p_3\ p_4\ p_5]
\end{aligned} \tag{4.5.11}$$

4.5.2 New coordinate variables

The half-plane or bimaterial interface considered in this section is different from those reported in the literature [43, 44]. The half-plane boundary (or bimaterial) is in the vertical ($x_1 = 0$ on the boundary in our analysis) rather than the horizontal direction (see Fig. 4.1). It is obvious that $z_k = x_1 + p_k x_2$ becomes a real number on the horizontal boundary $x_2 = 0$. However, z_k is, in general, neither a real number nor a pure imaginary number on the vertical boundary $x_1 = 0$, which complicates the related mathematical derivation. To bypass this problem, a new coordinate variable is introduced [38]

$$z_k^* = z_k / p_k \tag{4.5.12}$$

In this case z_k^* is a real number on the vertical boundary $x_1 = 0$. This coordinate transformation is used for both the half-plane and the bimaterial problem below.

4.5.3 Green's function for full space

For an infinite magneto-electro-elastic solid subjected to a line force q_0 and a line dislocation b both located at $z_0(x_{10}, x_{20})$ (see Fig. 4.1), the solution in the form of Eqs.(4.5.5) and (4.5.10) is [45]

$$U = \frac{1}{\pi}\,\mathrm{Im}\!\left[A\langle \ln(z_\alpha^* - z_{\alpha 0}^*)\rangle q\right], \quad \varphi = \frac{1}{\pi}\,\mathrm{Im}\!\left[B\langle \ln(z_\alpha^* - z_{\alpha 0}^*)\rangle q\right] \tag{4.5.13}$$

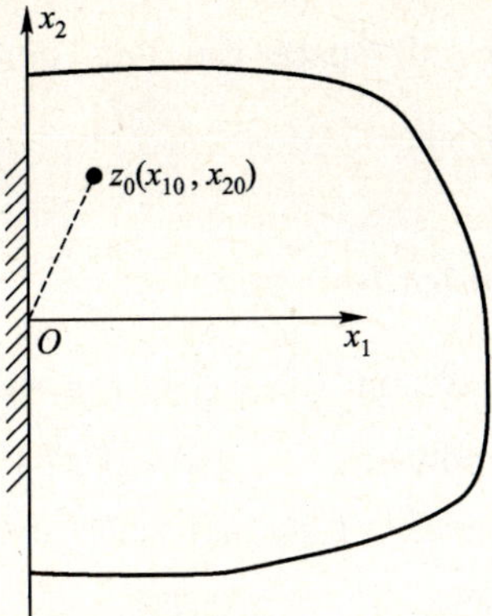

Fig.4.1   Magneto-electro-elastic half-plane

where $\boldsymbol{q}$ is a complex vector to be determined. Since $\ln(z_\alpha^* - z_{\alpha 0}^*)$ is a multi-valued function we introduce a cut along the line defined by $x_2 = x_{20}$ and $x_1 \leqslant x_{10}$. Using the polar coordinate system (r, θ) with its origin at $z_0(x_{10}, x_{20})$ and with $\theta = 0$ being parallel to the x_1-axis, the solution (4.5.13) applies to

$$-\pi < \theta < \pi, \qquad r > 0 \tag{4.5.14}$$

Therefore

$$\ln(z_\alpha^* - z_{\alpha 0}^*) = \ln r \pm i\pi, \qquad \theta = \pm\pi, \quad \alpha = 1\sim 5 \tag{4.5.15}$$

Owing to this relation, Eq.(4.5.13) must satisfy the conditions

$$\boldsymbol{U}(\pi) - \boldsymbol{U}(-\pi) = \boldsymbol{b}, \qquad \boldsymbol{\varphi}(\pi) - \boldsymbol{\varphi}(-\pi) = \boldsymbol{q}_0 \tag{4.5.16}$$

which lead to

$$2\operatorname{Re}(\boldsymbol{Aq}) = \boldsymbol{b}, \qquad 2\operatorname{Re}(\boldsymbol{Bq}) = \boldsymbol{q}_0 \tag{4.5.17}$$

This can be written as

$$\begin{bmatrix} \boldsymbol{A} & \bar{\boldsymbol{A}} \\ \boldsymbol{B} & \bar{\boldsymbol{B}} \end{bmatrix} \begin{bmatrix} \boldsymbol{q} \\ \bar{\boldsymbol{q}} \end{bmatrix} = \begin{bmatrix} \boldsymbol{b} \\ \boldsymbol{q}_0 \end{bmatrix} \tag{4.5.18}$$

It follows from the relation

$$\begin{bmatrix} \boldsymbol{B}^{\mathrm{T}} & \boldsymbol{A}^{\mathrm{T}} \\ \bar{\boldsymbol{B}}^{\mathrm{T}} & \bar{\boldsymbol{A}}^{\mathrm{T}} \end{bmatrix} \begin{bmatrix} \boldsymbol{A} & \bar{\boldsymbol{A}} \\ \boldsymbol{B} & \bar{\boldsymbol{B}} \end{bmatrix} = \begin{bmatrix} \boldsymbol{I} & 0 \\ 0 & \boldsymbol{I} \end{bmatrix} \tag{4.5.19}$$

that

$$\begin{bmatrix} \boldsymbol{q} \\ \bar{\boldsymbol{q}} \end{bmatrix} = \begin{bmatrix} \boldsymbol{B}^{\mathrm{T}} & \boldsymbol{A}^{\mathrm{T}} \\ \bar{\boldsymbol{B}}^{\mathrm{T}} & \bar{\boldsymbol{A}}^{\mathrm{T}} \end{bmatrix} \begin{bmatrix} \boldsymbol{b} \\ \boldsymbol{q}_0 \end{bmatrix} \tag{4.5.20}$$

Hence

$$\boldsymbol{q} = \boldsymbol{A}^{\mathrm{T}} \boldsymbol{q}_0 + \boldsymbol{B}^{\mathrm{T}} \boldsymbol{b} \tag{4.5.21}$$

The Green's function for full space can thus be obtained by substituting Eq.(4.5.21) into Eq.(4.5.13).

4.5.4 Green's function for half-space

Let the material occupy the region $x_1 > 0$ (see Fig. 4.1), and a line force-charge $\boldsymbol{q}_0$ and a line dislocation $\boldsymbol{b}$ apply at $z_0\,(x_{10}, x_{20})$. To satisfy the boundary conditions on the infinite straight boundary of the half-plane, the general solution (4.5.13) should be modified as follows:

$$\boldsymbol{U} = \frac{1}{\pi}\,\mathrm{Im}\!\left[\boldsymbol{A}\big\langle \ln(z_\alpha^* - z_{\alpha 0}^*)\big\rangle \boldsymbol{q}\right] + \sum_{\beta=1}^{5} \frac{1}{\pi}\,\mathrm{Im}\!\left[\boldsymbol{A}\big\langle \ln(z_{\alpha 0}^* - \bar{z}_{\beta 0}^*)\big\rangle \boldsymbol{q}_\beta\right] \quad (4.5.22)$$

$$\boldsymbol{\varphi} = \frac{1}{\pi}\,\mathrm{Im}\!\left[\boldsymbol{B}\big\langle \ln(z_\alpha^* - z_{\alpha 0}^*)\big\rangle \boldsymbol{q}\right] + \sum_{\beta=1}^{5} \frac{1}{\pi}\,\mathrm{Im}\!\left[\boldsymbol{B}\big\langle \ln(z_{\alpha 0}^* - \bar{z}_{\beta 0}^*)\big\rangle \boldsymbol{q}_\beta\right] \quad (4.5.23)$$

where $\boldsymbol{q}$ is given in Eq.(4.5.21) and $\boldsymbol{q}_\beta$ are unknown constants to be determined. To determine constants $\boldsymbol{q}_\beta$, the following two kinds of boundary conditions are considered.

Consider first the case in which the surface $x_1 = 0$ is traction-free, so that [45]

$$\boldsymbol{\varphi} = 0, \qquad x = 0 \quad (4.5.24)$$

Substituting Eq.(4.5.23) into Eq.(4.5.24) yields

$$\boldsymbol{\varphi} = \frac{1}{\pi}\,\mathrm{Im}\!\left[\boldsymbol{B}\big\langle \ln(x_2 - z_{\alpha 0}^*)\big\rangle \boldsymbol{q}\right] + \sum_{\beta=1}^{5} \frac{1}{\pi}\,\mathrm{Im}\!\left[\boldsymbol{B}\big\langle \ln(x_2 - \bar{z}_{\beta 0}^*)\big\rangle \boldsymbol{q}_\beta\right] = 0 \quad (4.5.25)$$

Noting that $\mathrm{Im}(f) = -\mathrm{Im}(\bar{f})$, we have

$$\mathrm{Im}\!\left[\boldsymbol{B}\big\langle \ln(x_2 - z_{\alpha 0}^*)\big\rangle \boldsymbol{q}\right] = -\mathrm{Im}\!\left[\bar{\boldsymbol{B}}\big\langle \ln(x_2 - \bar{z}_{\alpha 0}^*)\big\rangle \bar{\boldsymbol{q}}\right] \quad (4.5.26)$$

and

$$\big\langle \ln(x_2 - z_{\alpha 0}^*)\big\rangle = \sum_{\beta=1}^{5} \ln(x_2 - z_{\beta 0}^*)\boldsymbol{I}_\beta \quad (4.5.27)$$

where

$$\boldsymbol{I}_\beta = \big\langle \delta_{\beta\alpha}\big\rangle = \mathrm{diag}[\delta_{\beta 1} \quad \delta_{\beta 2} \quad \delta_{\beta 3} \quad \delta_{\beta 4} \quad \delta_{\beta 5}] \quad (4.5.28)$$

Eq.(4.5.25) now yields

$$\boldsymbol{q}_\beta = \boldsymbol{B}^{-1}\bar{\boldsymbol{B}}\boldsymbol{I}_\beta\bar{\boldsymbol{q}} = \boldsymbol{B}^{-1}\bar{\boldsymbol{B}}\boldsymbol{I}_\beta(\bar{\boldsymbol{A}}^{\mathrm{T}}\boldsymbol{q}_0 + \bar{\boldsymbol{B}}^{\mathrm{T}}\boldsymbol{b}) \quad (4.5.29)$$

If the boundary $x_1 = 0$ is a rigid surface, then

$$\boldsymbol{U} = 0, \qquad x = 0 \quad (4.5.30)$$

The same procedure shows that the solution is given by Eqs.(4.5.22) and (4.5.23) with

$$q_\beta = A^{-1}\overline{A}I_\beta(\overline{A}^T q_0 + \overline{B}^T b) \qquad (4.5.31)$$

Therefore the final version of the Green's function can be written in terms of z_k as

$$U = \frac{1}{\pi}\mathrm{Im}\left[A\left\langle \ln(z_\alpha - z_{\alpha 0})/p_\alpha\right\rangle q\right] + \sum_{\beta=1}^{5}\frac{1}{\pi}\mathrm{Im}\left[A\left\langle \ln(z_\alpha/p_\alpha - \overline{z}_{\beta 0}/\overline{p}_\beta)\right\rangle q_\beta\right] \qquad (4.5.32)$$

$$\varphi = \frac{1}{\pi}\mathrm{Im}\left[B\left\langle \ln(z_\alpha - z_{\alpha 0})/p_\alpha\right\rangle q\right] + \sum_{\beta=1}^{5}\frac{1}{\pi}\mathrm{Im}\left[B\left\langle \ln(z_\alpha/p_\alpha - \overline{z}_{\beta 0}/\overline{p}_\beta)\right\rangle q_\beta\right] \qquad (4.5.33)$$

4.5.5   Green's function for a bimaterial problem

We now consider a bimaterial solid whose interface is on the x_2-axis$(x_1 = 0)$. It is assumed that the left half-plane $(x_1 < 0)$ is occupied by material 1, and the right half-plane $(x_1 > 0)$ by material 2 (see Fig. 4.2). They are rigidly bonded together so that

$$U^{(1)} = U^{(2)}, \qquad \varphi^{(1)} = \varphi^{(2)}, \qquad x = 0 \qquad (4.5.34)$$

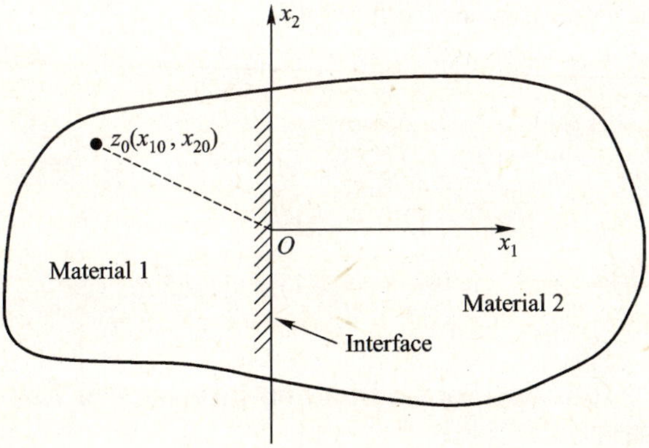

Fig.4.2   Magneto-electro-elastic bimaterial plate

where the superscripts (1) and (2) label the quantities relating to materials 1 and 2 respectively. The equality of traction continuity comes from the relations $t = \partial\varphi/\partial s$. When points along the interface are considered, integration of $t^{(1)} = t^{(2)}$ provides Eq.(4.5.34) since the integration constants corresponding to rigid motion can be ignored.

For a magneto-electro-elastic bimaterial plate subjected to a line force-charge q_0 and a line dislocation b both located in the left half-plane at $z_0(x_{10}, x_{20})$ (Fig.4.2), the solution may be assumed, using a similar treatment to that for the half-plane problem, in the form [38]

$$U^{(1)} = \frac{1}{\pi} \mathrm{Im}\left[A^{(1)} \left\langle \ln(z_\alpha^{*(1)} - z_{\alpha 0}^{*(1)}) \right\rangle q \right] + \sum_{\beta=1}^{5} \frac{1}{\pi} \mathrm{Im}\left[A^{(1)} \left\langle \ln(z_\alpha^{*(1)} - \overline{z}_{\beta 0}^{*(1)}) \right\rangle q_\beta^{(1)} \right]$$

$$(4.5.35)$$

$$\varphi^{(1)} = \frac{1}{\pi} \mathrm{Im}\left[B^{(1)} \left\langle \ln(z_\alpha^{*(1)} - z_{\alpha 0}^{*(1)}) \right\rangle q \right] + \sum_{\beta=1}^{5} \frac{1}{\pi} \mathrm{Im}\left[B^{(1)} \left\langle \ln(z_\alpha^{*(1)} - \overline{z}_{\beta 0}^{*(1)}) \right\rangle q_\beta^{(1)} \right]$$

$$(4.5.36)$$

for material 1 in $x_1 < 0$ and

$$U^{(2)} = \sum_{\beta=1}^{5} \frac{1}{\pi} \mathrm{Im}\left[A^{(2)} \left\langle \ln(z_\alpha^{*(2)} - z_{\beta 0}^{*(1)}) \right\rangle q_\beta^{(2)} \right] \qquad (4.5.37)$$

$$\varphi^{(2)} = \sum_{\beta=1}^{4} \frac{1}{\pi} \mathrm{Im}\left[B^{(2)} \left\langle \ln(z_\alpha^{*(2)} - z_{\beta 0}^{*(1)}) \right\rangle q_\beta^{(2)} \right] \qquad (4.5.38)$$

for material 2 in $x_1 > 0$, where $z_{\beta 0}^{*(1)} = z_{\beta 0}^{(1)} / p_\beta^{(1)}$, $z_\alpha^{*(i)} = z_\alpha^{(i)} / p_\alpha^{(i)}$ $(i = 1,2)$. The value of q is again given in Eq.(4.5.21), and $q_\beta^{(1)}$, $q_\beta^{(2)}$ are unknown constants which are determined by substituting Eqs.(4.5.35)~(4.5.38) into Eq.(4.5.34). Following the derivation in subsection 4.5.3, we obtain

$$A^{(1)} q_\beta^{(1)} + \overline{A}^{(2)} \overline{q}_\beta^{(2)} = \overline{A}^{(1)} I_\beta \overline{q}, \quad B^{(1)} q_\beta^{(1)} + \overline{B}^{(2)} \overline{q}_\beta^{(2)} = \overline{B}^{(1)} I_\beta \overline{q} \quad (4.5.39)$$

Solving Eq.(4.5.39) yields

$$q_\beta^{(1)} = B^{(1)-1}[I - 2(M^{(1)-1} + \overline{M}^{(2)-1})^{-1} L^{(1)-1}] \overline{B}^{(1)} I_\beta \overline{q} \qquad (4.5.40)$$

$$q_\beta^{(2)} = 2 B^{(2)-1} (\overline{M}^{(1)-1} + M^{(2)-1})^{-1} L^{(1)-1} B^{(1)} I_\beta q \qquad (4.5.41)$$

where $M^{(j)} = -i B^{(j)} A^{(j)-1}$ is the surface impedance matrix.

4.5.6   Green's function for an inclined interface or half-plane boundary

If the half-boundary is in an angle θ_0 $(\theta_0 \neq 0)$ (see Fig. 4.3) with positive x-axis, the corresponding Green's function can be obtained by introducing a new mapping function:

$$z = \zeta^{\theta_0 / \pi} \qquad (\theta_0 \neq 0) \qquad (4.5.42)$$

which maps the boundary $\theta = \theta_0$ in the z-plane onto the real axis in the ζ-plane $(\zeta + i\eta)$.

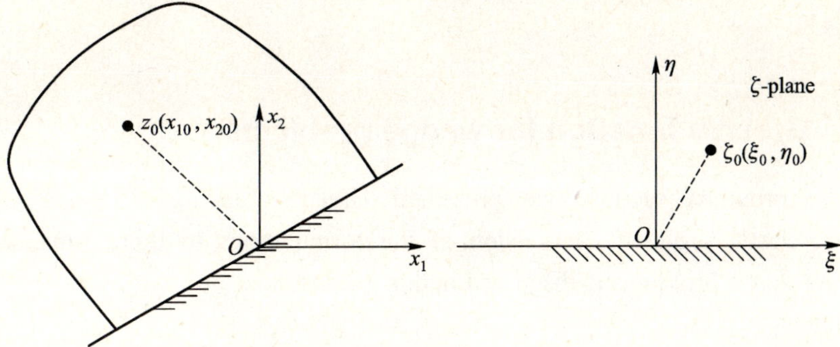

Fig.4.3   Magneto-electro-elastic solid with arbitrarily oriented half-plane

Following the procedure in Subsections 4.5.3 and 4.5.4 it can be shown that the resulting Green's functions can be expressed as

$$U = \frac{1}{\pi}\mathrm{Im}\Big[\,A\big\langle \ln(z_\alpha^{\pi/\theta_0} - z_{\alpha 0}^{\pi/\theta_0})\big\rangle q\,\Big] + \sum_{\beta=1}^{5}\frac{1}{\pi}\mathrm{Im}\Big[\,A\big\langle \ln(z_\alpha^{\pi/\theta_0} - \overline{z}_{\beta 0}^{\pi/\theta_0})\big\rangle q_\beta\,\Big]$$

$$(4.5.43)$$

$$\varphi = \frac{1}{\pi}\mathrm{Im}\Big[\,B\big\langle \ln(z_\alpha^{\pi/\theta_0} - z_{\alpha 0}^{\pi/\theta_0})\big\rangle q\,\Big] + \sum_{\beta=1}^{5}\frac{1}{\pi}\mathrm{Im}\Big[\,B\big\langle \ln(z_\alpha^{\pi/\theta_0} - \overline{z}_{\beta 0}^{\pi/\theta_0})\big\rangle q_\beta\,\Big]$$

$$(4.5.44)$$

for the half-plane problem, and

$$U^{(1)} = \frac{1}{\pi}\mathrm{Im}\Big[\,A^{(1)}\big\langle \ln(z_\alpha^{(1)\pi/\theta_0} - z_{\alpha 0}^{(1)\pi/\theta_0})\big\rangle q\,\Big] +$$

$$\sum_{\beta=1}^{5}\frac{1}{\pi}\mathrm{Im}\Big[\,A^{(1)}\big\langle \ln(z_\alpha^{(1)\pi/\theta_0} - \overline{z}_{\beta 0}^{(1)\pi/\theta_0})\big\rangle q_\beta^{(1)}\,\Big] \qquad (4.5.45)$$

$$\varphi^{(1)} = \frac{1}{\pi}\mathrm{Im}\Big[\,B^{(1)}\big\langle \ln(z_\alpha^{(1)\pi/\theta_0} - z_{\alpha 0}^{(1)\pi/\theta_0})\big\rangle q\,\Big] +$$

$$\sum_{\beta=1}^{5}\frac{1}{\pi}\mathrm{Im}\Big[\,B^{(1)}\big\langle \ln(z_\alpha^{(1)\pi/\theta_0} - \overline{z}_{\beta 0}^{(1)\pi/\theta_0})\big\rangle q_\beta^{(1)}\,\Big] \qquad (4.5.46)$$

for material 1 in $x_1 < 0$ and

$$U^{(2)} = \sum_{\beta=1}^{5}\frac{1}{\pi}\mathrm{Im}\Big[\,A^{(2)}\big\langle \ln(z_\alpha^{(2)\pi/\theta_0} - z_{\beta 0}^{(1)\pi/\theta_0})\big\rangle q_\beta^{(2)}\,\Big] \qquad (4.5.47)$$

$$\varphi^{(2)} = \sum_{\beta=1}^{5}\frac{1}{\pi}\mathrm{Im}\Big[\,B^{(2)}\big\langle \ln(z_\alpha^{(2)\pi/\theta_0} - z_{\beta 0}^{(1)\pi/\theta_0})\big\rangle q_\beta^{(2)}\,\Big] \qquad (4.5.48)$$

for material 2 in a biomaterial problem, where q_β, $q_\beta^{(1)}$, and $q_\beta^{(2)}$ have, respectively, the same forms as those given in Eqs.(4.5.31), (4.5.40), and

(4.5.41).

4.6   Green's function for wedge problems

In the previous section, we presented Green's functions for magneto-electro-elastic problems. Extension of the formulations to thermo-magneto-electro-elastic problems is discussed in this section.

4.6.1   Basic formulations

With the shorthand notation used in the last section and in the stationary case where no free electric charge, electric current, body force, and heat source are assumed to exist, the complete set of governing equations for coupled thermo-magneto-electro-elastic problems are [24]

$$h_{i,i} = 0, \qquad \Pi_{iJ,i} = 0 \tag{4.6.1}$$

together with

$$h_i = -k_{ij}T_{,j}, \qquad \Pi_{iJ} = E_{iJMn}U_{M,n} - \lambda_{iJ}T \tag{4.6.2}$$

where

$$\lambda_{iJ} = \begin{cases} \lambda_{ij}, & J \leqslant 3 \\ \rho_i, & J = 4 \\ \upsilon_i, & J = 5 \end{cases} \tag{4.6.3}$$

and where Π_{iJ}, U_M, and E_{iJMn} are defined by Eqs.(4.5.3) and (4.5.4).

A general solution to Eq.(4.6.1) can be expressed as [45]

$$T = 2\operatorname{Re}\left[g'(z_t)\right], \quad U = 2\operatorname{Re}\left[Af(z)q + cg(z_t)\right] \tag{4.6.4}$$

with

$$z_t = x_1 + \tau x_2 \tag{4.6.5}$$

in which the prime (') denotes differentiation with the argument, q represents unknown constants to be found by boundary conditions, g and f are arbitrary functions to be determined, τ and c are constants determined by [45]

$$k_{22}\tau^2 + (k_{12} + k_{21})\tau + k_{11} = 0$$

$$\left[Q + (R + R^{\mathrm{T}})\tau + T\tau^2\right]c = \lambda_1 + \tau\lambda_2 \tag{4.6.6}$$

where λ_i are 5×1 vectors defined by

$$\lambda_i = \begin{bmatrix} \lambda_{i1} & \lambda_{i2} & \lambda_{i3} & \rho_i & \upsilon_i \end{bmatrix}^{\mathrm{T}} \tag{4.6.7}$$

The heat flux, h, and the SEDMI, Π, obtained from Eq.(4.6.2) can be written as

$$h_i = -2\,\mathrm{Re}\big[(k_{i1} + \tau k_{i2})g''(z_t)\big], \quad \Pi_{1J} = -\varphi_{J,2}, \quad \Pi_{2J} = \varphi_{J,1} \qquad (4.6.8)$$

where φ is the SEDMI function given as

$$\varphi = 2\,\mathrm{Re}\big[\mathbf{B}f(z)\mathbf{q} + \mathbf{d}g(z_t)\big] \qquad (4.6.9)$$

with

$$\mathbf{d} = (\mathbf{R}^{\mathrm{T}} + \tau\mathbf{T})\mathbf{c} - \boldsymbol{\lambda}_2 = -(\mathbf{Q} + \tau\mathbf{R})\mathbf{c}/\tau + \boldsymbol{\lambda}_1/\tau \qquad (4.6.10)$$

Introducing a heat flow function [45]

$$\vartheta = 2k\,\mathrm{Im}\big[g'(z_t)\big] \qquad (4.6.11)$$

where $k = (k_{11}k_{22} - k_{12}^2)^{1/2}$, and "Im" stands for the imaginary part of the complex number, we have

$$h_1 = -\vartheta_{,2}, \qquad h_2 = \vartheta_{,1} \qquad (4.6.12)$$

which has the same form as those for SEDMI function [see Eq.(4.6.8)]. Thus we may use the same method as that in magneto-electro-elastic problems to derive the thermal solutions.

4.6.2   Green's function for a wedge or a semi-infinite crack

Consider an infinite magneto-electro-elastic wedge whose symmetric line extends infinitely in the negative direction of the x_1-axis (Fig.4.4). The wedge angle is denoted by $2\theta_0$. The solid is subjected to a temperature discontinuity $\hat{T}$ and a heat source h^*, both at a point z_0 (x_{10},x_{20}) as shown in Fig.4.4. The wedge faces are assumed to be thermal-insulated, and free of force, external electric current and charge. The boundary condition along the two wedge faces can thus be written as

$$\vartheta = \varphi = 0 \qquad (4.6.13)$$

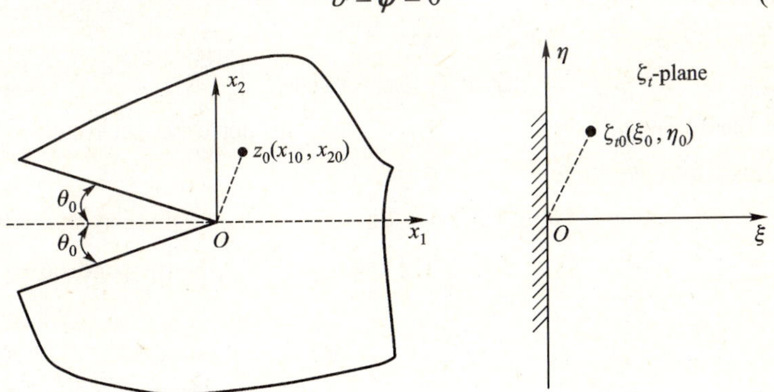

Fig.4.4   Wedge-shaped magneto-electro-elastic plate and its mapping in the ζ-plane

1. General solution for thermal field

Based on the concept of perturbation given by Stagni [47], the general solution for temperature and heat-flow function can be assumed in the form

$$T = 2\operatorname{Re}\left[g'(z_t)\right] = 2\operatorname{Re}\left[f_0(\zeta_t) + f_1(\zeta_t)\right] \qquad (4.6.14)$$

$$\vartheta = 2k\operatorname{Im}\left[g'(z_t)\right] = 2k\operatorname{Im}\left[f_0(\zeta_t) + f_1(\zeta_t)\right] \qquad (4.6.15)$$

where f_0 can be chosen to represent the solutions associated with the unperturbed thermal fields which are holomorphic in the entire domain except at some singular points such as the point at which a point heat source is applied, and f_1 is a function corresponding to the perturbed field due to the wedge. Here ζ_t and ζ_{t0} are related to z_t and $z_{t0} (= x_{10} + \tau x_{20})$ by the mapping functions [48]

$$z_t = \zeta_t^{1/\lambda} \quad \text{and} \quad z_{t0} = \zeta_{t0}^{1/\lambda} \qquad (4.6.16)$$

where $\lambda = \pi/(2\pi - 2\theta_0)$ and $\zeta_t = \xi + i\eta$ maps the wedge boundary $\theta = \pm(\pi - \theta_0)$ in the z_t-plane into the imaginary axis in the ζ_t-plane (Fig.4.4). Therefore the solution domain is mapped into the right half plane axis in the ζ_t-plane.

For a given loading condition, the function f_0 can be obtained easily since it is related to the solution of homogeneous media. When an infinite space is subjected to a line heat source h^* and the thermal analog of a line temperature discontinuity T_0 both located at (x_{10}, x_{20}), the function f_0 can be chosen in the form

$$f_0(\zeta_t) = q_0 \ln(\zeta_t - \zeta_{t0}) \qquad (4.6.17)$$

where q_0 is a complex number which can be determined from the conditions

$$\int_C dT = \hat{T}, \text{ for any closed curve } C \text{ enclosing the point } \zeta_{t0} \qquad (4.6.18)$$

$$\int_C d\vartheta = -h^*, \text{ for any closed curve } C \text{ enclosing the point } \zeta_{t0} \qquad (4.6.19)$$

With the substitution of Eq.(4.6.17) into Eqs.(4.6.14) and (4.6.15), the conditions (4.6.18) and (4.6.19) yield

$$q_0 = \hat{T}/4\pi i - h^*/4\pi k \qquad (4.6.20)$$

For the half plane in the $\zeta_t = \xi + i\eta$ system, the perturbation function can be assumed in the form [45]

$$f_1(\zeta_t) = q_1 \ln(-\zeta_t - \bar{\zeta}_{t0}) \qquad (4.6.21)$$

Substituting Eqs.(4.6.17) and (4.6.21) into Eq.(4.6.15), the condition (4.6.13) yields

$$\operatorname{Im}\left[q_0 \ln(i\eta - \zeta_{t0}) + q_1 \ln(-i\eta - \bar{\zeta}_{t0})\right] = 0 \qquad (4.6.22)$$

Noting that $\operatorname{Im}(f) = -\operatorname{Im}(\bar{f})$, we have

$$\operatorname{Im}\left[q_0 \ln(i\eta - \zeta_{t0})\right] = -\operatorname{Im}\left[\bar{q}_0 \ln(-i\eta - \bar{\zeta}_{t0})\right] \qquad (4.6.23)$$

Equation (4.6.22) now yields

$$q_1 = \overline{q}_0 \tag{4.6.24}$$

Having obtained the solution of f_0 and f_1, the function $g'(z_t)$ can now be written as

$$g'(z_t) = q_0 \ln(z_t^\lambda - z_{t0}^\lambda) + \overline{q}_0 \ln(-z_t^\lambda - \overline{z}_{t0}^\lambda) \tag{4.6.25}$$

Substituting Eq.(4.6.25) into Eqs.(4.6.14) and (4.6.15) yields

$$T = 2\,\mathrm{Re}\left[q_0 \ln(z_t^\lambda - z_{t0}^\lambda) + \overline{q}_0 \ln(-z_t^\lambda - \overline{z}_{t0}^\lambda) \right] \tag{4.6.26}$$

$$\vartheta = 2k\,\mathrm{Im}\left[q_0 \ln(z_t^\lambda - z_{t0}^\lambda) + \overline{q}_0 \ln(-z_t^\lambda - \overline{z}_{t0}^\lambda) \right] \tag{4.6.27}$$

The function g in Eq.(4.6.14) can thus be obtained by integrating the functions of f_0 and f_1 with respect to z_t, which leads to

$$g(z_t) = q_0 f_1(z_t) + \overline{q}_0 f_2(z_t) \tag{4.6.28}$$

where

$$f_1(z_t) = \lambda z_t\left[-1 + {}_2F_1(1/\lambda,1,1+1/\lambda, z_t^\lambda / z_{t0}^\lambda) \right] + z_t \ln(z_t^\lambda - z_{t0}^\lambda)$$

$$f_2(z_t) = \lambda z_t\left[-1 + {}_2F_1(1/\lambda,1,1+1/\lambda, -z_t^\lambda / \overline{z}_{t0}^\lambda) \right] + z_t \ln(-z_t^\lambda - \overline{z}_{t0}^\lambda) \tag{4.6.29}$$

with ${}_2F_1(a,b,c,z)$ being a hypergeometric function defined in [49]

$${}_2F_1(a,b,c,z) = \frac{\Gamma(c)}{\Gamma(b)\Gamma(c-b)} \int_0^1 \frac{t^{b-1}(1-t)^{c-b-1}}{(1-tz)^a}\,\mathrm{d}t \tag{4.6.30}$$

or series expansion

$${}_2F_1(a,b,c,z) = 1 + \frac{ab}{1!c}z + \frac{a(a+1)b(b+1)}{2!c(c+1)}z^2 + \cdots = \sum_{n=0}^{\infty} \frac{(a)_n (b)_n}{n!(c)_n}z^n \tag{4.6.31}$$

where $\Gamma(x)$ is a gamma function.

2. Green's function for magneto-electro-elastic field

The general solution of the thermo-magneto-electro-elastic problem can be written as

$$U = U_p + U_h, \quad \varphi = \varphi_p + \varphi_h \tag{4.6.32}$$

where subscripts "p" and "h" refer, respectively, to the particular and homogeneous solution.

From Eqs.(4.6.5) and (4.6.9) the particular solution of a magneto-electro-elastic field induced by thermal loading can be written as

$$U_p = 2\,\mathrm{Re}\left[cg(z_t) \right], \quad \varphi_p = 2\,\mathrm{Re}\left[dg(z_t) \right] \tag{4.6.33}$$

The particular solutions (4.6.33) do not generally satisfy the boundary condition (4.6.13) along the wedge boundary. We therefore need to seek a corrective isothermal solution for a given problem so that when it is superimposed on the particular thermo-magneto-electro-elastic solution, the surface conditions (4.6.13) will be satisfied. Owing to the fact that $f(z_k)$ and $g(z_t)$ have the same order of effect on stress and electric displacement in Eqs.(4.6.5) and (4.6.9) [note that the term $\boldsymbol{B}f(z)\boldsymbol{q}$ in Eq.(4.6.9) is now replaced by $\boldsymbol{B}(\langle f_1(z_k)\rangle\boldsymbol{q}_1 + \langle f_2(z_k)\rangle\boldsymbol{q}_2)$], possible function forms come from the partition of $g(z_t)$. They are

$$f_1(z_k) = \lambda z_k\left[-1 + {}_2F_1(1/\lambda,1,1+1/\lambda,z_k^\lambda/z_{t0}^\lambda)\right] + z_k\ln(z_k^\lambda - \overline{z}_{t0}^\lambda)$$
$$f_2(z_k) = \lambda z_k\left[-1 + {}_2F_1(1/\lambda,1,1+1/\lambda,-z_k^\lambda/\overline{z}_{t0}^\lambda)\right] + z_k\ln(-z_k^\lambda - \overline{z}_{t0}^\lambda)$$

$$(4.6.34)$$

The substitution of Eqs.(4.6.28) and (4.6.34) into Eq.(4.6.9), and then into Eq.(4.6.13), leads to

$$\boldsymbol{q}_1 = -\boldsymbol{B}^{-1}d\boldsymbol{q}_0, \quad \boldsymbol{q}_2 = -\boldsymbol{B}^{-1}d\overline{\boldsymbol{q}}_0 \tag{4.6.35}$$

Substituting Eq.(4.6.35) into Eqs.(4.6.5) and (4.6.9), the Green's functions can then be written as

$$\boldsymbol{U} = 2\operatorname{Re}\left[-\boldsymbol{A}(\langle f_1(z_k)\rangle q_0 + \langle f_2(z_k)\rangle\overline{q}_0)\boldsymbol{B}^{-1}d + cg(z_t)\right]$$
$$\boldsymbol{\varphi} = 2\operatorname{Re}\left[-\boldsymbol{B}(\langle f_1(z_k)\rangle q_0 + \langle f_2(z_k)\rangle\overline{q}_0)\boldsymbol{B}^{-1}d + dg(z_t)\right]$$

$$(4.6.36)$$

When $\theta_0=0$, i.e, $\lambda=1/2$, Eq.(4.6.36) represents the Green's functions for the case of a semi-infinite crack in an infinite magneto-electro-elastic solid.

4.7 Antiplane shear crack in a magneto-electro-elastic layer

In this section, the crack problem of a magnetoelectroelastic layer bonded to dissimilar half spaces under antiplane shear and inplane electric and magnetic loads is considered based on the formulation described in the previous sections of this chapter. In the analysis, Fourier transforms are used to reduce the mixed boundary value problems of the crack, which is assumed to be permeable, to simultaneous dual integral equations, and then expressed in terms of Fredholm integral equations of the second kind. The discussion follows the development in [50].

4.7.1 Statement of the problem

Consider a Griffith crack of length $2c$ situated in the mid-plane of a magneto-electro-elastic layer that is sandwiched between two elastic half planes with an elastic stiffness constant c_{44}^{E}, as shown in Fig. 4.5. Quantities in the elastic half plane will subsequently be designated by superscript E. A coordinate system (x, y, z) is set at the center of the crack for reference. Due to the assumed symmetry in geometry and loading conditions, it is sufficient to consider the problem for $0 \leqslant x < \infty$, $0 \leqslant y < \infty$ only.

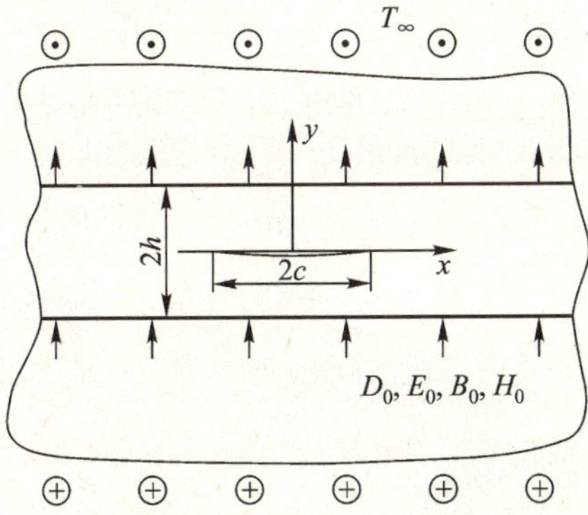

Fig.4.5   A magneto-electro-elastic laminate with a finite crack

The magneto-electro-elastic boundary value problem is simplified considerably if we consider only the out-of-plane displacement, the in-plane electric fields and in-plane magnetic fields, i.e.,

$$u_x = u_y = 0, \quad u_z = u_z(x, y) \tag{4.7.1}$$

$$E_x = E_x(x, y), \quad E_y = E_y(x, y), \quad E_z = 0 \tag{4.7.2}$$

$$H_x = H_x(x, y), \quad H_y = H_y(x, y), \quad H_z = 0 \tag{4.7.3}$$

$$u_x^E = u_y^E = 0, \quad u_z^E = u_z^E(x, y) \tag{4.7.4}$$

where (u_x, u_y, u_z), (E_x, E_y, E_z) and (H_x, H_y, H_z) are the components of displacement, electric field and magnetic field vectors, respectively. The constitutive equations for anti-plane magneto-electro-elastic material take the form of [51]

$$\begin{bmatrix} \sigma_{zx} \\ D_y \\ B_y \end{bmatrix} = \begin{bmatrix} c_{44} & -e_{15} & -\tilde{e}_{15} \\ e_{15} & \kappa_{11} & \alpha_{11} \\ \tilde{e}_{15} & \alpha_{11} & \mu_{11} \end{bmatrix} \begin{bmatrix} u_{z,y} \\ E_y \\ H_y \end{bmatrix}, \quad \begin{bmatrix} \sigma_{zx} \\ D_x \\ B_x \end{bmatrix} = \begin{bmatrix} c_{44} & -e_{15} & -\tilde{e}_{15} \\ e_{15} & \kappa_{11} & \alpha_{11} \\ \tilde{e}_{15} & \alpha_{11} & \mu_{11} \end{bmatrix} \begin{bmatrix} u_{z,x} \\ E_x \\ H_x \end{bmatrix}$$

$$\tag{4.7.5}$$

and the governing equations are

$$c_{44}\nabla^2 u_z + e_{15}\nabla^2\phi + \tilde{e}_{15}\nabla^2\psi = 0$$

$$e_{15}\nabla^2 u_z - \kappa_{11}\nabla^2\phi - \alpha_{11}\nabla^2\psi = 0 \tag{4.7.6}$$

$$\tilde{e}_{15}\nabla^2 u_z - \alpha_{11}\nabla^2\phi - \mu_{11}\nabla^2\psi = 0$$

$$\nabla^2 u_z^E = 0 \tag{4.7.7}$$

where $\nabla^2 = \partial^2/\partial x^2 + \partial^2/\partial y^2$ is the two-dimensional Laplacian operator in the variables x and y.

We consider four possible cases of electrical and magnetic boundary conditions on the edges of the magneto-electro-elastic layer

$$\text{Case 1:}\qquad D_y(x,h) = D_0, \quad B_y(x,h) = B_0 \tag{4.7.8}$$

$$\text{Case 2:}\qquad E_y(x,h) = E_0, \quad B_y(x,h) = B_0 \tag{4.7.9}$$

$$\text{Case 3:}\qquad D_y(x,h) = D_0, \quad H_y(x,h) = H_0 \tag{4.7.10}$$

$$\text{Case 4:}\qquad E_y(x,h) = E_0, \quad H_y(x,h) = H_0 \tag{4.7.11}$$

The mechanical conditions are

$$\sigma_{zy}(x,0) = 0, \quad 0 \leqslant x < c \tag{4.7.12}$$

$$u_z(x,0) = 0, \quad c \leqslant x < \infty \tag{4.7.13}$$

$$\sigma_{zy}^E(x,y) = T_\infty, \quad x^2 + y^2 \to \infty \tag{4.7.14}$$

$$\sigma_{zy}(x,h) = \sigma_{zy}^E(x,h) \tag{4.7.15}$$

$$u_z(x,h) = u_z^E(x,h) \tag{4.7.16}$$

The shear stress T_∞ can be expressed as

$$T_\infty = \begin{cases} \dfrac{\lambda_1}{c_{44}}T_0 + \dfrac{(e_{15}\alpha_{11} - \tilde{e}_{15}\kappa_{11})B_0 + (\tilde{e}_{15}\alpha_{11} - e_{15}\mu_{11})D_0}{\kappa_{11}\mu_{11} - \alpha_{11}^2}, & \text{Case 1} \\[4mm] \dfrac{\lambda_2}{c_{44}}T_0 + \dfrac{(\tilde{e}_{15}\alpha_{11} - e_{15}\mu_{11})E_0 - \tilde{e}_{15}B_0}{\mu_{11}}, & \text{Case 2} \\[4mm] \dfrac{\lambda_3}{c_{44}}T_0 + \dfrac{(e_{15}\alpha_{11} - \tilde{e}_{15}\kappa_{11})H_0 - e_{15}D_0}{\kappa_{11}}, & \text{Case 3} \\[4mm] \dfrac{\lambda_4}{c_{44}}T_0 - e_{15}E_0 - \tilde{e}_{15}H_0, & \text{Case 4} \end{cases} \tag{4.7.17}$$

where T_0 is a uniform shear stress at zero electrical and magnetic loads, and $\lambda_j\,(j=1,\ 2,\ 3,\ 4)$ are the magneto-electro-elastic stiffened elastic constants defined as

$$\lambda_1 = c_{44} + \frac{\mu_{11}e_{15}^2 + \kappa_{11}\tilde{e}_{15}^2 - 2\alpha_{11}\tilde{e}_{15}e_{15}}{\kappa_{11}\mu_{11} - \alpha_{11}^2}$$

$$\lambda_2 = (c_{44}\mu_{11} + \tilde{e}_{15}^2)/\mu_{11}$$

$$\lambda_3 = (c_{44}\kappa_{11} + e_{15}^2)/\kappa_{11} \tag{4.7.18}$$

$$\lambda_4 = c_{44}$$

The electrical and magnetic conditions for the permeable crack case can be expressed as [50]

$$D_y(x,0^+) = D_y(x,0^-), \quad E_x(x,0^+) = E_x(x,0^-), \qquad 0 \leqslant x < c$$

$$B_y(x,0^+) = B_y(x,0^-), \quad H_x(x,0^+) = H_x(x,0^-), \qquad 0 \leqslant x < c \tag{4.7.19}$$

$$\phi(x,0) = 0, \quad \psi(x,0) = 0, \quad c \leqslant x < \infty \tag{4.7.20}$$

4.7.2　Solution procedure

Fourier transforms are applied to Eqs.(4.7.6) and (4.7.7), and we obtain the results as

$$u_z(x,y) = \frac{2}{\pi}\int_0^\infty \left[A_1(\alpha)\exp(\alpha y) + A_2(\alpha)\exp(-\alpha y)\right]\cos(\alpha x)\mathrm{d}\alpha + a_0 y \tag{4.7.21}$$

$$\phi(x,y) = \frac{2}{\pi}\int_0^\infty \left[B_1(\alpha)\exp(\alpha y) + B_2(\alpha)\exp(-\alpha y)\right]\cos(\alpha x)\mathrm{d}\alpha - b_0 y \tag{4.7.22}$$

$$\psi(x,y) = \frac{2}{\pi}\int_0^\infty \left[C_1(\alpha)\exp(\alpha y) + C_2(\alpha)\exp(-\alpha y)\right]\cos(\alpha x)\mathrm{d}\alpha - c_0 y \tag{4.7.23}$$

$$u_z^E(x,y) = \frac{2}{\pi}\int_0^\infty A_3(\alpha)\exp(-\alpha y)\cos(\alpha x)\mathrm{d}\alpha + d_0 y + e_0 \tag{4.7.24}$$

where $A_j(\alpha)$ $(j=1,\ 2,\ 3)$ and $B_j(\alpha)$ $(i=1,\ 2)$ are the unknowns to be solved and a_0, b_0, c_0, d_0 and e_0 are real constants which can be determined by considering the far-field and interface conditions as

Case 1:
$$\begin{bmatrix} a_0 \\ b_0 \\ c_0 \end{bmatrix} = \begin{bmatrix} c_{44} & -e_{15} & -\tilde{e}_{15} \\ e_{15} & \kappa_{11} & \alpha_{11} \\ \tilde{e}_{15} & \alpha_{11} & \mu_{11} \end{bmatrix}^{-1} \begin{bmatrix} T_\infty \\ D_0 \\ B_0 \end{bmatrix} \tag{4.7.25}$$

Case 2:
$$\begin{bmatrix} a_0 \\ c_0 \end{bmatrix} = \begin{bmatrix} c_{44} & -\tilde{e}_{15} \\ \tilde{e}_{15} & \gamma_{11} \end{bmatrix}^{-1} \begin{bmatrix} T_\infty + e_{15}E_0 \\ B_0 - \alpha_{11}E_0 \end{bmatrix} \tag{4.7.26}$$

$$b_0 = E_0$$

$$\text{Case 3:} \qquad \begin{bmatrix} a_0 \\ b_0 \end{bmatrix} = \begin{bmatrix} c_{44} & -e_{15} \\ e_{15} & \kappa_{11} \end{bmatrix}^{-1} \begin{bmatrix} T_\infty + \tilde{e}_{15} H_0 \\ D_0 - \alpha_{11} H_0 \end{bmatrix} \qquad (4.7.27)$$

$$c_0 = H_0$$

$$\text{Case 4:} \quad a_0 = \frac{1}{c_{44}} (T_\infty + e_{15} E_0 + \tilde{e}_{15} H_0), \quad b_0 = E_0, \quad c_0 = H_0 \quad (4.7.28)$$

$$d_0 = T_\infty / c_{44}^E, \quad e_0 = h(a_0 - T_\infty / c_{44}^E) \qquad (4.7.29)$$

Then, a simple calculation leads to the stress, electric displacement and magnetic induction expressions

$$\sigma_{zy} = \frac{2}{\pi} \int_0^\infty \alpha \begin{Bmatrix} [c_{44} A_1(\alpha) + e_{15} B_1(\alpha) + \tilde{e}_{15} C_1(\alpha)] \exp(\alpha y) - \\ [c_{44} A_2(\alpha) + e_{15} B_2(\alpha) + \tilde{e}_{15} C_2(\alpha)] \exp(-\alpha y) \end{Bmatrix} \cos(\alpha x) d\alpha +$$

$$c_{44} a_0 - e_{15} b_0 - \tilde{e}_{15} c_0 \qquad (4.7.30)$$

$$D_y = \frac{2}{\pi} \int_0^\infty \alpha \begin{Bmatrix} [e_{15} A_1(\alpha) - \kappa_{11} B_1(\alpha) - \alpha_{11} C_1(\alpha)] \exp(\alpha y) - \\ [e_{15} A_2(\alpha) - \kappa_{11} B_2(\alpha) - \alpha_{11} C_2(\alpha)] \exp(-\alpha y) \end{Bmatrix} \cos(\alpha x) d\alpha +$$

$$e_{15} a_0 + \kappa_{11} b_0 + \alpha_{11} c_0 \qquad (4.7.31)$$

$$B_y = \frac{2}{\pi} \int_0^\infty \alpha \begin{Bmatrix} [\tilde{e}_{15} A_1(\alpha) - \alpha_{11} B_1(\alpha) - \mu_{11} C_1(\alpha)] \exp(\alpha y) - \\ [\tilde{e}_{15} A_2(\alpha) - \alpha_{11} B_2(\alpha) - \mu_{11} C_2(\alpha)] \exp(-\alpha y) \end{Bmatrix} \cos(\alpha x) d\alpha +$$

$$\tilde{e}_{15} a_0 + \alpha_{11} b_0 + \mu_{11} c_0 \qquad (4.7.32)$$

$$\sigma_{zy}^E = -\frac{2}{\pi} c_{44}^E \int_0^\infty \alpha A_3(\alpha) \exp(-\alpha y) \cos(\alpha x) d\alpha + c_{44}^E d_0 \qquad (4.7.33)$$

Satisfaction of the boundary conditions (4.7.8)~(4.7.11), (4.7.15) and (4.7.16) leads to the result that

$$A_1(\alpha) = \frac{(\lambda_i - c_{44}^E) \exp(-2\alpha h) F(\alpha)}{\Omega_i} \qquad (4.7.34)$$

$$A_2(\alpha) = \frac{(\lambda_i + c_{44}^E) F(\alpha)}{\Omega_i} \qquad (4.7.35)$$

$$A_3(\alpha) = 2\lambda_i F(\alpha) / \Omega_i \qquad (4.7.36)$$

$$B_1(\alpha) = \frac{\exp(-2\alpha h)[G(\alpha) + m_i F(\alpha)]}{1 + \exp(-2\alpha h)} \qquad (4.7.37)$$

$$B_2(\alpha) = \frac{G(\alpha) - m_i \exp(-2\alpha h) F(\alpha)}{1 + \exp(-2\alpha h)} \qquad (4.7.38)$$

$$C_1(\alpha) = \frac{\exp(-2\alpha h)\left[H(\alpha)+n_iF(\alpha)\right]}{1+\exp(-2\alpha h)} \tag{4.7.39}$$

$$C_2(\alpha) = \frac{H(\alpha)-n_i\exp(-2\alpha h)F(\alpha)}{1+\exp(-2\alpha h)} \tag{4.7.40}$$

where $F(\alpha)$, $G(\alpha)$ and $H(\alpha)$ are the only unknown functions, and m_i, n_i and Ω_i ($i=1,\ 2,\ 3,\ 4$) are defined for Case i ($i=1,\ 2,\ 3,\ 4$), respectively, as

$$m_1 = \frac{2c_{44}^E(\tilde{e}_{15}\alpha_{11}-e_{15}\mu_{11})}{\Omega_1(\kappa_{11}\mu_{11}-\alpha_{11}^2)}, \quad m_2 = m_4 = 0, \quad m_3 = -\frac{2c_{44}^E e_{15}}{\kappa_{11}\Omega_3} \tag{4.7.41}$$

$$n_1 = \frac{2c_{44}^E(e_{15}\alpha_{11}-\tilde{e}_{15}\kappa_{11})}{\Omega_1(\kappa_{11}\mu_{11}-\alpha_{11}^2)}, \quad n_2 = -\frac{2c_{44}^E\tilde{e}_{15}}{\mu_{11}\Omega_2}, \quad n_3 = n_4 = 0 \tag{4.7.42}$$

$$\Omega_i = \lambda_i + c_{44}^E + (\lambda_i - c_{44}^E)\exp(-2\alpha h), \quad i=1,\ 2,\ 3,\ 4 \tag{4.7.43}$$

By applying the mixed boundary conditions (4.7.13), (4.7.19) and (4.7.20), we can reduce the problem to the unknowns $F(\alpha)$, $G(\alpha)$ and $H(\alpha)$ that satisfy the following simultaneous dual integral equations:

$$\int_0^\infty \alpha M_i(\alpha)F(\alpha)\cos(\alpha x)\mathrm{d}\alpha = \frac{\pi T_\infty}{2c_{44}}, \quad 0 \leqslant x < c, \quad i=1,\ 2,\ 3,\ 4 \tag{4.7.44}$$

$$\int_0^\infty F(\alpha)\cos(\alpha x)\mathrm{d}\alpha = 0, \quad x \geqslant c$$

$$\int_0^\infty \alpha G(\alpha)\sin(\alpha x)\mathrm{d}\alpha = 0, \quad \int_0^\infty \alpha H(\alpha)\cos(\alpha x)\mathrm{d}\alpha = 0, \quad 0 \leqslant x < c \tag{4.7.45}$$

$$\int_0^\infty G(\alpha)\cos(\alpha x)\mathrm{d}\alpha = 0, \quad \int_0^\infty H(\alpha)\cos(\alpha x)\mathrm{d}\alpha = 0, \quad x \geqslant c \tag{4.7.46}$$

where $M_i(\alpha)$ are defined as

$$M_1(\alpha) = \frac{2}{\Omega_1}\left\{\lambda_1 + c_{44}^E + \frac{2c_{44}^E\exp(-2\alpha h)(\mu_{11}e_{15}^2 + \kappa_{11}\tilde{e}_{15}^2 - 2\alpha_{11}\tilde{e}_{15}e_{15})}{c_{44}(\kappa_{11}\mu_{11}-\alpha_{11}^2)[1+\exp(-2\alpha h)]}\right\} - 1 \tag{4.7.47}$$

$$M_2(\alpha) = \frac{2}{\Omega_2}\left\{\lambda_2 + c_{44}^E + \frac{2c_{44}^E\tilde{e}_{15}^2\exp(-2\alpha h)}{c_{44}\mu_{11}[1+\exp(-2\alpha h)]}\right\} - 1 \tag{4.7.48}$$

$$M_3(\alpha) = \frac{2}{\Omega_3}\left\{\lambda_3 + c_{44}^E + \frac{2c_{44}^E e_{15}^2\exp(-2\alpha h)}{c_{44}\kappa_{11}[1+\exp(-2\alpha h)]}\right\} - 1 \tag{4.7.49}$$

$$M_4(\alpha) = \frac{2(\lambda_4 + c_{44}^E)}{\Omega_4} - 1 \tag{4.7.50}$$

Eqs.(4.7.44)~(4.7.46) can be solved by using the method of Copson [52], and the solutions are as follows:

$$F(\alpha) = \frac{\pi T_\infty c^2}{2c_{44}} \int_0^\infty \xi \Phi_i(\xi) J_0(\alpha c \xi) \mathrm{d}\xi$$

$$G(\alpha) = \frac{\pi T_\infty c^2}{2c_{44}} \int_0^\infty \xi \Psi_1(\xi) J_0(\alpha c \xi) \mathrm{d}\xi \qquad (4.7.51)$$

$$H(\alpha) = \frac{\pi T_\infty c^2}{2c_{44}} \int_0^\infty \xi \Psi_2(\xi) J_0(\alpha c \xi) \mathrm{d}\xi$$

where $J_0(\ )$ is the zero order Bessel function of the first kind. The function $\Phi_i(\xi)$ should satisfy the Fredholm integral equations of the second kind in the form

$$\Phi_i(t) + \int_0^1 \Phi_i(\eta) K_i(\eta,t) \mathrm{d}\eta = 1 , \quad i = 1,\ 2,\ 3,\ 4 \qquad (4.7.52)$$

where

$$K_i(\eta,t) = \eta \int_0^\infty s\left[M_i(s/c) - 1 \right] J_0(st) J_0(s\eta) \mathrm{d}\eta \qquad (4.7\ 53)$$

The functions $\Psi_j(\xi)$ $(j = 1,\ 2)$ are $\Psi_j(\xi) = 0$.

The stress intensity factor (SIF), the electric displacement intensity factor (EDIF), and the magnetic induction intensity factor (MIIF) are defined and determined respectively as

$$K^{\mathrm{T}} = \lim_{x \to c^+} \sqrt{2\pi(x-c)}\, \sigma_{zy}(x,0) = T_\infty \Phi_i(1)\sqrt{\pi c} , \quad i = 1,\ 2,\ 3,\ 4 \qquad (4.7.54)$$

$$K^D = \lim_{x \to c^+} \sqrt{2\pi(x-c)}\, D_y(x,0) = \frac{e_{15}}{c_{44}} K^{\mathrm{T}} \qquad (4.7.55)$$

$$K^B = \lim_{x \to c^+} \sqrt{2\pi(x-c)}\, B_y(x,0) = \frac{\tilde{e}_{15}}{c_{44}} K^{\mathrm{T}} \qquad (4.7.56)$$

For this particular problem, the stresses, electric displacements and magnetic inductions at the crack tip show inverse square root singularities. It is clear that the SIF, EDIF and MIIF are dependent on the geometry size of the magneto-electro-elastic layer, the mechanical load conditions and the material constants.

In the case of $\beta_{11} = 0$ and $h_{15} = 0$, the results are reduced exactly to the solution of a cracked piezoelectric layer bonded to dissimilar half spaces given by Narita et al[53].

References

[1] Soh A K, Liu J X. On the constitutive equations of magnetoelectroelastic solids. J.

Intelligent Mat. Sys. Struc., 2005, 16:597-602.

[2] Van Suchtelen J. Product properties: a new application of composite materials. Philips Research Reports, 1972, 27:28-37.

[3] Van den Boomgaard J, Terrell D R, Born R A J, et al. An in situ grown eutectic magnetoelectric composite material: part 1, composition and unidirectional solidification. J. Mat. Sci., 1974, 9:1705-1709.

[4] Lee P C Y. A variational principle for the equations of piezoelectromagnetism in elastic dielectric crystals. J. Appl. Phys., 1991, 6:7470-7473.

[5] He J H. Variational theory for linear magnetoelectroelasticity. Int. J. Non. Sci. Numer. Simulation, 2001, 2:309-316.

[6] Qing G H, Qiu J J and Liu Y H. Modified H-R mixed variational principle for magnetoelectroelastic bodies and state-vector equation. Appl. Mathe. Mech., 2005, 26:722-728.

[7] Alshits V I, Darinskii A N and Lothe J, On the existence of surface waves in half-infinite anisotropic elastic media with piezoelectric and piezomagnetic properties. Wave Motion, 1992, 16:265-283.

[8] Lee P C Y, Yang J S, Yu J D, et al. Stress sensitivity of electromagnetic resonances in circular dielectric disks. J. Appl. Phys., 1996, 79:1224-1232.

[9] Li J Y, Dunn M L. Micromechanics of magnetoelectroelastic composite materials: average fields and effective behaviour. J. Intelligent Mat. Sys. Struct., 1998, 9:404-416.

[10] Li J Y. Magnetoelectroelastic multi-inclusion and inhomogeneity problems and their applications in composite materials. Int. J. Eng. Sci., 2000, 38:1993-2011.

[11] Pan E. Exact solution for simply supported and multilayered magneto-electro-elastic plates. J. Appl. Mech., 2001, 68: 608-613.

[12] Pan E, Heyliger P R. Free vibrations of simply supported and multilayered magnetoelectroelastic plates. J. Sound Vib., 2002, 252:429-442.

[13] Jiang X, Pan E. Exact solution for 2D polygonal inclusion problem in anisotropic magnetoelectroelastic full-, half- and biomaterial-planes. Int. J. Solids Struc., 2004, 41:4361-4382.

[14] Pan E, Han F. Exact solution for functionally graded and layered magneto-electro-elastic plates. Int. J. Eng. Sci., 2005, 43:321-339.

[15] Wang X, Shen Y P. The general solution of three dimensional problems in magnetoelectroelastic media. Int. J. Eng. Sci., 2002, 40:1069-1080.

[16] Chen W Q, Lee K Y, Ding H J. General solution for transversely isotropic magneto-electro-thermo- elasticity and the potential theory method. Int. J. Eng. Sci., 2004, 42: 1361-1379.

[17] Wu X H, Shen Y P, Wang X. Stress concentration around a hyperboloidal notch under tension in a magnetoelectroelastic material. ZAMM, 2004, 84:818-824.

[18] Liu J X, Soh A K, Fang D N, A moving dislocation in a magneto-electro-elastic solid. Mech. Res. Communi., 2005, 32:504-513.

[19] Ootao Y, Tanigawa Y. Transient analysis of multilayered magneto-electro-thermo-elastic strip due to ununiform heat supply. Composite Struc., 2005, 68:471-480.

[20] Soh A K, Liu J X. Interfacial debonding of a circular inhomogeneity in piezoelectric-piezomagnetic composites under anti-plane mechanical and in-plane electromagnetic loading. Composites Sci. Tech., 2005, 65:1347-1353.

[21] Gao C F, Tong P, Zhang T Y. Interfacial crack problems in magneto-electroelastic solids. Int. J. Eng. Sci., 2003, 41:2105-2121.

[22] Spyropoulos C P, Sih G C, Song Z F. Magnetoelectroelastic composite with poling parallel to plane of plane crack under out-of-plane deformation. Theore. Appl. Frac. Mech., 2003, 40:281-289.

[23] Wang B L, Mai Y W. Crack tip field in piezoelectric/piezomagnetic media. Euro. J. Mech. A/Solids, 2003, 22:591-602.

[24] Gao C F, Kessler H, Balke H. Fracture analysis of electromagnetic thermoelastic solids. Euro. J. Mech. A/Solids, 2003, 22:433-442.

[25] Tian W Y, Gabbert U. Parallel crack near the interface of magnetoelectroelastic bi-materials. Compu. Mat. Sci., 2005, 32:562-567.

[26] Hu K Q, Li G Q. Constant moving crack in a magnetoelectroelastic material under anti-plane shear loading. Int. J. Solids Struc., 2005, 42:2823-2835.

[27] Tian W Y, Gabbert U. Microcrack-microcrack interaction problem in magnetoelectroelastic solids. Mech. Mat., 2005, 37:565-592.

[28] Li X F. Dynamic analysis of a cracked magnetoelectroelastic medium under antiplane mechanical and inplane electric and magnetic impacts. Int. J. Solids Struc., 2005, 42:3185-3205.

[29] Tian W Y, Rajapakse R K N D. Fracture analysis of magnetoelectroelastic solids by path independent integrals. Int. J. Frac., 2005, 131:311-335.

[30] Zhou Z G, Wu L Z, Wang B. The dynamic behaviour of two collinear interface cracks in magneto-electro-elastic materials. Euro. J. Mech. A/Solids, 2005, 24:253-262.

[31] Hu K Q, Li G Q. Electro-magneto-elastic analysis of a piezoelectromagnetic strip with a finite crack under longitudinal shear. Mech. Mat., 2005, 37:925-934.

[32] Chung M Y, Ting T C T. The Green function for a piezoelectric piezomagnetic magnetoelectric anisotropic elastic medium with an elliptic hole or rigid inclusion. Philos. Mag. Letters, 1995, 72:405-410.

[33] Liu J X, Liu X L, Zhao Y B. Green's functions for anisotropic magnetoelectroelastic solids with an elliptical cavity or a crack. Int. J. Eng. Sci., 2001, 39: 1405-1418.

[34] Pan E. Three-dimensional Green's functions in anisotropic magneto-electro-elastic bi-materials. ZAMP, 2002, 53:815-838.

[35] Hou P F, Ding H J, Chen J Y. Green's function for transversely isotropic magnetoelectroelastic media. Int. J. Eng. Sci., 2005, 43:826-858.

[36] Li J Y. Magnetoelectric Green's functions and their application to the inclusion and inhomogeneity problems. Int. J. Solids Struc., 2002, 39:4201-4213.

[37] Ding H J, Jiang A M, Hou P F, et al. Green's functions for two-phase transversely isotropic magneto-electro-elastic media. Eng. Analysis with Boundary Elements, 2005, 29:551-561.

[38] Qin Qinghua. Green's functions of magnetoelectroelastic solids with a half-plane boundary or bimaterial interface. Philos. Mag. Letters, 2004, 84:771-779.

[39] Qin Qinghua. 2D Green's functions of defective magnetoelectroelastic solids under thermal loading. Eng. Analysis Boundary with Elements, 2005, 29:577-585.

[40] Yao W A. Generalized variational principles of three dimensional problems in magnetoelectroelastic solids. Chinese J. Compu. Mech., 2003, 20:487-489.

[41] Liu J X, Wang X Q, Wang B. General solution for the coupled equations of transversely isotropic magnetoelectroelastic solids. Appl. Mathe. Mech., 2003, 24:684-690.

[42] Gao C F, Noda N. Thermal-induced interfacial cracking of magnetoelectroelastic materials. Int. J. Eng. Sci., 2004, 42, 1347-1360.

[43] Gao C F, Fan W X. Green's functions for the plane problem in a half-infinite piezoelectric medium. Mech. Res. Commun., 1998,25:69-74.

[44] Qin Qinghua, Mai Y W. Thermoelectroelastic Green's function and its application for bi-material of piezoelectric materials. Arch. App. Mech., 1998,68:433-444.

[45] Qin Qinghua. Fracture mechanics of piezoelectric materials, Southampton: WIT Press, 2001.

[46] Qin Qinghua. Green's functions of magnetoelectroelastic solids and applications to fracture analysis//Proc. of 9th International Conference on Inspection, Appraisal, Repairs & Maintenance of Structures, October 20-21, Fuzhou, China, 2005:93-106.

[47] Stagni L. On the elastic field perturbation by inhomogeneous in plane elasticity. ZAMP, 1982, 33:313-325.

[48] Chen B J, Xiao Z M, Liew K M. A screw dislocation in a piezoelectric bi-material wedge. Int. J. Eng. Sci., 2002, 40:1665-1685.

[49] Seaborn J B. Hypergeometric functions and their applications. New York: Springer-Verlag, 1991.

[50] Hu K Q, Qin Qinghua, Kang Y L. Antiplane shear crack in a magnetoelectroelastic layer sandwiched between dissimilar half spaces. Eng. Frac. Mech. (in press)

[51] Hu K Q, Li G Q. Constant moving crack in a magnetoelectroelastic material under anti-plane shear loading. Int. J. Solids Struct., 2005, 42: 2823-2835.

[52] Copson E T. On certain dual integral equations. Proc. Glasgow Math. Assoc., 1961, 5:19-24.

[53] Narita F, Shindo Y, Watanabe K. Anti-plane shear crack in a piezoelectric layer bonded to dissimilar half spaces. JSME Int. J. Series A, 1999, 42: 66-72.

Chapter 5   Thermo-electro-chemo-mechanical coupling

Many synthetic and natural media which are often described as multifunctional smart materials demonstrate thermo-electro-chemo-mechanical coupling behavior and are sensitive to external environmental stimuli. This chapter presents a set of basic equations, a variational principle and a finite element procedure for investigating the coupled behavior of thermo-electro-chemo-elastic media. The emphasis here is placed on introducing chemical effects into the coupled equation system. Using the governing equations of thermal conduction, electric field, ionic diffusion and momentum balance, a variational principle is deduced for a linearly coupled system by means of the extended Gibb's free energy function. The variational principle is then used to derive a fully coupled multifield finite element formulation for simulating the coupled thermo-electro-chemo-elastic behavior of such media. Numerical examples are considered to illustrate the coupled phenomena of the materials and to verify the proposed variational theory and numerical procedure.

5.1   Introduction

With the rapid development of material sciences and technologies, many new multifunction materials have been created and applied to industrial engineering, including materials that exhibit coupled multifield behavior and interaction among fields. For example, conducting polymers have been widely used as artificial muscles and biosensors [1-4], because the conducting polymers can accomplish the transformation of electrical, chemical and mechanical energy and demonstrate response to external environmental variables including temperature, pH and electrical and mechanical loadings. Such versatile polymers, which seem to be increasingly manufactured, include hydrogel and various advanced polymers [5-7]. In general, these multifield materials consist of a solid network and interval fluid, and deform in volume and shape. They can be applied in practical engineering as biosensors, artificial skin of robots, artificial muscles, and actuators of adaptive structures [2, 4].

On the other hand, natural materials such as biological tissues, clays and shales exhibit strong swelling and contractive properties under chemical, electrical and mechanical stimuli. For example, articular cartilage is a porous medium bathed in an electrolyte and its electro-chemo-mechanical coupling behavior cannot be ignored. This cartilage consists of hydrated proteoglycans and collagen fibers which form fibrillar structures that trap their own water. To sustain external loads, including mechanical and electrochemical loads, cartilage modifies its internal configuration by means of water and ion exchanges [8-10]. The performance of saturated porous media has also attracted the attention of researchers and scientists over the past decades [11, 12]. Early studies in this field focused on the interactions between the solid and the fluid in the saturated porous media. A poroelastic theory was developed [13, 14] and used for deriving various numerical algorithms. The concept of effective stress and the equations of mass balance and momentum balance form the primary framework of poroelastic theory.

Using poroelastic theory [14], a triphasic mixture model of porous media was proposed to consider the electric and diffusion effect induced by ions in the fluid [15]. In the triphasic theoretical model, the porous medium was assumed to be composed of solid, fluid and ions. Modified mass balance and momentum balance equations, in addition to the ionic diffusion equation, were introduced. These equations were then used to describe the deformation and stress of biological soft tissues like cartilage and to derive corresponding finite element (FE) formulations [16]. Later, a quadriphasic model was presented, to investigate quasi-static finite deformation of swelling of incompressible porous media, where the ions in the fluid are decompounded as anions and cations [17]. In this model, balance laws are derived for each phase and for the mixture as a whole. The quadriphasic model, considering electric-osmosis and streaming current effects, can be applied to the analysis of intervertebral disk tissue [18]. More recently, a thermo-electro-chemo-mechanical formulation based on the quadriphasic mixture model has been developed for quasi-static finite deformation of swelling incompressible porous media [19]. It is noted that the triphasic and quadriphasic models belong to the category of mixture theory based on the poroelasitic framework of porous media. However, these two models provide balance equations of the mixture only, and an explicit form of constitutive law does not appear in the related literature. Therefore the solution of these theoretical models largely depends on the form of constitutive law used.

In addition to the mixture methods discussed above, other types of multifield approache have been presented in recent years [20-27] to reveal the electro-chemo- mechanical coupling behavior of porous media and to try to explain interactions among the fields. On the basis of these theories some numerical methods have been developed to solve the coupled multifield differential equations, including direct iteration procedures [22, 23] and the FE method [24-27].

In this chapter, a theoretical model and the corresponding FE formulation for thermo-electro-chemo-mechanical coupled problems are presented, developed by redefining linearly coupled constitutive relations and extending the traditional Gibb's free energy to include chemical effects. In contrast to previous work, the theoretical model proposed is based on a newly introduced linear constitutive chemical law instead of the balance laws that were used in the triphasic and quadriphasic mixture models [15-19]. As existing chemical governing equations are not suitable for FE analysis, we start by deriving a modified form of basic equation for the chemical field (Section 5.2). By extending the traditional Gibb's free energy to include the chemical field, we obtain linear forms of coupled constitutive laws and a variational principle including chemical effect (Sections 5.3 and 5.4). The variational principle is then used to derive the FE formulation (Section 5.5). As a special case of the coupled system, coupling between chemical and mechanical fields is discussed in detail, and the determination of some coupled property parameters is also demonstrated (Section 5.6).

5.2   Governing equations of fields

Consider a thermo-electro-chemo-mechanical body of volume Ω bounded by surface S. The governing equations, including heat conduction equation, Maxwell's equation of electrostatics, equilibrium equations of stresses, and diffusion equations of ions, are as follows:

(1) Equilibrium equations of stresses momentum balance equations.

$$\sigma_{ij,j} + f_i = 0, \text{ in } \Omega \tag{5.2.1}$$

(2) Boundary conditions.

$$u_i = \bar{u}_i, \text{ on } S_u \tag{5.2.2a}$$

$$\sigma_{ij}n_j = \bar{t}_i, \text{ on } S_t \tag{5.2.2b}$$

(3) Maxwell's equation of electrostatics.

$$D_{i,i} = q_b, \text{ in } \Omega \tag{5.2.3}$$

(4) Electric boundary conditions.

$$\phi = \bar{\phi}, \text{ on } S_\phi \tag{5.2.4a}$$

$$D_i n_i = -\bar{q}_s, \text{ on } S_D \tag{5.2.4b}$$

(5) Heat conduction equation.

$$h_{i,i} = -T_0\dot{\eta}, \text{ in } \Omega \tag{5.2.5}$$

(6) Thermal boundary conditions.

$$T = \bar{T}, \text{ on } S_T \tag{5.2.6a}$$

$$h_i n_i = \bar{h}_n, \text{ on } S_h \tag{5.2.6b}$$

In these equations, σ_{ij}, D_i, and h_i are respectively stress tensor, electric displacement vector, and heat flux vector; f_i and q_b are the mechanical body force and body electric charge density; $\bar{u}_i$ and $\bar{t}_i$ are the prescribed surface displacements and tractions; $\bar{\phi}$ and $\bar{q}_s$ are the prescribed electric potential and surface electric charge; $\bar{T}$ and $\bar{h}_n$ are the prescribed temperature change and heat flux on the surface S; n_i is the unit outward normal vector on the surface S; η is the entropy density; T_0 is reference temperature; and $S = S_u + S_t = S_\phi + S_D = S_T + S_h$.

(7) Basic equations of chemical field.

Fick's law shows that the mass flux $\xi^\pm$ is proportional to the gradient of the ionic concentrations $\nabla c^\pm$ by [28]

$$\xi^\pm = -\varphi^\pm \nabla c^\pm \text{ or } \xi_i^\pm = -\varphi_{ij}^\pm c_{,j}^\pm \tag{5.2.7}$$

where "+" and "−" denote anion and cation, respectively. The proportional coefficients $\varphi_{ij}^\pm$ denote the diffusion coefficients of anions and cations, depending on the intrinsic features of the medium. For an isotropic medium, $\varphi_{ij}^\pm = \varphi^\pm \delta_{ij}$. $c^\pm$ are increments of concentrations for the anion and cation. Therefore $c^\pm$ can be related to the current concentrations $\bar{c}^\pm$ by $\bar{c}^\pm = c_0^\pm + c^\pm$, where $c_0^\pm$ are the reference concentrations. $\nabla = \left[\dfrac{\partial}{\partial x} \quad \dfrac{\partial}{\partial y}\right]$ is the gradient

operator. The convection-diffusion equations of the ions can thus be written as

$$\frac{\partial c^\pm}{\partial t} + \nabla \bullet (c^\pm v) - \nabla \bullet \left(\frac{\varphi^\pm c^\pm}{RT^*}\nabla \mu^\pm\right) = 0 \tag{5.2.8}$$

where v is the velocity of the ions, R is the universal gas constant and T^* is the absolute temperature. The first term in Eq. (5.2.8) represents the change rate of the concentrations with respect to the time. The second term stands for the convection effect that describes the macroscopic motion of the ions. The third term is the diffusion of the ions. The electric potential produced by ion is very small compared with that produced by the applied electric field and is therefore ignored.

For the motion of ions in fluid, the primary mechanism is ionic diffusion. By ignoring the convection effect in Eq. (5.2.8), which means that the macroscopic motion of the ions is not considered, we have

$$\frac{\partial c^\pm}{\partial t} - \nabla \bullet \left(\frac{\varphi^\pm c^\pm}{RT^*}\nabla \mu^\pm\right) = 0 \tag{5.2.9}$$

where $\mu^\pm$ is chemical potential. In classical physical chemistry, the chemical potential and concentration of the ions have the following relations:

$$\mu^\pm = \mu_0^\pm + RT^* \ln \overline{c}^\pm \tag{5.2.10}$$

where $\mu_0^\pm$ is a reference potential of anion and cation in the standard state.

Substituting Eq. (5.2.10) into Eq. (5.2.9) leads to

$$\frac{\partial c^\pm}{\partial t} - \nabla \bullet \varphi^\pm \nabla c^\pm = 0 \tag{5.2.11}$$

It is noted that Eqs. (5.2.8) and (5.2.11) agree with those applied in [20] for the motion of ions. For an isotropic medium, we have

$$\frac{\partial c^\pm}{\partial t} + \nabla \xi^\pm = 0, \quad \text{in } \Omega \tag{5.2.12}$$

The corresponding natural boundary condition is

$$\xi_i^\pm n_i = \xi_n^\pm, \quad \text{on } S \tag{5.2.13}$$

where $\xi_n^\pm$ is the ionic flux on the surface of the domain.

Eq. (5.2.12) and natural boundary condition (5.2.13) are the governing equations of ionic diffusion. For the multifield coupling case, we must modify

Eq. (5.2.12) to consider the coupling effect. For this purpose, taking differentiation in respect to time t leads to

$$\frac{\partial \mu^{\pm}}{\partial t} = \frac{RT^{*}}{\overline{c}^{\pm}} \frac{\partial c^{\pm}}{\partial t} \qquad (5.2.14a)$$

Consider a small concentration increment, thus $\overline{c}^{\pm} = c_0^{\pm} + c^{\pm} \approx c_0^{\pm}$. We can obtain

$$\frac{\partial \mu^{\pm}}{\partial t} = \frac{RT^{*}}{c_0^{\pm}} \frac{\partial c^{\pm}}{\partial t} \qquad (5.2.14b)$$

Employing Eq. (5.2.12) and Eq.(5.2.14b), we obtain the diffusion equations of ions in the form

$$\frac{\partial \mu^{\pm}}{\partial t} + \frac{c_0^{\pm}}{RT^{*}} \nabla \xi^{\pm} = 0 \qquad (5.2.15)$$

The relations between displacements u_i and strains ε_{ij} for elastic field, electric potential ϕ and electric fields E_i for electrostatics, temperature change T and heat flux h_i for heat conduction, ionic flux $\xi_i^{\pm}$ and concentration change $c^{\pm}$ for the chemical field are as follows:

$$\varepsilon_{ij} = \frac{1}{2}(u_{i,j} + u_{j,i}) \qquad (5.2.16)$$

$$E_i = -\phi_{,i} \qquad (5.2.17)$$

$$h_i = -k_{ij}T_{,j} \qquad (5.2.18)$$

$$\xi_i^{\pm} = -\varphi_{ij}^{\pm}c_{,j}^{\pm} \qquad (5.2.19)$$

where k_{ij} are heat conduction coefficients. It is noted that the theory outlined here is restricted to small deformations as linear equations are used.

5.3　Free energy and constitutive laws

In this section, Gibb's free energy function in [29,30] is extended to include chemical effect and is used to derive the linear constitutive law for thermo-electro-chemo-mechanical systems. For a system that includes thermal, electrical, chemical and mechanical interaction, the extended Gibb's free energy per volume can be written by adding chemical energy in the form

$$g = U - E_i D_i - \eta T + \sum_{\alpha=+,-} \mu^{\alpha} c^{\alpha} \qquad (5.3.1)$$

where U denotes internal energy, the second and third terms stand for the energy contributions of electric and temperature fields, respectively, and the last term is the chemical energy. Gibb's free energy containing the first three terms in Eq. (5.3.1) has been discussed elsewhere [29, 30]. The last term should be added to the Gibb's free energy when the chemical effect is considered [28]. An exact differential of Gibb's free energy function (5.3.1) with respect to its independent variables leads to

$$\mathrm{d}g = \sigma_{ij}\mathrm{d}\varepsilon_{ij} - D_m\mathrm{d}E_m - \eta\mathrm{d}T + \sum_{\alpha=+,-} \mu^\alpha \mathrm{d}c^\alpha \qquad (5.3.2)$$

Thus, we obtain

$$\sigma_{ij} = \frac{\partial g}{\partial \varepsilon_{ij}}, \quad D_i = -\frac{\partial g}{\partial E_i}, \quad \eta = -\frac{\partial g}{\partial T}, \quad \mu^\pm = \frac{\partial g}{\partial c^\pm} \qquad (5.3.3)$$

When the function g is expanded with respect to T, ε_{ij}, E_m and $c^\pm$ within the scope of linear interactions, we have

$$g = \frac{1}{2}\left(T\frac{\partial}{\partial T} + \varepsilon_{ij}\frac{\partial}{\partial \varepsilon_{ij}} + E_m\frac{\partial}{\partial E_m} + c^\pm\frac{\partial}{\partial c^\pm}\right)\left(T\frac{\partial}{\partial T} + \varepsilon_{kl}\frac{\partial}{\partial \varepsilon_{kl}} + E_n\frac{\partial}{\partial E_n} + c^\pm\frac{\partial}{\partial c^\pm}\right)g$$

$$(5.3.4)$$

The following constants can then be defined:

$$c_{ijkl} = \left[\frac{\partial^2 g}{\partial \varepsilon_{ij}\partial \varepsilon_{kl}}\right], \quad \kappa_{nm} = -\left[\frac{\partial^2 g}{\partial E_n\partial E_m}\right]$$

$$R_{ij}^\pm = -\left[\frac{\partial^2 g}{\partial \varepsilon_{ij}\partial c^\pm}\right], \quad \nu^\pm = -\left[\frac{\partial^2 g}{\partial T\partial c^\pm}\right]$$

$$\alpha = \frac{\rho C_v}{T_0} = -\left[\frac{\partial^2 g}{\partial T^2}\right], \quad e_{mij} = -\left[\frac{\partial^2 g}{\partial \varepsilon_{ij}\partial E_m}\right], \quad w_m^\pm = -\left[\frac{\partial^2 g}{\partial E_m\partial c^\pm}\right]$$

$$\lambda_{ij} = -\left[\frac{\partial^2 g}{\partial \varepsilon_{ij}\partial T}\right], \quad \chi_m = -\left[\frac{\partial^2 g}{\partial T\partial E_m}\right], \quad s^\pm = \left[\frac{\partial^2 g}{\partial c^\pm\partial c^\pm}\right]$$

$$(5.3.5)$$

where c_{ijkl} are the elastic moduli, κ_{nm} the dielectric constants, c_v the specific heat per unit mass, e_{mij} the piezoelectric coefficients, λ_{ij} the thermal-stress coefficients, and χ_m the pyroelectric coefficients. The newly introduced constants $R_{ij}^\pm$, $\nu^\pm$, $w_m^\pm$, and $s^\pm$ are respectively the mechanical-chemical coefficients measured at a constant temperature and electric field, the thermo-chemical coefficients measured at constant strain and electric field, the electrical-chemical coefficients measured at a constant strain and temperature, and the chemical potential constant measured at a constant strain and temperature and electric field for anion and cation.

When the function g is differentiated according to Eq. (5.3.2), and the above constants are used, we find

$$\sigma_{ij} = C_{ijkl}\varepsilon_{ij} - \lambda_{ij}T - e_{ijn}E_n - \sum_{\alpha=+,-} R_{ij}^{\alpha}c^{\alpha} \tag{5.3.6a}$$

$$\eta = \lambda_{ij}\varepsilon_{ij} + \alpha T + \chi_n E_n + \sum_{\alpha=+,-} \nu^{\alpha}c^{\alpha} \tag{5.3.6b}$$

$$D_m = e_{klm}\varepsilon_{kl} + \chi_m T + \kappa_{mn}E_n + \sum_{\alpha=+,-} w_m^{\alpha}c^{\alpha} \tag{5.3.6c}$$

$$\mu^{\pm} = -R_{kl}^{\pm}\varepsilon_{kl} - \nu^{\pm}T - w_n^{\pm}E_n + s^{\pm}c^{\pm} \tag{5.3.6d}$$

A set of these equations is the constitutive relation in the coupled system.

It is noted that the constitutive equations (5.3.6a~d) are extensions of known thermo-electro-mechanical coupling [29, 30] to include a chemical field. In classical physical chemistry, the relation between chemical potential and ionic concentrations is expressed by a logarithm function [see Eq. (5.2.10)], but here we assume a linear relationship between the potential and the concentration changes, which means that Eqs. (5.3.6a~d) apply to a small change of ionic concentrations only. This assumption allows us to develop the corresponding numerical model for FE formulation in a simple way. Actually, the classical logarithm relation is applied to determine the present linear coefficients, as shown in the following section.

Finally, by means of the material parameters defined in Eqs. (5.3.6a~d) we can rewrite the Gibb's free energy function as

$$g = \frac{1}{2}C_{ijkl}\varepsilon_{kl}\varepsilon_{ij} - \lambda_{ij}T\varepsilon_{ij} - e_{ijn}E_n\varepsilon_{ij} - \sum_{\alpha=+,-} R_{ij}^{\alpha}c^{\alpha}\varepsilon_{ij} - \frac{1}{2}\alpha T^2 -$$

$$\chi_n E_n T - \sum_{\alpha=+,-} \nu^{\alpha}c^{\alpha}T - \frac{1}{2}\in_{mn} E_n E_m - \sum_{\alpha=+,-} w_m^{\alpha}c^{\alpha}E_m + \frac{1}{2}\sum_{\alpha=+,-} s^{\alpha}c^{\alpha}c^{\alpha} \tag{5.3.7}$$

This energy function is used as a basis for developing FE formulations in the following sections.

5.4   Variational principle

The variational functional plays a central role in the formulation of the fundamental governing equations in finite element method (FEM). For the boundary value problem described in Section 5.2 and the linear constitutive equations (5.3.6), the variational functional used for deriving FE formulation of thermo-electro-chemo-mechanical coupling system can be constructed in the form

$$\Pi = \int_{\Omega} (\dot{B} + F + J)\mathrm{d}\Omega - \int_{\Omega} (f_i \dot{u}_i - q_b \dot{\phi})\mathrm{d}\Omega$$

$$- \int_{S} (\overline{t_i} \dot{u}_i + \overline{q}_s \dot{\phi} - \frac{T}{T_0}\overline{h}_n - \tau \sum_{\alpha=+,-} \overline{\xi}_n^{\alpha} c^{\alpha})\mathrm{d}S \tag{5.4.1}$$

where $\tau = \dfrac{RT^*}{c_0^{\pm}}$. The vanishing variation of functional (5.4.1) leads to

$$\delta\Pi = \int_{\Omega} \delta\dot{B}\mathrm{d}\Omega + \delta\int_{\Omega} (F + J)\mathrm{d}\Omega - \int_{\Omega} (f_i \delta u_i - q_b \delta\phi)\mathrm{d}\Omega -$$

$$\int_{S} (\overline{t_i}\delta u_i + \overline{q}_s \delta\phi - \frac{\overline{h}_n}{T_0}\delta T - \tau \sum_{\alpha=+,-} \overline{\xi}_n^{\alpha}\delta c^{\alpha})\mathrm{d}S = 0 \tag{5.4.2}$$

where B is the generalized Biot's free energy [30] of the coupled four-field system. The function B should satisfy the following conditions:

$$\frac{\partial B}{\partial \varepsilon_{ij}} = \sigma_{ij}, \quad \frac{\partial B}{\partial T} = \eta, \quad \frac{\partial B}{\partial E_i} = D_i, \quad \frac{\partial B}{\partial c^{\pm}} = \mu^{\pm} \tag{5.4.3}$$

It should be noted that the minus sign in Eq. (5.4.3) disappears here. Thus the function B has the following differential form

$$\delta\dot{B} = \delta\dot{g} + 2D_n\delta\dot{E}_n + 2\dot{\eta}\delta T \tag{5.4.4}$$

The function J in Eq. (5.4.1) is the dissipation energy caused by ionic diffusion and F is the heat dissipation. They are defined by

$$J = -\frac{1}{2}\tau \sum_{\alpha=+,-} \xi_i^{\alpha} c_{,i}^{\alpha} \tag{5.4.5a}$$

$$F = -\frac{1}{2T_0} h_i T_{,i} \tag{5.4.5b}$$

Substituting Eqs. (5.4.4) and (5.4.5) into Eq. (5.4.2) yields

$$\delta\Pi = \int_{\Omega} (\sigma_{ij}\delta\dot{\varepsilon}_{ij} + \dot{\eta}\delta T + D_n\delta\dot{E}_n + \sum_{\alpha=+,-} \dot{\mu}^{\alpha}\delta c^{\alpha})\mathrm{d}\Omega -$$

$$\frac{1}{2}\delta\int_{\Omega}\left(\frac{1}{T_0} h_i T_{,i} + \tau \sum_{\alpha=+,-} \xi_i^{\alpha} c_{,i}^{\alpha}\right)\mathrm{d}\Omega - \int_{\Omega} (f_i\delta u_i - q_b\delta\phi)\mathrm{d}\Omega -$$

$$\int_{S}\left(\overline{t_i}\delta u_i + \overline{q}_s\delta\phi - \frac{\overline{h}_n}{T_0}\delta T - \tau \sum_{\alpha=+,-} \overline{\xi}_n^{\alpha}\delta c^{\alpha}\right)\mathrm{d}S = 0 \tag{5.4.6}$$

Through a series of integrating by parts and mathematical operations, Eq. (5.4.6) can be further written as

$$\delta\Pi = \int_{\Omega}\left[-(\sigma_{ij,j}+f_i)\delta\dot{u}_i+\left(\frac{h_{i,i}}{T_0}+\dot{\eta}\right)\delta T+(D_{i,i}-q_b)\delta\dot{\phi}+\right.$$

$$\left.\sum_{\alpha=+,-}(\tau\xi_{i,i}^{\alpha}+\dot{\mu}^{\alpha})\delta c^{\alpha}\right]\mathrm{d}\Omega+\int_{S}\left[(\sigma_{ij}n_j-\overline{t}_i)\delta\dot{u}_i+(D_in_i+\overline{q}_s)\delta\dot{\phi}+\right.$$

$$\left.\left(\frac{\overline{h}_n}{T_0}-\frac{h_in_i}{T_0}\right)\delta T+\sum_{\alpha=+,-}\tau(\overline{\xi}_n^{\alpha}-\xi_i^{\alpha}n_i)\delta c^{\alpha}\right]\mathrm{d}S=0 \qquad (5.4.7)$$

Due to the arbitrariness of δu_i, $\delta\phi$, δT and $\delta c^{\pm}$, the variational equation (5.4.7) leads to the following governing equations and natural boundary conditions:

$$\sigma_{ij,j}+f_i=0, \quad \sigma_{ij}n_j=\overline{t}_i$$

$$D_{i,i}=q_b, \quad D_in_i=-\overline{q}_s$$

$$h_{i,i}=-T_0\dot{\eta}, \quad h_in_i=\overline{h}_n$$

$$\xi_{i,i}^{\pm}+\frac{1}{\tau}\dot{\mu}^{\pm}=0, \quad \xi_i^{\pm}n_i=\overline{\xi}_n^{\pm}$$

Obviously, they are the Euler equations of functional (5.4.1) and represent the governing equations (5.2.1), (5.2.3), (5.2.5) and (5.2.15) of the mechanical, thermal, electrical and chemical fields, respectively, as well as the corresponding natural boundary conditions (5.2.2b), (5.2.4b), (5.2.6b) and (5.2.14), respectively.

It is interesting to note that the governing equations of the chemical field have the same forms as those of heat conduction problems. In other words, there is an analogous relation between ionic diffusion and heat conduction. Table 5.1 lists the analogous relations between chemical problems and heat conduction problems.

Table 5.1   Analogies of ionic diffusion and heat conduction

Heat conduction	Ionic diffusion
Temperature T	Concentration c
Heat flux $h_i=-\kappa_{ij}T_{,j}$	Mass flux $\xi_i=-\varphi_{ij}c_{,j}$
Entropy change rate $\dot{\eta}$	Chemical potential change rate $\dot{\mu}$
Energy form ηT	Energy form μc
Governing equation $h_{i,i}+\dfrac{1}{T_0}\dot{\eta}=0$	Governing equation $\xi_{i,i}+\dfrac{c_0}{RT^*}\dot{\mu}=0$
Boundary condition $h_in_i=\overline{h}_n$	Boundary condition $\xi_in_i=\overline{\xi}_n$

5.5 Finite element formulation

Substituting the constitutive equations (5.3.6a~d) and the gradient equations(5.2.16)~(5.2.19) into the variational equation (5.4.6), we obtain

$$\int_{\Omega}\left(C_{ijkl}\varepsilon_{kl} - \lambda_{ij}T - e_{ijn}E_n + \sum_{\alpha=+,-} R_{ij}^{\alpha}c^{\alpha} \right)\delta\varepsilon_{ij}\,d\Omega +$$

$$\int_{\Omega}\left(\lambda_{ij}\varepsilon_{ij} + \alpha T + \chi_n E_n + \sum_{\alpha=+,-} \nu^{\alpha}c^{\alpha} \right)\delta T d\Omega +$$

$$\int_{\Omega}\left(e_{klm}\varepsilon_{kl} + \chi_m T + \in_{mn} E_n + \sum_{\alpha=+,-} w_m^{\alpha}c^{\alpha} \right)\delta E_m d\Omega +$$

$$\int_{\Omega}\left[\sum_{\alpha=+,-} (-R_{kl}^{\alpha}\varepsilon_{kl} - \nu^{\alpha}T - w_n^{\alpha}E_n + s^{\alpha}c^{\alpha}) \right]\delta c^{\alpha} d\Omega +$$

$$\int_{\Omega}\left[\frac{1}{T_0}\kappa_{ij}T_{,i}\delta T_{,j} + \tau \sum_{\alpha=+,-} \varphi_{ij}^{\alpha}c_{,i}^{\alpha}\delta c_{,i}^{\alpha} \right] d\Omega -$$

$$\int_{\Omega}(f_i\delta u_i - q_b\delta\phi)d\Omega - \int_{S}\left(\overline{t}_i\delta u_i + \overline{q}_s\delta\phi - \frac{\overline{h}_n}{T_0}\delta T - \tau \sum_{\alpha=+,-} \overline{\xi}_n^{\alpha}\delta c^{\alpha}\right) dS = 0$$

$$(5.5.1)$$

This is a variational equation in terms of the independent variables u, ϕ, T and $c^{\pm}$. It is a basis for establishing the FE formulation. As in the conventional FEM, the boundary S and the domain Ω are divided into a series of boundary elements and internal cells. Over each internal element, the independent variables u, ϕ, T and $c^{\pm}$ are interpolated by the nodal discrete values in the form

$$u = \sum_{i=1}^{n} N_i u_i = Nu^e, \quad \varepsilon = Lu = LNu^e = Bu^e \qquad (5.5.2a)$$

$$T = \sum_{i=1}^{n} \overline{N}_i T_i = \overline{N}T^e, \quad (T_{,j}) = \sum_{i=1}^{n} \overline{N}_{i,j}T_j = \overline{B}T^e \qquad (5.5.2b)$$

$$\phi = \sum_{i=1}^{n} \overline{N}_i \phi_i = \overline{N}\Phi^e, \quad (E_i) = -\sum_{i=1}^{n} \overline{N}_{i,i}\phi_i = -\overline{B}\Phi^e \qquad (5.5.2c)$$

$$c^{\pm} = \sum_{i=1}^{n} \tilde{N}_i c_i^{\pm} = \tilde{N}c^e, \quad (c_{,i}^{\pm}) = \sum_{i=1}^{n} \tilde{N}_{i,i}c_i^{\pm} = \tilde{B}c^e \qquad (5.5.2d)$$

where N, $\overline{N}$ and $\tilde{N}$ are the matrices of shape functions, n is the nodal number of the element, and $c^e = [c_1^+ \quad c_1^- \quad \cdots \quad c_n^+ \quad c_n^-]^{\mathrm{T}}$ is the nodal ionic concentration vector of element e. Let $c_i = [c_i^+ \quad c_i^-]^\mathrm{T}$ denote the degree of freedom for ionic concentrations at node i. L is a matrix of differential operators and depends on the forms of the gradient equation.

To simplify the notations, the compact form of matrix is applied in the following formulations. When the domain is discretized, the potential energy of the whole system can be obtained by summation of the energy over each element

$$\delta\Pi = \delta\sum_e \Pi^e \tag{5.5.3}$$

Thus we have the variational equation for element e

$$\delta\Pi^e =$$

$$\int_{\Omega^e} \delta u^{e\mathrm{T}}(B^\mathrm{T}CBu^e - B^\mathrm{T}\lambda\bar{N}T^e + B^\mathrm{T}e\bar{B}\Phi^e - B^\mathrm{T}R\tilde{N}c^e)\,\mathrm{d}\Omega +$$

$$\int_{\Omega^e} \delta T^{e\mathrm{T}}(\bar{N}^\mathrm{T}\lambda Bu^e + \bar{N}^\mathrm{T}\alpha\bar{N}T^e - \bar{N}^\mathrm{T}\chi\bar{B}\Phi^e + \bar{N}^\mathrm{T}\nu\tilde{N}c^e)\,\mathrm{d}\Omega -$$

$$\int_{\Omega^e} \delta\Phi^{e\mathrm{T}}(\bar{B}^\mathrm{T}eBu^e + \bar{B}^\mathrm{T}\chi\bar{N}T^e - \bar{B}^\mathrm{T}\in\bar{B}\Phi^e + \bar{B}^\mathrm{T}w\tilde{N}c^e)\,\mathrm{d}\Omega +$$

$$\int_{\Omega^e} \delta c^{e\mathrm{T}}(-\tilde{N}^\mathrm{T}R^\mathrm{T}Bu^e - \tilde{N}^\mathrm{T}\nu\bar{N}T^e + \tilde{N}^\mathrm{T}w\bar{B}\Phi^e + \tilde{N}^\mathrm{T}S\tilde{N}c^e)\,\mathrm{d}\Omega -$$

$$\int_{\Omega^e} \left(\frac{-\delta T^{e\mathrm{T}}\bar{B}^\mathrm{T}\kappa}{T_0}\bar{B}T^e - \tau\delta c^{e\mathrm{T}}\tilde{B}^\mathrm{T}\varphi\tilde{B}c^e + \delta u^{e\mathrm{T}}N^\mathrm{T}f - \delta\Phi^{e\mathrm{T}}\bar{N}^\mathrm{T}q_b\right)\mathrm{d}\Omega -$$

$$\int_{S_t^e} \delta u^{e\mathrm{T}}N^\mathrm{T}\bar{t}\,\mathrm{d}S - \int_{S_D^e} \delta\Phi^{e\mathrm{T}}\bar{N}^\mathrm{T}\bar{q}_s\,\mathrm{d}S + \int_{S_h^e} \frac{\delta T^{e\mathrm{T}}\bar{N}^\mathrm{T}}{T_0}\bar{h}_n\,\mathrm{d}S + \int_{S_\xi^e} \tau\delta c^{e\mathrm{T}}\tilde{N}^\mathrm{T}\bar{\xi}_n\,\mathrm{d}S = 0$$

$$\tag{5.5.4}$$

where u^e, T^e, Φ^e, c^e are unknowns to be solved and $\delta u^{e\mathrm{T}}, \delta\Phi^{e\mathrm{T}}, \delta T^{e\mathrm{T}}, \delta c^{e\mathrm{T}}$ are arbitrary variational variables. The equivalent forms of Eq. (5.5.4) are

$$\int_{\Omega^e} (B^\mathrm{T}CBu^e - B^\mathrm{T}\lambda\bar{N}T^e + B^\mathrm{T}e\bar{B}\Phi^e + B^\mathrm{T}R\tilde{N}c^e - N^\mathrm{T}f)\,\mathrm{d}\Omega - \int_{S_t^e} N^\mathrm{T}\bar{t}\,\mathrm{d}S = 0$$

$$\tag{5.5.5a}$$

$$\int_{\Omega^e} (\bar{N}^\mathrm{T}\lambda^\mathrm{T}Bu^e + \bar{N}^\mathrm{T}\alpha\bar{N}T^e - \bar{N}^\mathrm{T}\chi\bar{B}\Phi^e + \bar{N}^\mathrm{T}\nu\tilde{N}c^e)\,\mathrm{d}\Omega +$$

$$\int_{\Omega^e} \frac{\bar{B}^\mathrm{T}\kappa}{T_0}\bar{B}T^e\mathrm{d}\Omega + \int_{S_h^e} \frac{\bar{N}^\mathrm{T}}{T_0}\bar{h}_n\mathrm{d}S = 0 \tag{5.5.5b}$$

$$\int_{\Omega^e} (\bar{B}^\mathrm{T}e^\mathrm{T}Bu^e + \bar{B}^\mathrm{T}\chi^\mathrm{T}\bar{N}T^e - \bar{B}^\mathrm{T}\kappa\bar{B}\Phi^e + \bar{B}^\mathrm{T}w^\mathrm{T}\tilde{N}c^e)\,\mathrm{d}\Omega -$$

$$\int_{\Omega^e} \bar{N}^\mathrm{T}q_b\mathrm{d}\Omega - \int_{S_D^e} \bar{N}^\mathrm{T}\bar{q}_s\mathrm{d}S = 0 \tag{5.5.5c}$$

$$\int_{\Omega^e} (-\tilde{N}^{\mathrm{T}} R^{\mathrm{T}} B u^e - \tilde{N}^{\mathrm{T}} \nu^{\mathrm{T}} \bar{N} T^e + \tilde{N}^{\mathrm{T}} w \bar{B} \Phi^e + \tilde{N}^{\mathrm{T}} S \tilde{N} c^e) \, \mathrm{d}\Omega +$$

$$\int_{\Omega^e} \tau \tilde{B}^{\mathrm{T}} \varphi \tilde{B} c^e \mathrm{d}\Omega + \int_{S_\xi^e} \tau \tilde{N}^{\mathrm{T}} \bar{\xi}_n \mathrm{d}S = 0 \qquad (5.5.5\mathrm{d})$$

or in matrix form

$$
\begin{bmatrix}
K_{mm} & K_{mt} & K_{me} & K_{mc} \\
K_{tm} & K_{tt} & K_{te} & K_{tc} \\
K_{em} & K_{et} & K_{ee} & K_{ec} \\
K_{cm} & K_{ct} & K_{ce} & K_{cc}
\end{bmatrix}
\begin{bmatrix}
u^e \\
T^e \\
\Phi^e \\
c^e
\end{bmatrix}
=
\begin{bmatrix}
F_m^e \\
F_t^e \\
F_e^e \\
F_c^e
\end{bmatrix}
\qquad (5.5.6)
$$

in which

$$K_{mm} = \int_{\Omega^e} B^{\mathrm{T}} C B \mathrm{d}\Omega, \quad K_{mt} = -\int_{\Omega^e} B^{\mathrm{T}} \lambda \bar{N} \mathrm{d}\Omega, \quad K_{me} = \int_{\Omega^e} B^{\mathrm{T}} e \bar{B} \mathrm{d}\Omega$$

$$K_{mc} = \int_{\Omega^e} B^{\mathrm{T}} R \tilde{N}^{\mathrm{T}} \mathrm{d}\Omega, \quad F_m^e = \int_{\Omega^e} N^{\mathrm{T}} f \mathrm{d}\Omega + \int_{S_t^e} N^{\mathrm{T}} \bar{t} \mathrm{d}S$$

$$K_{tm} = \int_{\Omega^e} \bar{N}^{\mathrm{T}} \lambda^{\mathrm{T}} B \mathrm{d}\Omega, \quad K_{tt} = \int_{\Omega^e} \left(\bar{N}^{\mathrm{T}} \alpha \bar{N} + B^{\mathrm{T}} \frac{\kappa \bar{B}}{T_0} \right) \mathrm{d}\Omega, \quad K_{te} = \int_{\Omega^e} \bar{N}^{\mathrm{T}} \chi \bar{B} \mathrm{d}\Omega$$

$$K_{tc} = \int_{\Omega^e} \bar{N}^{\mathrm{T}} \nu \tilde{N} \mathrm{d}\Omega, \quad F_t^e = -\int_{S_h^e} \frac{\bar{N}^{\mathrm{T}}}{T_0} \bar{h}_n \mathrm{d}S$$

$$K_{em} = \int_{\Omega^e} \bar{B}^{\mathrm{T}} e^{\mathrm{T}} B \mathrm{d}\Omega, \quad K_{et} = \int_{\Omega^e} \bar{B}^{\mathrm{T}} \chi^{\mathrm{T}} \bar{N} \mathrm{d}\Omega, \quad K_{ee} = -\int_{\Omega^e} \bar{B}^{\mathrm{T}} \kappa \bar{B} \mathrm{d}\Omega,$$

$$K_{ec} = \int_{\Omega^e} \bar{B}^{\mathrm{T}} w^{\mathrm{T}} \tilde{N} \mathrm{d}\Omega, \quad F_e^e = \int_{\Omega^e} \bar{N}^{\mathrm{T}} q_b \mathrm{d}\Omega + \int_{S_D^e} \bar{N}^{\mathrm{T}} \bar{q}_s \mathrm{d}S$$

$$K_{cm} = \int_{\Omega^e} \tilde{N}^{\mathrm{T}} R^{\mathrm{T}} B \mathrm{d}\Omega, \quad K_{ct} = -\int_{\Omega^e} \tilde{N}^{\mathrm{T}} \nu^{\mathrm{T}} \bar{N} \mathrm{d}\Omega, \quad K_{ce} = \int_{\Omega^e} \tilde{N}^{\mathrm{T}} w \bar{B} \mathrm{d}\Omega$$

$$K_{cc} = \int_{\Omega^e} (\tilde{N}^{\mathrm{T}} S \tilde{N} + \tau \tilde{B}^{\mathrm{T}} \varphi \tilde{B}) \mathrm{d}\Omega, \quad F_c^e = -\int_{S_\xi^e} \tau \tilde{N}^{\mathrm{T}} \bar{\xi}_n \mathrm{d}S$$

The FE equation (5.5.6) is for element e only. The whole linear system can be obtained by a regular assembly process of elemental stiffness equations.

5.6   Chemo-mechanical coupling

When no external electric field is applied and the electric potential induced by ions is ignored, the problem can be significantly simplified by omitting the electrical effect. In addition, when the heat conduction of the system under consideration is not induced by stress and ionic diffusion, or in other words when the temperature field is not affected by stress and a chemical field, the temperature field can affect other fields through the constitutive relation only, and conversely, the other fields have no effect on the temperature field. In this section, a coupled chemical and mechanical problem is considered to illustrate more clearly how the chemical field can be coupled with the other fields. For this purpose, consider a time-dependent coupled chemical and mechanical problem. The governing equations are

$$\varphi^{\pm}c_{,ii}^{\pm} - \frac{1}{\tau}\dot{\mu}^{\pm} = 0, \quad \sigma_{ij,j} + f_i = 0 \tag{5.6.1}$$

where the inertial effect is neglected. The natural boundary conditions are

$$\varphi^{\pm}c_{,i}^{\pm}n_i = -\overline{\xi}_n^{\pm}, \quad \sigma_{ij}n_j - \overline{t}_i = 0 \tag{5.6.2}$$

The generalized variational principle for the chemo-mechanical coupling is then given by

$$\delta\int_{\Omega}(\dot{g} + J)\,\mathrm{d}\Omega - \int_{\Omega}f_i\delta\dot{u}_i\mathrm{d}\Omega -$$
$$\int_{S_t}\overline{t}_i\delta\dot{u}_i\mathrm{d}S - \sum_{\alpha=+,-}\int_{S_\xi}\tau\overline{\xi}_n^{\alpha}\delta c^{\alpha}\mathrm{d}S = 0 \tag{5.6.3}$$

where g is Gibb's free energy for chemo-mechanical coupling problems, and J is chemical dissipation energy

$$\delta g = \sigma_{ij}\delta\varepsilon_{ij} + \mu\delta c \tag{5.6.4a}$$

$$J = -\frac{1}{2}\tau\sum_{\alpha=+,-}\xi_i^{\alpha}c_{,i}^{\alpha} = \frac{1}{2}\tau\sum_{\alpha=+,-}\varphi^{\alpha}c_{,i}^{\alpha}c_{,i}^{\alpha} \tag{5.6.4b}$$

Substituting Eq. (5.6.4) into Eq. (5.6.3), we have

$$\delta\Pi = \int_{\Omega}\left(\sigma_{ij}\delta\dot{\varepsilon}_{ij} + \mu\delta c + \tau\sum_{\alpha=+,-}\varphi^{\alpha}c_{,i}^{\alpha}\delta c_{,i}^{\alpha}\right)\mathrm{d}\Omega - \int_{\Omega}f_i\delta\dot{u}_i\mathrm{d}\Omega -$$
$$\int_S\overline{t}_i\delta\dot{u}_i\mathrm{d}S + \sum_{\alpha=+,-}\int_S\tau\overline{\xi}_n^{\alpha}\delta c^{\alpha}\mathrm{d}S = 0 \tag{5.6.5}$$

After some mathematical operations, the equation becomes

$$\delta\Pi = \int_{\Omega}(\sigma_{ij} + f_i)\delta\dot{u}_i\mathrm{d}\Omega + \int_S(\sigma_{ij}n_j - \overline{t}_i)\delta\dot{u}_i\mathrm{d}S +$$
$$\sum_{\alpha=+,-}\int_{\Omega}(-\tau\varphi^{\alpha}c_{,ii}^{\alpha} + \dot{\mu})\delta c^{\alpha}\mathrm{d}\Omega + \sum_{\alpha=+,-}\int_S\tau(\varphi^{\alpha}c_{,i}^{\alpha}n_i + \xi_n^{\alpha})\delta c^{\alpha}\mathrm{d}S = 0 \tag{5.6.6}$$

Equation (5.6.6) is equivalent to Eqs. (5.6.1) and (5.6.2). Substitution of constitutive laws into Equation (5.6.5) yields

$$\delta\Pi = \int_{\Omega}\left(C_{ijkl}\varepsilon_{ij}\delta\dot{\varepsilon}_{kl} - \sum_{\alpha=+,-}R_{ij}^{\alpha}c^{\alpha}\delta\dot{\varepsilon}_{kl} - \sum_{\alpha=+,-}R_{kl}^{\alpha}\dot{\varepsilon}_{kl}\delta c^{\alpha} + \sum_{\alpha=+,-}s^{\alpha}\dot{c}^{\alpha}\delta c^{\alpha}\right)\mathrm{d}\Omega +$$
$$\int_{\Omega}\tau\sum_{\alpha=+,-}\varphi^{\alpha}c_{,i}^{\alpha}\delta c_{,i}^{\alpha}\mathrm{d}\Omega - \int_{\Omega}f_i\delta\dot{u}_i\mathrm{d}\Omega - \int_S\overline{t}_i\delta\dot{u}_i\mathrm{d}S + \sum_{\alpha=+,-}\int_S\tau\overline{\xi}_n^{\alpha}\delta c^{\alpha}\mathrm{d}S = 0$$
$$\tag{5.6.7}$$

The FE discrete form of independent variables u and c can read

$$u = \sum_{i=1}^{n} N_i u_i = N u^e, \quad \varepsilon = L u = L N u^e = B u^e \tag{5.6.8a}$$

$$c = \sum_{i=1}^{n} \tilde{N}_i c_i = \tilde{N} c^e, \quad c_{,i} = \sum_{i=1}^{n} \tilde{N}_{i,i} c_i = \tilde{B} c^e \tag{5.6.8b}$$

Thus we obtain

$$\delta \Pi_e = \int_{\Omega_e} (\delta \dot{u}^{eT} B^T C B u^e - \delta \dot{u}^{eT} B^T R \tilde{N} c^e - \delta c^{eT} \tilde{N}^T R B \dot{u}^e + \delta c^{eT} \tilde{N}^T S \tilde{N} \dot{c}^e) \, \mathrm{d}\Omega +$$

$$\int_{\Omega_e} \tau \delta c^{eT} \tilde{B}^T \varphi \tilde{B} c^e \mathrm{d}\Omega - \int_{\Omega_e} \delta \dot{u}^{eT} N^T f \mathrm{d}\Omega - \int_{S_e} \delta \dot{u}^{eT} N^T \bar{t} \mathrm{d}S + \int_{S_e} \tau \delta c^{eT} \tilde{N}^T \bar{\xi} \, \mathrm{d}S = 0$$

$$\tag{5.6.9a}$$

That is

$$\delta \Pi = \delta \dot{u}^{eT} \left[\int_{\Omega} \left(B^T C B u^e - B^T R \tilde{N} c^e \right) \mathrm{d}\Omega - \int_{\Omega} N^T f \mathrm{d}\Omega - \int_{S} N^T \bar{t} \mathrm{d}S \right] +$$

$$\delta c^{eT} \left(\int_{\Omega} \tilde{N}^T R B \dot{u}^e + \tilde{N}^T S \tilde{N} \dot{c}^e + \tau \tilde{B}^T \varphi \tilde{B} c^e + \int_{S} \tau \tilde{N}^T \bar{\xi} \, \mathrm{d}S \right) = 0$$

$$\tag{5.6.9b}$$

The above equation leads to the following FE formulations

$$\begin{bmatrix} 0 & 0 \\ K_{mc}^T & M_c \end{bmatrix} \begin{bmatrix} \dot{u} \\ \dot{c} \end{bmatrix} + \begin{bmatrix} K_{mm} & K_{mc} \\ 0 & K_{cc} \end{bmatrix} \begin{bmatrix} u \\ c \end{bmatrix} = \begin{bmatrix} F_m \\ F_c \end{bmatrix} \tag{5.6.10}$$

in which the coefficient matrices have the following forms

$$M_c = \sum_e \int_{\Omega_e} \tilde{N}^T S \tilde{N} \mathrm{d}\Omega \tag{5.6.11}$$

$$K_{mm} = \sum_e \int_{\Omega_e} B^T C B \mathrm{d}\Omega$$

$$K_{mc} = K_{cm}^T = -\sum_e \int_{\Omega_e} B^T R \tilde{N} \mathrm{d}\Omega$$

$$K_{cc} = \sum_e \int_{\Omega_e} \tau \tilde{B}^T \varphi \tilde{B} \mathrm{d}\Omega \tag{5.6.12}$$

The equivalent nodal force vectors are

$$F_m^e = \sum_e \int_{\Omega_e} N^T f \mathrm{d}\Omega + \int_{S_e} N^T \bar{t} \mathrm{d}S, \quad F_c^e = -\sum_e \int_{S_e} \tau \tilde{N}^T \xi_n^e \mathrm{d}S \tag{5.6.13}$$

It should be mentioned that the material parameters used in coupled constitutive equations are classified into two sets. One set covers "stiffness" coefficients which reflect the strength of the fields such as Young's modulus, dielectric coefficient, etc. The another set consists of the coupled coefficients between fields, representing the interaction between fields such as piezoelectric coefficients, chemo-mechanical coefficients, etc. In principle, all the material parameters should be determined by experiments. But we can roughly estimate the ranges of some parameters by a theoretical method. Since the material properties of thermo-electro-mechanical media are well known [28], the following discussion focuses on the chemo-mechanical coupling parameters.

For material parameters of chemo-mechanical media, the mechanical stiffness matrix can be calculated through Young's modulus and Poisson ratio, and the diffusion coefficient can be measured by the physicochemical method [28]. Here the new parameters to be determined are the proportional coefficients of chemical potential and concentration, and the coupled coefficient of the chemical and mechanical fields.

At first, we consider the linearly proportional coefficients of chemical potential and concentration.

In our study the linear relation $\mu^{\pm} = s^{\pm} c^{\pm}$ is assumed. According to the physico-chemical theory, $c^{\pm}$ has the dimension of $\mathrm{mol} \cdot \mathrm{m}^{-3}$ and $\mu^{\pm}$ has the dimension of $\mathrm{J} \cdot \mathrm{m}^{-3}$, thus $s^{\pm}$ should have the dimension of $\mathrm{J} \cdot \mathrm{mol}^{-1}$.

We use the same notation to write Eq. (5.2.10) as

$$\mu^{\pm} = \frac{RT^{*}}{\overline{V}^{\pm}} \ln \overline{c}^{\pm} = \frac{RT^{*}}{\overline{V}^{\pm}} \ln(c_0^{\pm} + c^{\pm}) \qquad (5.6.14)$$

Then we have

$$\mathrm{d}\mu^{\pm} = s^{\pm} \mathrm{d}c^{\pm} = \frac{RT^{*}}{\overline{V}^{\pm}} \frac{\mathrm{d}c^{\pm}}{c_0^{\pm} + c^{\pm}} \qquad (5.6.15)$$

Considering an infinitesimal change in the concentration, we have

$$s^{\pm} = \frac{RT^{*}}{\overline{V}^{\pm}} \lim_{c^{\pm} \to 0} \frac{1}{c_0^{\pm} + c^{\pm}} = \frac{RT^{*}}{\overline{V}^{\pm} c_0^{\pm}} \qquad (5.6.16)$$

If the temperature change is very small, i.e., $T^{*} = T_0 + T \approx T_0$, we have

$$s^{\pm} = \frac{RT_0}{\overline{V}^{\pm} c_0^{\pm}} \qquad (5.6.17)$$

Since the universal gas constant R has the dimension of $\mathrm{J} \cdot \mathrm{mol}^{-1} \cdot \mathrm{K}^{-1}$, T has the dimension of K, and $\overline{V}^{\pm}$ has the dimension of $\mathrm{m}^3 \cdot \mathrm{mol}^{-1}$, the dimension of $s^{\pm}$ should be $\mathrm{J} \cdot \mathrm{mol}^{-1}$, which is consistent with the dimension in the present linear relation. The value of $s^{\pm}$ depends on the material used.

For the coupled coefficient of chemical and mechanical effects, we can use the analogous relation of chemical and heat conduction to predict its value. Considering an isotropic material and supposing that concentration changes induce only swelling and contraction, with no shear deformation of the medium, the constitutive laws for chemo-mechanical coupling can be written as

$$\sigma_{ij} = C_{ijkl}\varepsilon_{ij} - \sum_{\alpha=+,-} R_{ij}^{\alpha} c^{\alpha} = \frac{\nu E}{(1+\nu)(1-2\nu)}\varepsilon_{kk}\delta_{ij} + \frac{E}{1+\nu}\varepsilon_{ij} - \sum_{\alpha=+,-} R_0^{\alpha}\delta_{ij}c^{\alpha}$$

(5.6.18a)

$$\mu^{\pm} = R_{kl}^{\pm}\varepsilon_{kl} + s^{\pm}c^{\pm} = R_0^{\pm}\delta_{kl}\varepsilon_{kl} + s^{\pm}c^{\pm}$$

(5.6.18b)

In order to estimate the coupled coefficients $R_0^{\pm}$, setting $i = j$ in the above equations and denoting $\varepsilon_{ii} = \varepsilon_{11} + \varepsilon_{22} + \varepsilon_{33}$ as the volume strain, the volume stress is given by

$$\sigma_{ii} = \frac{E}{1-2\nu}\varepsilon_{ii} - 3R_0^{+}c^{+} - 3R_0^{-}c^{-}$$

(5.6.19)

It is assumed that the swelling and contraction of the material induced by concentration change begins from the free stress state, e.g. $\sigma_{ii} = 0$. Thus we can obtain the volume expansion coefficient

$$K^{+} = \left(\frac{\varepsilon_{ii}}{c^{+}}\right)_{\sigma=0} = \frac{3(1-2\nu)}{E}R_0^{+}$$

(5.6.20a)

$$K^{-} = \left(\frac{\varepsilon_{ij}}{c^{-}}\right)_{\sigma=0} = \frac{3(1-2\nu)}{E}R_0^{-}$$

(5.6.20b)

Then the coupled coefficient is

$$R_0^{\pm} = \frac{E}{3(1-2\nu)}K^{\pm}$$

(5.6.21)

It is noted that the dimension of E is $\mathrm{N \bullet m^{-2}}$ and dimension of $K^{\pm}$ is $\mathrm{m^3 \bullet mol^{-1}}$, the dimension of $R_0^{\pm}$ is thus $\mathrm{N \bullet m \bullet mol^{-1}}$.

5.7   FE procedure and numerical examples

The basic governing equations and FE formulations for the thermo-electric-chemo-mechanical coupling have been given in previous sections. In this section, the FE implementation and numerical examples are presented. Discussion focuses on the chemo-mechanical coupling problem.

The FE program is developed on the base of elastic FE procedure. The nodal degrees of freedom for an elastic problem are replaced by generalized degrees of freedom, including elastic displacements, electric potential and magnetic potential at the nodes, for the present coupling problem. Fig.5.1 shows the steps for analysis of coupling problems by a FE method.

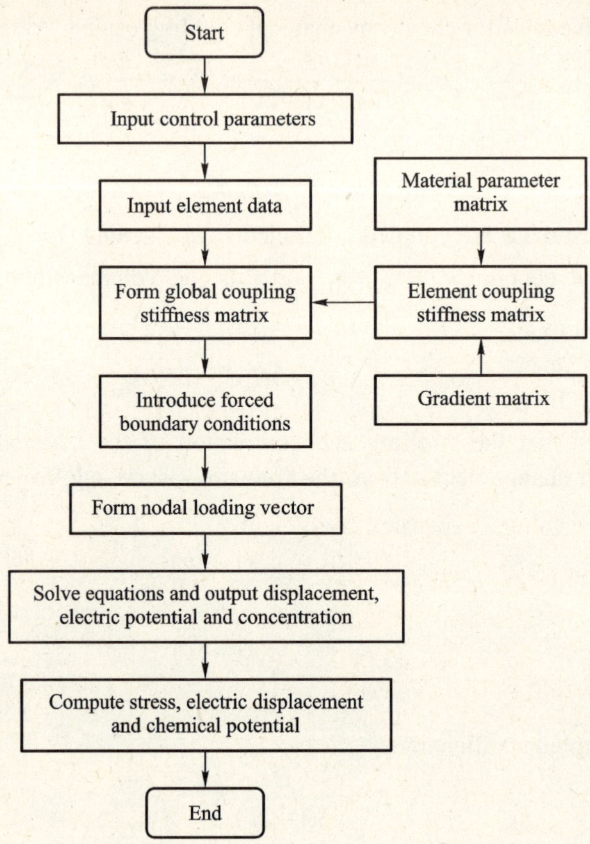

Fig.5.1   Flow sequence for the coupled finite element method

Some examples are now given to illustrate the analysis process and coupling behavior of multifield problems. Numerical simulation is carried out for the biological tissues and polymer gel.

To delineate the basic principles of the proposed multifield approach including a chemical effect, the assessment is limited to the swelling of a rectangular plate subjected to chemical load and a rectangular strip subjected to a chemical stimulus on its longer side. The two examples illustrate the coupling between chemical and mechanical fields.

The material parameters used in the calculations are as follows. The Young's Modulus is $E = 3.5 \times 10^5$ Pa ; Poisson ratio is 0.45 . The diffusion coefficients of the ions are $\phi^+ = 4.8 \times 10^{-10}$ m^2/s , $\phi^- = 7.8 \times 10^{-10}$ m^2/s . The linearly proportional coefficients of the chemical potential and ionic concentration are $s^+ = s^- = 1.0 \times 10^3$ N•m•mol^{-1}. The coupling coefficients of mechanical and chemical effect are $R_0^+ = R_0^- = 1.75 \times 10^4$ N•m•mol^{-1}.

Example 1   A square plate subjected to chemical load.

Consider the swelling of a square plate of 0.01 m$\times 0.01$ m caused by a chemical load. It is assumed that the plate is body force free and traction free. The lower boundary is completely constricted (no displacement on this side) and the other sides are free of surface traction. An increment linear distribution of ionic concentration from maximum on the upper boundary to zero on the lower boundary is applied to the plate. Thus the distribution of the ions can be completely determined. The concentration of the ions varies linearly vary in the x direction and is evenly distributed in the y directions. Obviously, the same distributions are assumed here for the anions and cations.

The deformation of the plate obtained from the proposed formulation is shown in Fig. 5.2. It is evident that a volume swelling of the plate occurs under the chemical stimuli, while the degree of swelling varies from point to point. Due to the linear distribution of ionic concentration, the maximum swelling occurs on the upper boundary and no swelling occurs on the lower boundary. It is evident that chemical swelling of the medium is very similar to heat expansion where pure volume expansion is produced under thermal load. The variations of maximum displacement versus ionic concentration are shown in Fig.5.3. The displacements increase linearly along with an increase in the ionic concentration. It is noted that the maxima of displacements u_x and u_y do not appear at the same point. The maximum of u_x occurs at the ends of the upper boundary, while the maximum of u_y is at the center of the upper boundary. It is shown that the maxima of displacements u_x and u_y are very different

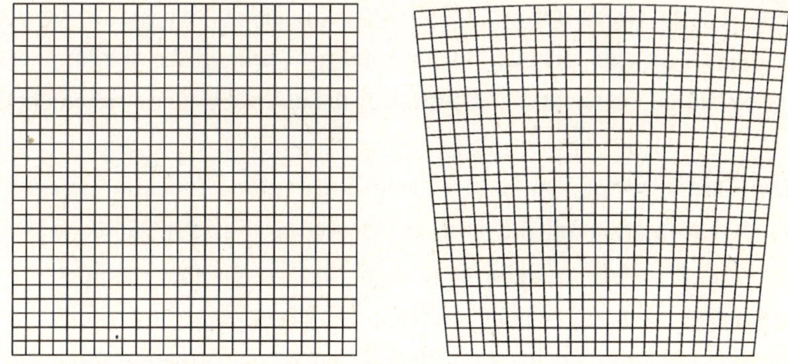

Fig.5.2   Swelling of a rectangular plate under chemical load

because of the difference of ionic distribution in two directions. On the upper boundary, even distribution of the ions with maximum values occurs in the x direction and consequently relatively large displacement in the x direction is found. Symmetry of the deformation is also exhibited in the x direction. In contrast, the ionic distribution in the y direction varies linearly, and the maximum of u_y is the accumulation of displacement in the y direction. Therefore, a comparatively small maximum displacement occurs in the y direction.

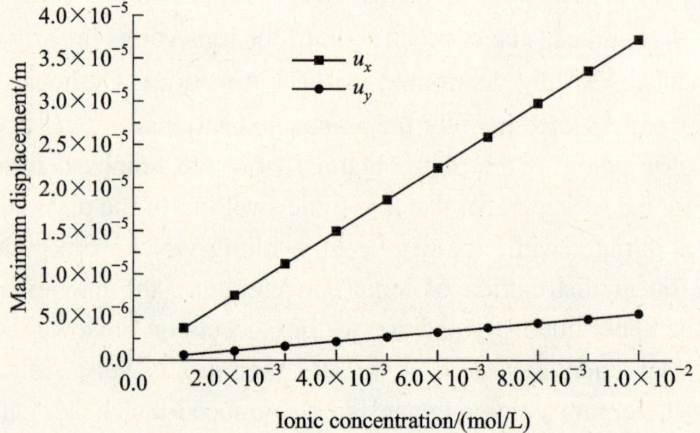

Fig.5.3   Variations of maximum displacement versus ionic concentrations

Example 2   A rectangular strip subjected to a chemical stimulus on its longer side.

Consider a 0.01 m×0.004 m rectangular strip with a displacement-free boundary on the lower side and traction-free boundaries on the remaining sides. A chemical stimulus is applied to the strip on one of its longer sides. The unique interest in this problem is that deformation of the medium can be affected by the geometry and constraints of the sample under applied chemical load.

The ionic diffusion results in a linear distribution of ions in the strip. The calculated deformation of the strip is shown in Fig. 5.4. In contrast with Example 1, a bending deformation of the strip appears, due to the non-symmetric swelling which leads to different expansions on the two opposite sides of the strip. It is concluded that the deformation model for the whole of the sample can be controlled by applying proper constraints on the boundaries. The same phenomenon for a gel fiber has already been demonstrated in [23]. This example illustrates the capability of the present theory for modelling the deformation of coupled media under chemical stimulus.

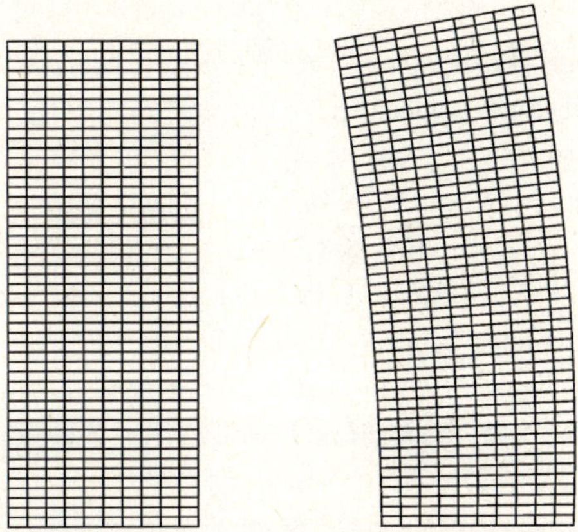

Fig.5.4   Bending deformation of a strip under chemical load

A theoretical model and FE formulation were developed in this chapter, based on the proposed governing equations of coupled thermal, electrical, chemical and mechanical fields. Using the proposed four-field equations a variational principle for deriving the FE formulation can be easily be constructed. Coupling between the chemical field and the other fields is enforced. Thus the resulting FE procedure is fully coupled in terms of the four fields. Two numerical examples were considered to illustrate the application of the FEM and to verify the proposed theory.

A linearly constitutive relation was obtained using the concept of extended Gibb's free energy. The materials parameters for the chemo-mechanical coupling problem were discussed from theoretical estimations and their dimensions. It was shown that material parameters in the present linear model are coordinated with the physical constants in classical physical chemistry.

References

[1]   Baughman R H. Conducting polymer artificial muscles. Synthetic Metals, 1996, 78: 339-353.

[2]   Garard M, Chaubey A, Malhotra B D. Application of conducting   polymers to

biosensors. Biosensor & Bioelectronics, 2002, 17(5): 345-359.

[3] MacDiarmid A G. Synthetic metals: a novel role for organic polymers. Synthetic Metals, 2002, 125: 11-22.

[4] Yoseph B C. Electroactive polymers as artificial muscles-capabilities, potentials and challenges[EB/OL]. http://trs-new.jpl.nasa.gov/dspace/bitstream/2014/37531/1/05-0308.pdf.

[5] Otero T F, Grande H, Rodriguez J. A new model for electrochemical oxidation of polypyrrole under conformational relaxation control. J. Electroanalytical Chemistry, 1995, 394: 211-216.

[6] Doi M, Matsumoto M, Hirose Y. Deformation of ionic polymer gels by electric fields. Macromolecules, 1992, 25, 5504-5511.

[7] Santa A D, Rossi D D, Mazzoldi A. Performance and work capacity of a polypyrrole conducting polymer linear actuator. Synthetic Metals, 1997, 90: 93-100.

[8] Simom B R, Kaufman M V, Liu J, et al. Porohyperelatic-transport-swelling theory, material properties and finite element models for large arteries. Int. J. Solides Structures, 1998, 35(34-35): 5021-5031.

[9] Garikipatia K, Arrudab E M, Groshc K, et al. A continuum treatment of growth in biological tissue: the coupling of mass transport and mechanics. Journal of the Mechanics and Physics of Solids, 2004, 52: 1595-1625.

[10] Loret B, Simões F M F. A framework for deformation, generalized diffusion, mass transfer and growth in multi-species multi-phase biological tissues. European Journal of Mechanics A/Solids, 2005, 24: 757-781.

[11] de Boer R. Reflections on the development of the theory of porous media. Appl. Mech. Rev., 2003, 56(6): 27-42.

[12] de Boer, R. Theory of Porous Media: Highlights in the Historical Development and Current State. Berlin: Springer, 2000.

[13] Biot M A. Theory of elasticity and consolidation for a porous anisotropic solid. J. Appl Phys., 1995, 26: 182-185.

[14] Bowen R M. Compressible porous media models by use of the theory of mixtures. Int. J. Eng. Sci., 1982, 20: 697-735.

[15] Lai W M, Hou J S, Mow V C A. A triphasic theory for the swelling and deformation behaviors of articular cartilage. ASME J. Biomech. Eng, 1991, 113: 245-258.

[16] Snijders H, Huyghe J M, Janssen J D. Triphasic finite element model for swelling porous media.Int. J. Num. Meth. Fluids, 1995, 20: 1039-1046.

[17] Huyghe J M, Janssen J D. Quadriphasic mechanics of swelling incompressible porous media. Int. J. Engng Sci, 1997, 18: 793-802.

[18] Frijns A J H, Huyghe J M, Janssen J D. A validation of the quadriphasic mixture theory for intervertebral disc tissue. Int. J. Engng Sci, 1997, 35(15): 1419-1429.

[19] Huyghe J M, Janssen J D. Thermo-chemo-electro-mechanical formulation of saturated charged porous solid. Transport in Porous Media, 1999, 34: 129-141.

[20] Moyne C, Murad M A. Electro-chemo-mechanical coupling in swelling clays derived from a micro/macro-homogenization procedure. Int. J. Solids and Strut., 2002, 39: 6159-6190.

[21] Loret B, Hueckel T, Gajo A. Chemo-mechanical coupling in saturated porous media: elasto-plastic behavior of homoionic expansive clays. Int. J. Solids and Strut., 2002, 39: 2773-2806.

[22] Sudipto K D,Aluru N R. A chemo-electro-mechanical mathematical model for simulation of pH sensitive hydrogels. Mechanics of materials, 2004, 36: 395-410.

[23] Wallmersperger T, Kroplin B, and Gulch R. Coupled chemo-electro-mechanical formulation for ionic polymer gels numerical and experimental investigations. Mechanics of Materials, 2004, 36: 411-420.

[24] Gajo A, Loret B. Finite element simulations of chemo-mechanical coupling in elastic-plastic homoionic expansive clays. Comput. Methods Appl. Mech. Engrg, 2003, 192: 3489-3530.

[25] Yang Qingsheng, Cui C Q, Lu X Z. A general procedure for modeling physicochemical coupling behavior of advanced materials—part I: theory. MMMS—Multidisciplinary Modeling in Materials and Structures, 2005, 1(3): 223-230.

[26] Yang Qingsheng, Cui C Q, Lu X Z. A MFE model for thermo-electro-chemo-mechanical coupled problem//Proc. WCCM6, Sept. 5-10, Beijing, China, 2004.

[27] Zohdi T I. Modeling and simulation of a class of coupled thermo-chemo-mechanical processes in multiphase solids. Comput. Methods Appl. Mech. Engrg, 2004, 193: 679-699.

[28] Levine I N. Physical Chemistry. 5th ed. New York: McGraw Hill, 2002.

[29] Qin Qinghua. Fracture mechanics of piezoelectric materials. Southampton: WIT press, 2001.

[30] Mason W P. Piezoelectric crystals and their application to ultrasonics. Toronto: D. Van Nostrand Company, Inc., 1950.

Chapter 6 Thermo-electro-elastic bone remodelling

6.1 Introduction

In Chapters 3 and 4, the multi-field theories of thermo-electro-elastic and thermomagneto-electro-elastic problems were presented. Applications of the theory to bone remodelling are described in this chapter. Bone is a kind of dynamically adaptable material. Like any other living system, it has mechanisms for repair and growth or remodelling, and mechanisms to feed its constituent parts and ensure that any materials needed for structural work are supplied to the correct area as and when required. These bone functions are performed via three types of bone cell: osteoblast, osteoclast, and osteocyte. Osteoblasts are cells that form new bone and are typically found lining bone surfaces that are undergoing extensive remodelling. Osteoclasts are large, multinucleated, bone-removing cells. Their function is to break down and remove bone material that is no longer needed or that has been damaged in some way. The third cell type is the osteocyte. Osteocytes, called the bone "sensor cells", are responsible for sensing the physical environment to which the skeleton is subjected. Osteocytes are characterized by many protoplasmic processes, or dendrites, emanating from the cell body. These cell dendrites form a communication network with surrounding cells, other osteocytes, osteoblasts, and possibly osteoclasts, which passes the signals from the osteocytes that control the action of osteoblasts osteoclasts. The activities of these three cell populations, and numerous other biological and biochemical factors, are coordinated in a continuous process throughout our lives to maintain a strong, healthy skeleton system.

It should be noted that applications of the multi-field theory to bone remodelling have been the subject of fruitful scientific attention by many distinguished researchers (e.g. Fukada and Yasuda [1,2], Kryszewski [3], Robiony [4], Qin and Ye [5] and others). Early in the 1950s, Fukada and Yasuda [1, 2] found that some living bone and collagen exhibit piezoelectric behaviour. Later, Gjelsvik [6] presented a physical description of the remodelling of bone tissue, in terms of a very simplified form of linear theory of piezoelectricity. Williams and Breger [7] explored the applicability of stress gradient theory for explaining the experimental data for a cantilever bone beam subjected to constant end load, showing that the approximate gradient theory was in good agreement with the experimental data. Guzelsu [8] presented a piezoelectric model for analysing a cantilever dry bone beam subjected to a vertical end load. Johnson [9] et al. further addressed the problem of a dry bone beam by presenting some theoretical expressions for the piezoelectric response to cantilever bending of the beam. Demiray [10] provided some theoretical descriptions of electro-mechanical remodelling models of bones. Aschero [11] et al. investigated the converse piezoelectric effect of fresh bone using a highly sensitive dilatometer. They further investigated the piezoelectric properties of bone and presented a set of repeated measurements of coefficient d_{23} on 25 cow bone samples [12]. Fotiadis [13] et al. studied wave propagation in a long cortical piezoelectric bone with arbitrary cross-section. El-Naggar and Abd-Alla [14], and Ahmed and Abd-Alla [15] further obtained an analytical solution for wave propagation in long cylindrical bones with and without cavity. Silva [16] et al. explored the physicochemical, dielectric and piezoelectric properties of anionic collagen and collagen-hydroxyapatite composites. Recently, Qin and Ye [5], and Qin [17] et al. presented a thermo-electro-elastic solution for internal and surface bone remodelling, respectively. Accounts of most of the developments in this area can also be found in [3, 18]. In this chapter, however, we restrict our discussion to the findings presented in [5, 17, 18, 27].

6.2   Thermo-electro-elastic internal bone remodelling

6.2.1   Linear theory of thermo-electro-elastic bone

Consider a hollow circular cylinder composed of linearly thermo- piezoelectric bone material subjected to axisymmetric loading. The axial, circumferential and normal to the middle-surface co-ordinate length parameters are denoted by z, θ and r, respectively. Using the cylindrical coordinate system, the constitutive equations (3.6.6) can be rewritten in the form [5,19]

$$\sigma_{rr} = c_{11}\varepsilon_{rr} + c_{12}\varepsilon_{\theta\theta} + c_{13}\varepsilon_{zz} - e_{31}E_z - \lambda_{11}T$$

$$\sigma_{\theta\theta} = c_{12}\varepsilon_{rr} + c_{11}\varepsilon_{\theta\theta} + c_{13}\varepsilon_{zz} - e_{31}E_z - \lambda_{11}T$$

$$\sigma_{zz} = c_{13}\varepsilon_{rr} + c_{13}\varepsilon_{\theta\theta} + c_{33}\varepsilon_{zz} - e_{33}E_z - \lambda_{33}T$$

$$\sigma_{zr} = c_{44}\varepsilon_{zr} - e_{15}E_r, \quad D_r = e_{15}\varepsilon_{zr} + \kappa_{11}E_r \qquad (6.2.1)$$

$$D_z = e_{31}(\varepsilon_{rr} + \varepsilon_{\theta\theta}) + e_{33}\varepsilon_{zz} + \kappa_{33}E_z - p_3 T$$

$$h_r = k_r W_r, \qquad h_z = k_z W_z$$

where W_i is the heat intensity. The associated strains, electric fields, and heat intensities are respectively related to displacements u_i, electric potential ϕ, and temperature change T as

$$\varepsilon_{rr} = u_{r,r}, \quad \varepsilon_{\theta\theta} = \frac{u_r}{r}, \quad \varepsilon_{zz} = u_{z,z} \quad \varepsilon_{zr} = u_{z,r} + u_{r,z} \qquad (6.2.2)$$

$$E_r = -\phi_{,r} \quad E_z = -\phi_{,z}, \quad W_r = -T_{,r}, \quad W_z = -T_{,z}$$

For quasi-stationary behaviour, in the absence of a heat source, free electric charge and body forces, the set of equations for thermo-piezoelectric theory of bones is completed by adding the following equations of equilibrium for heat flow, stress and electric displacements to Eqs. (6.2.1) and (6.2.2).

$$\frac{\partial \sigma_{rr}}{\partial r} + \frac{\partial \sigma_{zr}}{\partial z} + \frac{\sigma_{rr} - \sigma_{\theta\theta}}{r} = 0, \qquad \frac{\partial \sigma_{zr}}{\partial r} + \frac{\partial \sigma_{zz}}{\partial z} + \frac{\sigma_{zr}}{r} = 0$$

$$\frac{\partial D_r}{\partial r} + \frac{\partial D_z}{\partial z} + \frac{D_r}{r} = 0, \qquad \frac{\partial h_r}{\partial r} + \frac{\partial h_z}{\partial z} + \frac{h_r}{r} = 0 \qquad (6.2.3)$$

6.2.2 Adaptive elastic theory

Adaptive theory is used to model the normal adaptive processes that occur in bone remodelling as strain controlled mass deposition or resorbtion processes which modify the porosity of the porous bone material [20]. In the adaptive elastic constitutive equation presented in [20], the authors introduced an independent variable which is a measure of the volume fraction of the matrix structure. Let ξ denote the volume fraction of the matrix material in an unstrained reference state and assume that the density of the material composing the matrix is constant. Thus the conservation of mass will give the equation governing ξ. Then an important constitutive assumption was made [20] that, at constant temperature and zero body force, there exists a unique zero-strain reference state for all values of ξ. Thus ξ may change without changing the reference state for strain. One might imagine a block of porous elastic material with the four points, the vertices of a tetrahedron, marked on the block for the purpose of measuring the strain. When the porosity changes, material is added or taken away from the pores, but if the material is unstrained it remains so and the distance between the four vertices marked on the block do not change. Thus ξ can change while the zero-strain reference state remains the same. Keeping this in mind, a formal definition of the remodelling rate $\dot{e}$ (the rate at which mass per unit volume is added to or removed from the porous matrix structure) and free energy Ψ can be given [20].

$$\dot{e} = \dot{e}(\phi, F), \quad \Psi = \Psi(\phi, F) \tag{6.2.4}$$

where ϕ is the volume fraction of the matrix, and F stands for deformation gradient. More detailed discussion of this formulation (6.2.4) is found in [20]. Considering the adaptive property discussed above, the traditional elastic stress-strain relationship becomes [20]

$$\sigma_{ij} = (\phi_0 + e)C_{ijkl}(e)\varepsilon_{kl}, \quad \dot{e} = A^*(e) + A_{ij}(e)\varepsilon_{ij} \tag{6.2.5}$$

where ξ_0 is a reference volume fraction of bone matrix material, e is a change in the volume fraction of bone matrix material from its reference value ξ_0, $C_{ijkm}(e)$ is the stiffness matrix dependent upon the volume fraction change e, and $A^*(e)$ and $A_{ij}(e)$ are material constants also dependent upon the volume fraction change e. Eq. (6.2.5) is deduced from mass balance considerations. When e is very small, Eq. (6.2.5) can be approximated by a simple form

$$\dot{e} = C_0 + C_1 e + C_2 e^2 + (A_{ij}^0 + eA_{ij}^1)\varepsilon_{ij} \tag{6.2.6}$$

where C_0, C_1, C_2, A_{ij}^0, and A_{ij}^1 are material constants. When ξ_0 is one and e is zero the stress-strain relation (6.2.5) is reduced to Hooke's law for a solid elastic material. In this situation all the pores of the bone matrix would be completely filled with bone material.

The bone remodelling equation (6.2.5) can be extended to include the effect of thermal and electric fields by introducing some new terms as [5]

$$\dot{e} = A^*(e) + A_r^E(e)E_r + A_z^E(e)E_z + A_{rr}^\varepsilon(e)(\varepsilon_{rr} + \varepsilon_{\theta\theta}) + A_{zz}^\varepsilon(e)\varepsilon_{zz} + A_{rz}^\varepsilon(e)\varepsilon_{rz} \tag{6.2.7}$$

where $A_i^E(e)$ and $A_{ij}^\varepsilon(e)$ are material coefficients dependent upon the volume fraction e. Eqs.(6.2.1)~(6.2.3) together with Eq.(6.2.7) form the basic set of equations for the adaptive theory of internal piezoelectric bone remodelling.

6.2.3 Analytical solution of a homogeneous hollow circular cylindrical bone

We now consider a hollow circular cylinder of bone subjected to an external temperature change T_0, a quasi-static axial pressure load P, an external pressure p and an electric load ϕ_a (or/and ϕ_b). The boundary conditions are

$$T = 0, \quad \sigma_{rr} = \sigma_{r\theta} = \sigma_{rz} = 0, \quad \phi = \phi_a, \quad \text{at } r = a$$
$$T = T_0, \quad \sigma_{rr} = -p, \quad \sigma_{r\theta} = \sigma_{rz} = 0, \quad \phi = \phi_b, \quad \text{at } r = b \tag{6.2.8}$$

and

$$\int_S \sigma_{zz} \, dS = -P \tag{6.2.9}$$

where a and b denote, respectively, the inner and outer radii of the bone, and S is the cross-sectional area. For a long bone, it is assumed that all displacements, temperature and electrical potential except the axial displacement u_z are independent of the z coordinate and that u_z may have linear dependence on z. Using (6.2.1) and (6.2.2), differential equations (6.2.3) can be written as

$$\left(\frac{\partial^2}{\partial r^2} + \frac{1}{r} \frac{\partial}{\partial r} \right) T = 0, \quad c_{11} \left(\frac{\partial^2}{\partial r^2} + \frac{1}{r} \frac{\partial}{\partial r} - \frac{1}{r^2} \right) u_r = \lambda_{11} \frac{\partial T}{\partial r} \tag{6.2.10}$$

$$c_{44} \left(\frac{\partial^2}{\partial r^2} + \frac{1}{r} \frac{\partial}{\partial r} \right) u_z + e_{15} \left(\frac{\partial^2}{\partial r^2} + \frac{1}{r} \frac{\partial}{\partial r} \right) \phi = 0 \tag{6.2.11}$$

$$e_{15} \left(\frac{\partial^2}{\partial r^2} + \frac{1}{r} \frac{\partial}{\partial r} \right) u_z - \kappa_{11} \left(\frac{\partial^2}{\partial r^2} + \frac{1}{r} \frac{\partial}{\partial r} \right) \phi = 0 \tag{6.2.12}$$

The solution to the heat conduction equation (6.2.10) satisfying boundary conditions (6.2.8) can be written as

$$T = \frac{\ln(r/a)}{\ln(b/a)} T_0 \tag{6.2.13}$$

It is easy to prove that eqs.(6.2.10)~(6.2.12) will be satisfied if we assume

$$u_r = A(t)r + \frac{B(t)}{r} + \frac{\bar{\omega} r \left[\ln(r/a) - 1 \right]}{c_{11}} \tag{6.2.14}$$

$$u_z = zC(t) + D(t)\ln(r/a), \quad \phi = F(t)\ln(r/a) + \phi_a \tag{6.2.15}$$

where A, B, C, D and F are unknown variables to be determined by introducing boundary conditions, and $\bar{\omega} = \dfrac{\lambda_{11}T_0}{2\ln(b/a)}$. Substituting Eqs.(6.2.14) and (6.2.15)

into Eq. (6.2.2), and later into Eq.(6.2.1), we obtain

$$\sigma_{rr} = A(t)(c_{11}+c_{12}) - \frac{B(t)}{r^2}(c_{11}-c_{12}) + c_{13}C(t) + \bar{\omega}\left[\frac{c_{12}}{c_{11}}\left(\ln\frac{r}{a}-1\right)-\ln\frac{r}{a}\right]$$

$$(6.2.16)$$

$$\sigma_{\theta\theta} = A(t)(c_{11}+c_{12}) + \frac{B(t)}{r^2}(c_{11}-c_{12}) + c_{13}C(t) + \bar{\omega}\left[\frac{c_{12}}{c_{11}}\ln\frac{r}{a}-\ln\frac{r}{a}-1\right]$$

$$(6.2.17)$$

$$\sigma_{zz} = 2A(t)c_{13} + c_{33}C(t) + \bar{\omega}\frac{c_{13}}{c_{11}}\left[2\ln(r/a)-1\right] - \lambda_{33}T_0\frac{\ln(r/a)}{\ln(b/a)} \qquad (6.2.18)$$

$$\sigma_{zr} = \frac{1}{r}\left[c_{44}D(t)+e_{15}F(t)\right], \quad D_r = \frac{1}{r}[e_{15}D(t)-\kappa_{11}F(t)] \qquad (6.2.19)$$

$$D_z = 2A(t)e_{31} + C(t)e_{33} + \bar{\omega}\frac{e_{31}}{c_{11}}\left[2\ln(r/a)-1\right] - p_3T_0\frac{\ln(r/a)}{\ln(b/a)} \qquad (6.2.20)$$

The boundary conditions (6.2.8) and (6.2.9) of stresses and electric potential require that

$$c_{44}D(t)+e_{15}F(t)=0, \quad \phi_b = F(t)\ln(b/a)+\phi_a \qquad (6.2.21)$$

$$A(t)(c_{11}+c_{12}) - \frac{B(t)}{a^2}(c_{11}-c_{12}) + c_{13}C(t) - \frac{c_{12}}{c_{11}}\bar{\omega} = 0 \qquad (6.2.22)$$

$$A(t)(c_{11}+c_{12}) - \frac{B(t)}{b^2}(c_{11}-c_{12}) + c_{13}C(t) + \bar{\omega}\left[\frac{c_{12}}{c_{11}}\left(\ln\frac{b}{a}-1\right)-\ln\frac{b}{a}\right] = -p$$

$$(6.2.23)$$

$$\pi(b^2-a^2)\left[2A(t)c_{13}+C(t)c_{33}-F_1^*T_0\right]+F_2^*T_0 = -P \qquad (6.2.24)$$

where

$$F_1^* = \frac{1}{\ln(b/a)}\left(\frac{c_{13}\lambda_{11}}{c_{11}}-\frac{\lambda_{33}}{2}\right), \quad F_2^* = \pi b^2\left(\frac{c_{13}}{c_{11}}\lambda_{11}-\lambda_{33}\right) \qquad (6.2.25)$$

The unknown functions $A(t)$, $B(t)$, $C(t)$, $D(t)$ and $F(t)$ are readily found from Eqs.(6.2.21)~(6.2.24) as

$$A(t) = \frac{1}{F_3^*}\left\{c_{33}\beta_1^*[\beta_2^*T_0+p(t)]+\bar{\omega}\frac{c_{33}c_{12}}{c_{11}}+\frac{F_2^*T_0+P(t)}{\pi(b^2-a^2)}c_{13}-F_1^*T_0c_{13}\right\}$$

$$(6.2.26)$$

$$B(t) = \frac{a^2\beta_1^*[\beta_2^*T_0+p(t)]}{c_{11}-c_{12}} \qquad (6.2.27)$$

$$C(t) = \frac{1}{F_3^*}\left\{\left[F_1^*T_0 - \frac{F_2^*T_0 + P(t)}{\pi(b^2 - a^2)}\right](c_{11} + c_{12}) - 2c_{13}\beta_1^*[\beta_2^*T_0 + p(t)] - \frac{2c_{13}c_{12}\bar{\omega}}{c_{11}}\right\}$$

$$(6.2.28)$$

$$D(t) = -\frac{e_{15}(\phi_b - \phi_a)}{c_{44}\ln(b/a)} \qquad (6.2.29)$$

$$F(t) = \frac{\phi_b - \phi_a}{\ln(b/a)} \qquad (6.2.30)$$

where

$$F_3^* = c_{33}(c_{11} + c_{12}) - 2c_{13}^2, \quad \beta_1^* = \frac{b^2}{(a^2 - b^2)}, \quad \beta_2^* = \frac{\lambda_{11}}{2}\left(\frac{c_{12}}{c_{11}} + 1\right) \quad (6.2.31)$$

Using expressions (6.2.26)~(6.2.30), the displacements u_r, u_z and electrical potential ϕ are given by

$$u_r = \frac{r}{F_3^*}\left\{c_{33}\beta_1^*[\beta_2^*T_0 + p(t)] + \bar{\omega}\frac{c_{33}c_{12}}{c_{11}} + \frac{F_2^*T_0 + P(t)}{\pi(b^2 - a^2)}c_{13} - F_1^*T_0c_{13}\right\} +$$

$$\frac{a^2\beta_1^*[\beta_2^*T_0 + p(t)]}{r(c_{11} - c_{12})} + \frac{\bar{\omega}r[\ln(r/a) - 1]}{c_{11}} \qquad (6.2.32)$$

$$u_z = \frac{z}{F_3^*}\left\{\left[F_1^*T_0 - \frac{F_2^*T_0 + P(t)}{\pi(b^2 - a^2)}\right](c_{11} + c_{12}) - 2c_{13}\beta_1^*[\beta_2^*T_0 + p(t)] -\right.$$

$$\left.\frac{2c_{13}c_{12}\bar{\omega}}{c_{11}}\right\} - \frac{e_{15}(\phi_b - \phi_a)\ln(r/a)}{c_{44}\ln(b/a)} \qquad (6.2.33)$$

$$\phi = \frac{\ln(r/a)}{\ln(b/a)}(\phi_b - \phi_a) + \phi_a \qquad (6.2.34)$$

The strains and electric field intensity appearing in Eq.(6.2.7) can be found by substituting Eqs.(6.2.13) and (6.2.32)~(6.2.34) into Eq.(6.2.2). They are, respectively,

$$\varepsilon_{rr} = \frac{1}{F_3^*}\left\{c_{33}\beta_1^*[\beta_2^*T_0 + p(t)] + \bar{\omega}\frac{c_{33}c_{12}}{c_{11}} + \frac{F_2^*T_0 + P(t)}{\pi(b^2 - a^2)}c_{13} - F_1^*T_0c_{13}\right\} -$$

$$\frac{a^2\beta_1^*[\beta_2^*T_0 + p(t)]}{r^2(c_{11} - c_{12})} + \frac{\bar{\omega}\ln(r/a)}{c_{11}} \qquad (6.2.35)$$

$$\varepsilon_{\theta\theta} = \frac{1}{F_3^*}\left\{c_{33}\beta_1^*[\beta_2^*T_0 + p(t)] + \bar{\omega}\frac{c_{33}c_{12}}{c_{11}} + \frac{F_2^*T_0 + P(t)}{\pi(b^2 - a^2)}c_{13} - F_1^*T_0c_{13}\right\} +$$

$$\frac{a^2\beta_1^*[\beta_2^*T_0 + p(t)]}{r^2(c_{11} - c_{12})} + \frac{\bar{\omega}[\ln(r/a) - 1]}{c_{11}} \qquad (6.2.36)$$

$$\varepsilon_{zz} = \frac{1}{F_3^*}\left\{\left[F_1^*T_0 - \frac{F_2^*T_0 + P(t)}{\pi(b^2 - a^2)}\right](c_{11} + c_{12}) - 2c_{13}\beta_1^*[\beta_2^*T_0 + p(t)] - \right.$$

$$\left.\frac{2c_{13}c_{12}\bar{\omega}}{c_{11}}\right\} \tag{6.2.37}$$

$$\varepsilon_{rz} = -\frac{e_{15}(\phi_b - \phi_a)}{rc_{44}\ln(b/a)} \tag{6.2.38}$$

$$E_r = -\frac{(\phi_b - \phi_a)}{r\ln(b/a)} \tag{6.2.39}$$

Then substituting the solutions (6.2.35)~(6.2.39) into Eq.(6.2.7) yields

$$\dot{e} = A^*(e) + \frac{2A_{rr}^\varepsilon}{F_3^*}\left\{c_{33}\beta_1^*[\beta_2^*T_0 + p(t)] + \bar{\omega}\frac{c_{33}c_{12}}{c_{11}} + \frac{F_2^*T_0 + P(t)}{\pi(b^2 - a^2)}c_{13} - F_1^*T_0 c_{13}\right\} +$$

$$\frac{A_{rr}^\varepsilon\bar{\omega}[2\ln(r/a) - 1]}{c_{11}} + \frac{A_{zz}^\varepsilon}{F_3^*}\left\{\left[F_1^*T_0 - \frac{F_2^*T_0 + P(t)}{\pi(b^2 - a^2)}\right](c_{11} + c_{12}) - \right.$$

$$\left.2c_{13}\beta_1^*[\beta_2^*T_0 + p(t)] - \frac{2c_{13}c_{12}\bar{\omega}}{c_{11}}\right\} - \frac{\phi_b - \phi_a}{r\ln(b/a)}\left(A_r^E + \frac{e_{15}}{c_{44}}A_{zr}^\varepsilon\right) \tag{6.2.40}$$

Since we do not know the exact expressions of the material functions $A^*(e)$, $A_i^E(e)$, $A_{ij}^\varepsilon(e)$, c_{ij}, e_{ij}, λ_{jj}, κ_{jj} and χ_3, the following approximate forms of them, as proposed by Cowin and Van Buskirk [21] for small values of e, are used here

$$A^*(e) = C_0 + C_1 e + C_2 e^2, \quad A_i^E(e) = A_i^{E0} + eA_i^{E1}, \quad A_{ij}^\varepsilon(e) = A_{ij}^{\varepsilon 0} + eA_{ij}^{\varepsilon 1} \tag{6.2.41}$$

and

$$c_{ij}(e) = c_{ij}^0 + \frac{e}{\xi_0}(c_{ij}^1 - c_{ij}^0), \quad e_{ij}(e) = e_{ij}^0 + \frac{e}{\xi_0}(e_{ij}^1 - e_{ij}^0)$$

$$\lambda_{ii}(e) = \lambda_{ii}^0 + \frac{e}{\xi_0}(\lambda_{ii}^1 - \lambda_{ii}^0), \quad \kappa_{ii}(e) = \kappa_{ii}^0 + \frac{e}{\xi_0}(\kappa_{ii}^1 - \kappa_{ii}^0) \tag{6.2.42}$$

$$\rho_3(e) = \rho_3^0 + \frac{e}{\xi_0}(\rho_3^1 - \rho_3^0)$$

where C_0, C_1, C_2, A_i^{E0}, A_i^{E1}, $A_{ij}^{\varepsilon 0}$, $A_{ij}^{\varepsilon 1}$, c_{ij}^0, c_{ij}^1, e_{ij}^0, e_{ij}^1, λ_{ii}^0, λ_{ii}^1, κ_{ii}^0, κ_{ii}^1, ρ_3^0 and ρ_3^1 are material constants. Using these approximations the remodelling rate equation (6.2.40) can be simplified as

$$\dot{e} = \alpha(e^2 - 2\beta e + \gamma) \tag{6.2.43}$$

by neglecting terms of e^3 and the higher orders of e, where α, β and γ are constants. The solution to Eq. (6.2.43) is straightforward and has been discussed by Hegedus and Cowin [22]. For the reader's benefit, the solution process is briefly described here. Let e_1 and e_2 denote solutions to $e^2 - 2\beta e + \gamma = 0$, i.e.

$$e_{1,2} = \beta \pm (\beta^2 - \gamma)^{1/2} \tag{6.2.44}$$

When $\beta^2 < \gamma$, e_1 and e_2 are a pair of complex conjugate, the solution of Eq.(6.2.43) is

$$e(t) = \beta + \sqrt{(\gamma - \beta^2)}\,\tan\left[\alpha t\sqrt{(\gamma - \beta^2)} + \arctan\frac{\sqrt{(\gamma - \beta^2)}}{\beta - e_0}\right] \tag{6.2.45}$$

where $e=e_0$ is initial condition. When $\beta^2 = \gamma$, the solution is

$$e(t) = e_1 - \frac{e_1 - e_0}{1 + \alpha(e_1 - e_0)t} \tag{6.2.46}$$

Finally, when $\beta^2 > \gamma$, we have

$$e(t) = \frac{e_1(e_0 - e_2) + e_2(e_1 - e_0)\exp[\alpha(e_1 - e_2)t]}{(e_0 - e_2) + (e_1 - e_0)\exp[\alpha(e_1 - e_2)t]} \tag{6.2.47}$$

Since it has been proved that both solutions (6.2.45) and (6.2.46) are physically unlikely [21], we will use the solution (6.2.47) in our numerical analysis.

6.2.4   Semi-analytical solution for inhomogeneous cylindrical bone layers

The solution obtained in the previous section is suitable for analyzing bone cylinders if they are assumed to be homogeneous [21]. It can be useful if explicit expressions and a simple analysis are required. It is a fact, however, that all bone materials exhibit inhomogeneity. In particular, for a hollow bone cylinder, the volume fraction of bone matrix materials varies from the inner to the outer surface. To solve this problem we present here a semi-analytical model.

Considering Eqs.(6.2.1), (6.2.2) and (6.2.3) and assuming a constant longitudinal strain, the following first-order differential equations can be obtained [5]:

$$\frac{\partial}{\partial r}\begin{bmatrix} u_r \\ \sigma_{rr} \end{bmatrix} = \begin{bmatrix} -\dfrac{c_{12}}{c_{11}r} & \dfrac{1}{c_{11}} \\ \dfrac{\psi}{r^2} & \dfrac{c_{12}/c_{11}-1}{r} \end{bmatrix} \begin{bmatrix} u_r \\ \sigma_{rr} \end{bmatrix} +$$

$$\begin{bmatrix} -\dfrac{c_{13}}{c_{11}} \\ \dfrac{c_{13}(1-c_{12}/c_{11})}{r} \end{bmatrix} \varepsilon_{zz} + \begin{bmatrix} \dfrac{\lambda_{11}}{c_{11}} \\ \dfrac{(c_{12}/c_{11}-1)\lambda_{11}}{r} \end{bmatrix} T \qquad (6.2.48)$$

where $\psi = c_{11} - c_{12}^2/c_{11}$. In the above equation, the effect of electrical potential is absent. This is because it is independent of u_r and σ_r. The contribution of electrical field can be calculated separately as described in the previous section and then included in the remodeling rate equation.

Assuming that a bone layer is sufficiently thin, we can replace r with its mean value R, and let $r=a+s$, where $0 \leqslant s \leqslant h$, a and h are the inner radius and the thickness of the thin bone layer, respectively. Thus, Eq. (6.2.48) is reduced to

$$\frac{\partial}{\partial s}\begin{bmatrix} u_r \\ \sigma_{rr} \end{bmatrix} = \begin{bmatrix} -\dfrac{c_{12}}{c_{11}R} & \dfrac{1}{c_{11}} \\ \dfrac{\psi}{R^2} & \dfrac{c_{12}/c_{11}-1}{R} \end{bmatrix} \begin{bmatrix} u_r \\ \sigma_{rr} \end{bmatrix} +$$

$$\begin{bmatrix} -\dfrac{c_{13}}{c_{11}} \\ \dfrac{c_{13}(1-c_{12}/c_{11})}{R} \end{bmatrix} \varepsilon_{zz} + \begin{bmatrix} \dfrac{\lambda_{11}}{c_{11}} \\ \dfrac{(c_{12}/c_{11}-1)\lambda_{11}}{R} \end{bmatrix} T \qquad (6.2.49)$$

The above equation can be written symbolically as

$$\frac{\partial}{\partial s}[F] = [G][F] + [H_L] + [H_T] \qquad (6.2.50)$$

where $[G]$, $[H_L]$ and $[H_T]$ are all constant matrices.

Equation (6.2.50) can be solved analytically and the solution is [23]

$$\begin{bmatrix} u_r(s) \\ \sigma_{rr}(s) \end{bmatrix} = e^{[G]s} \begin{bmatrix} u_r(0) \\ \sigma_{rr}(0) \end{bmatrix} + \int_0^h e^{[G](s-\tau)}[\mathbf{H}_L]d\tau + \int_0^h e^{[G](s-\tau)}[\mathbf{H}_T]d\tau \qquad (6.2.51)$$

where $u_r(0)$ and $\sigma_{rr}(0)$ are, respectively, the displacement and stress at the bottom surface of the layer. Rewrite Eq.(6.2.51) as

$$[F(s)] = [D(s)][F(0)] + [D_L] + [D_T] \qquad (6.2.52)$$

The exponential matrix can be calculated as follows

$$[\boldsymbol{D}(s)] = e^{[\boldsymbol{G}]s} = \alpha_0(s)\boldsymbol{I} + \alpha_1(s)[\boldsymbol{G}] \tag{6.2.53}$$

where $\alpha_0(s)$ and $\alpha_1(s)$ can be solved from

$$\alpha_0(s) + \alpha_1(s)\beta_1 = e^{\beta_1 s}$$
$$\alpha_0(s) + \alpha_1(s)\beta_2 = e^{\beta_2 s} \tag{6.2.54}$$

In Eq.(6.2.54) β_1 and β_2 are two eigenvalues of $[\boldsymbol{G}]$, which are given by

$$\begin{bmatrix} \beta_1 \\ \beta_2 \end{bmatrix} = -\frac{1}{2R} \pm \frac{1}{2R}\sqrt{5 - 4\frac{c_{12}}{c_{11}}} \tag{6.2.55}$$

Considering now $s=h$, i.e. the external surface of the bone layer, we obtain

$$[\boldsymbol{F}(h)] = [\boldsymbol{D}(h)][\boldsymbol{F}(0)] + [\boldsymbol{D}_L] + [\boldsymbol{D}_T] \tag{6.2.56}$$

The axial stress applied at the end of the bone can be found as

$$\sigma_{zz} = c_{13}\left(1 - \frac{c_{12}}{c_{11}}\right)\frac{u}{R} + \left(c_{33} - \frac{c_{13}^2}{c_{11}}\right)\varepsilon_{zz} + \frac{c_{13}}{c_{11}}\sigma_{rr} + \left(\frac{c_{13}}{c_{11}}\lambda_{11} - \lambda_{33}\right)T \tag{6.2.57}$$

The stress problem can be solved by introducing the boundary conditions described on the top and bottom surfaces into Eq.(6.2.56) and

$$\iint_S \left[c_{13}\left(1 - \frac{c_{12}}{c_{11}}\right)\frac{u}{R} + \left(c_{33} - \frac{c_{13}^2}{c_{11}}\right)\varepsilon_{zz} + \frac{c_{13}}{c_{11}}\sigma_{rr} + \right.$$
$$\left. \left(\frac{c_{13}}{c_{11}}\lambda_{11} - \lambda_{33}\right)T \right] \mathrm{d}S = -P'(t) \tag{6.2.58}$$

where $P'(t)$ is the axial force excluding the effect of electric field.

For a thick-walled bone section or a section with variable volume fraction in the radial direction, we can divide the bone into a number of sub-layers, each of which is sufficiently thin and is assumed to be composed of a homogeneous material. Within a layer we take the mean value of volume fraction of the layer as the layer's volume fraction. As a consequence, the analysis described above for a thin and homogeneous bone can be applied here for the sub-layer in a straightforward manner. For instance, for the j-th layer, Eq. (6.2.56) becomes

$$[\boldsymbol{F}^{(j)}(h_j)] = \left[\boldsymbol{D}^{(j)}(h_j)\right][\boldsymbol{F}^{(j)}(0)]_j + [\boldsymbol{D}_L^{(j)}] + [\boldsymbol{D}_T^{(j)}] \tag{6.2.59}$$

where h_j denotes thickness of the j-th sub-layer.

Considering the continuity of displacements and transverse stresses across the interfaces between these fictitious sub-layers, we have

$$[\boldsymbol{F}^{(j)}(h_j)] = [\boldsymbol{F}^{(j+1)}(0)] \tag{6.2.60}$$

After establishing Eq.(6.2.59) for all sub-layers, the following equation can be obtained by using Eqs.(6.2.59) and (6.2.60) recursively:

$$[F(h_N)] = [D^{(N)}(h_N)][F(h_{N-1})] + [D_L^{(N)}] + [D_T^{(N)}]$$

$$= [D^{(N)}(h_N)]\{[D^{(N-1)}(h_{N-1})][F(h_{N-2})] + [D_L^{(N-1)}] + [D_T^{(N-1)}]\} + [D_L^{(N)}] + [D_T^{(N)}]$$

$$= [D^{(N)}(h_N)][D^{(N-1)}(h_{N-1})][F(h_{N-2})] +$$

$$[D^{(N)}(h_N)]\{[D_L^{(N-1)}] + [D_T^{(N-1)}]\} + [D_L^{(N)}] + [D_T^{(N)}]$$

$$= \cdots$$

$$= [D^{(N)}(h_N)][D^{(N-1)}(h_{N-1})][D^{(N-2)}(h_{N-2})]\cdots[D^{(N-j)}(h_{N-j})][F(h_{N-j-1})] +$$

$$[D^{(N)}(h_N)][D^{(N-1)}(h_{N-1})]\cdots[D^{N-j+1}(h_{N-j+1})]\{[D_L^{(N-j)}] + [D_T^{(N-j)}]\} +$$

$$[D^{(N)}(h_N)][D^{(N-1)}(h_{N-1})]\cdots[D^{N-j+2}(h_{N-j+2})]\{[D_L^{(N-j+1)}] +$$

$$[D_T^{(N-j+1)}]\} + \cdots + [D^{(N)}(h_N)]\{[D_L^{(N-1)}] + [D_T^{(N-1)}]\} + [D_L^{(N)}] + [D_T^{(N)}]$$

$$= \cdots$$

$$= [\Psi][F(0)] + [\Omega] \tag{6.2.61}$$

where

$$[\Psi] = \prod_{j=N}^{1} [D^{(j)}(h_j)]$$

$$\tag{6.2.62}$$

$$[\Omega] = \sum_{i=2}^{N} \left\{ \prod_{j=N}^{i} [D^{(j)}(h_j)] \right\} \{\{[D_L^{(N-1)}] + [D_T^{(N-1)}]\} + \{[D_L^{(N)}] + [D_T^{(N)}]\}\}$$

It can be seen that Eq.(6.2.61) has the same structure and dimension as those of Eq.(6.2.56). After introducing the boundary condition imposed on the two transverse surfaces and considering Eq.(6.2.58), the surface displacements and/or stresses can be obtained. Introducing these solutions back into the equations at sub-layer level, the displacements, stresses and then strains within each sub-layer can be further calculated.

6.2.5   Internal surface pressure induced by a medullar pin

Prosthetic devices often employ metallic pins fitted into the medulla of a long bone as a means of attachment. These medullar pins will cause the bone in the vicinity of the pin to change its internal structure and external shape. In this section we introduce the model presented in [17,18,21] for external changes in bone shape. The theory is applied here to the problem of determining the changes in external bone shape that result from a pin force-fitted into the medulla. The diaphysial region of a long bone is modelled here as a hollow circular cylinder, and external changes in shape are changes in the external and internal radii of the hollow circular cylinder.

The solution of this problem can be obtained by decomposing the problem into two separate sub-problems: the problem of the remodelling of a hollow circular cylinder of adaptive bone material subjected to external loads, and the problem of an isotropic solid elastic cylinder subjected to an external pressure. These two problems are illustrated in Fig. 6.1.

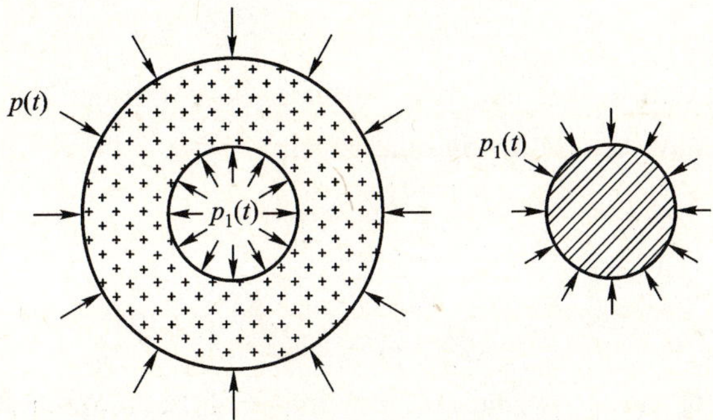

Fig. 6.1 Decomposition of the medullar pin problem into two separate sub-problems

For an isotropic solid elastic cylinder subjected to an external pressure $p(t)$, the displacement in the radial direction is given by

$$u = \frac{-(2\mu + \lambda)p(t)r}{2\mu(3\lambda + 2\mu)} \tag{6.2.63}$$

where λ and μ are Lamè's constants for an isotropic solid elastic cylinder.

In this problem we calculate the pressure of interaction $p(t)$ which occurs when an isotropic solid cylinder of radius $a_0 + \delta/2$ is forced into a hollow adaptive bone cylinder of radius a_0.

Let a and b denote the inner and outer radii, respectively, of the hollow bone cylinder at the instant after the solid isotropic cylinder has been forced into the hollow cylinder. Although the radii of the hollow cylinder will actually change during the adaptation process, the deviation of these quantities from a and b will be a small quantity negligible in small strain theory.

At an arbitrary time instant after the two cylinders have been forced together the pressure of the interaction is $p_1(t)$. The radial displacement of the solid cylinder at its surface is

$$u_1 = \frac{-(2\mu + \lambda)p_1(t)a}{2\mu(3\lambda + 2\mu)} \tag{6.2.64}$$

Using the expression (6.2.32), the radial displacement of the bone at its inner surface is obtained as

$$u_2 = \frac{a}{F_3^*}\left\{c_{33}\beta_1^*[\beta_2^*T_0 - p_1(t) + p(t)] + \bar{\omega}\frac{c_{33}c_{12}}{c_{11}} + \frac{F_2^*T_0 + P(t)}{\pi(b^2 - a^2)}c_{13} - F_1^*T_0c_{13}\right\} +$$

$$\frac{a\beta_1^*[\beta_2^*T_0 - p_1(t) + p(t)]}{(c_{11} - c_{12})} - \frac{\bar{\omega}a}{c_{11}} \tag{6.2.65}$$

Since it is assumed that the two surfaces have perfect contact, the two displacements have the following relationship:

$$a_0 + \frac{\delta}{2} + u_1 = a_0 + u_2 \tag{6.2.66}$$

Hence we find

$$\delta = 2(u_2 - u_1) \tag{6.2.67}$$

Substituting Eqs. (6.2.64) and (6.2.65) into Eq. (6.2.66), and then solving Eq.(6.2.66) for $p_1(t)$ we obtain

$$p_1(t) = -\frac{1}{H}\left[\frac{\delta}{a} - \left(H_1\frac{b^2}{b^2 - a^2} + H_2\frac{1}{b^2 - a^2} + H_3\frac{1}{\ln\left(\dfrac{b}{a}\right)}\right)\right] \tag{6.2.68}$$

$$H = \frac{2\mu + \lambda}{\mu(3\lambda + 2\mu)} - 2\left(\frac{c_{33}}{F_3^*} + \frac{1}{c_{11} - c_{12}}\right)\frac{b^2}{b^2 - a^2} \tag{6.2.69}$$

$$H_1 = \frac{2}{F_3^*}\left[c_{13}\left(\frac{c_{13}}{c_{11}}\lambda_{11} - \lambda_{33}\right)T_0\right] - 2\left(\frac{c_{33}}{F_3^*} + \frac{1}{c_{11} - c_{12}}\right)[\beta_2^*T_0 + p(t)] \tag{6.2.70}$$

$$H_2 = \frac{2}{F_3^*}\frac{c_{13}P(t)}{\pi} \tag{6.2.71}$$

$$H_3 = \frac{c_{33}c_{12}\lambda_{11}T_0}{F_3^*c_{11}} - \frac{2}{F_3^*}\left(\frac{c_{13}}{c_{11}}\lambda_{11} - \frac{\lambda_{33}}{2}\right)c_{13}T_0 - \frac{\lambda_{11}T_0}{c_{11}} \tag{6.2.72}$$

Eq.(6.2.68) is the solution of the internal surface pressure induced by an inserting medullar pin.

6.2.6   Numerical examples

As numerical illustration of the proposed analytical and semi-analytical solutions, we consider a femur with a=25 mm and b=35 mm. The material properties assumed for the bone are

$$c_{11} = 15(1+e) \text{ GPa}, \quad c_{12} = c_{13} = 6.6(1+e) \text{ GPa}, \quad c_{33} = 12(1+e) \text{ GPa}$$

$$c_{44} = 4.4(1+e) \text{ GPa}, \quad \lambda_{11} = 0.621(1+e) \times 10^5 \text{ NK}^{-1}\text{m}^{-2}$$

$$\lambda_{33} = 0.551(1+e) \times 10^5 \text{ NK}^{-1}\text{m}^{-2}, \quad \rho_3 = 0.0133(1+e) \text{ CK}^{-1}\text{m}^{-2} \qquad (6.2.73)$$

$$e_{31} = -0.435(1+e) \text{ C/m}^2, \quad e_{33} = 1.75(1+e) \text{ C/m}^2$$

$$e_{15} = 1.14(1+e)\text{C/m}^2, \quad \kappa_{11} = 111.5(1+e)\kappa_0, \quad \kappa_{33} = 126(1+e)\kappa_0$$

$$\kappa_0 = 8.85 \times 10^{-12} \text{C}^2/\text{Nm}^2 = \text{permittivity of free space}$$

The remodelling rate coefficients are assumed to be

$$C_0 = 3.09 \times 10^{-9} \text{ s}^{-1}, \quad C_1 = 2 \times 10^{-7} \text{ s}^{-1}, \quad C_2 = 10^{-6} \text{ s}^{-1}$$

and

$$A_{rr}^{\varepsilon 0} = A_{rr}^{\varepsilon 1} = A_{zz}^{\varepsilon 0} = A_{zz}^{\varepsilon 1} = A_{rz}^{\varepsilon 0} = A_{rz}^{\varepsilon 1} = 10^{-5} \text{ s}^{-1}$$

$$A_r^{E0} = A_r^{E1} = 10^{-15} \text{ m/(V}\bullet\text{s)}=10^{-15} \text{ C/(N}\bullet\text{s)}$$

The initial inner and outer radii are assumed to be

$$a_0 = 25 \text{ mm}, \quad b_0 = 35 \text{ mm}$$

and e_0=0 is assumed. In the calculation, $u_r(t) \ll a_0$ has been assumed for the sake of simplicity, i.e., $a(t)$ and $b(t)$ may be approximated by a_0 and b_0.

(1) A hollow, homogeneous circular cylindrical bone subjected to various external loads.

To analyse remodelling behaviour affected by various loading cases we distinguish the following five loading cases:

① $p(t) = n \times 2$ MPa (n=1, 2, 3, 4), $P(t)$=1500 N, with no other types of load applied.

Table 6.1 lists the results at some typical time instances obtained by both the analytical and semi-analytical solutions. The semi-analytical solution is obtained by dividing the bone into N (=10, 20, 40) sub-layers. It is evident from the table, and also from other extensive comparisons that are not shown here, that the solutions have excellent agreement on the change rate of porosity e. Hence, for the numerical results presented below, no references are given regarding which method is used to obtain the solution, unless otherwise stated. It is also evident from the table that the numerical results will gradually converge to the exact value as the layer number N increases.

Table 6.1   Comparison of porosity e obtained by analytical and semi-analytical solution (P=1500 N, p=2 Mpa)

	Time (sec)	500 000	1 000 000	1 500 000	2 000 000
	N=10	7.283×10^{-5}	1.533×10^{-4}	2.423×10^{-4}	3.406×10^{-4}
Semi-analytical	N=20	7.294×10^{-5}	1.536×10^{-4}	2.427×10^{-4}	3.412×10^{-4}
	N=40	7.297×10^{-5}	1.536×10^{-4}	2.428×10^{-4}	3.413×10^{-4}
Analytical		7.298×10^{-5}	1.536×10^{-4}	2.428×10^{-4}	3.414×10^{-4}

The extended results for this loading case are shown in Fig. 6.2 to demonstrate the effect of external pressure on the bone remodelling process. It is evident that there is a critical value p_{r0}, above which the porosity of the femur will be reduced. The critical value p_{r0} in this problem is approximately 2.95 MPa. It is also evident that the porosity of the femur increases along with the increase of external pressure p.

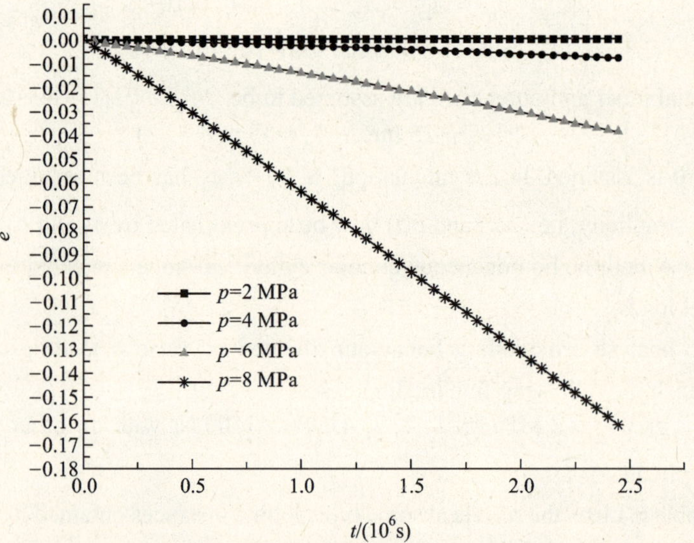

Fig. 6.2   Variation of e with time t ($\phi_b - \phi_a = T_0 = 0$ and $P = 1500$ N)

② P=1500 N and internal pressure is produced by inserting a rigid pin whose radius a^* is greater than a.

The values of e as a function of t for a^*-a=0.01 mm, 0.03 mm, and 0.05 mm are shown in Fig. 6.3. It is interesting to note that for the three cases, the bone structure at the pin-bone interface adapts itself initially to become less porous and then to a state with even less porosity. This is followed by a quick recovery of porosity, indicated by a sharply decreased value of e. As time approaches infinity, the bone structure stabilizes itself at a moderately reduced porosity. Although dramatic change of the remodelling constant is observed during the remodelling process, it is believed that the effect of the change on bone structures is limited by the fact that the duration of the change is very short compared to the entire remodelling process. This result coincides with Cowin and Van Buskirk's [21] theoretical observation which showed that a bone structure might tend to a physiologically impossible bone structure in finite time. Both of these have been observed clinically and classified as osteoporosis (excess density with the maximum value of e) and osteopetrosis (excess porosity with the minimum value of e), respectively. Fig.6.3 also shows the variation of e against the tightness of fit. It is evident that the tightness of fit has significant effects on the remodelling process, especially during the time period when the abrupt change of porosity occurs. It must be mentioned here is that the remodelling rate for this period can only serve as an indication of the modelling process, since equation (6.2.43) is only valid for predicting a low remodelling rate. Thus, detailed analysis of the equation will not provide any further reliable information. More sophisticated and advanced remodelling models are apparently needed. Nevertheless, the prediction does suggest that the possibility exists of loss of grip on the pin, or of high level tensile stresses in the bone layer surrounding the pin that may induce cracks.

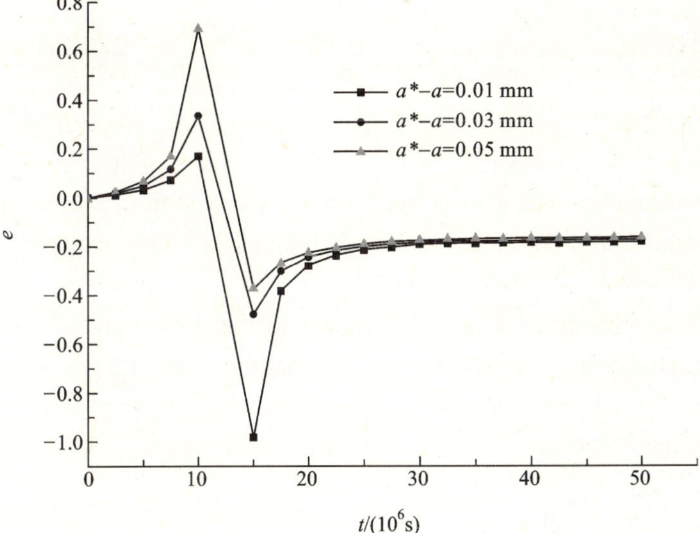

Fig.6.3   Variation of e with time induced by a solid pin

③ $T_0(t) = 10°C, 20°C, 30°C, 40°C$.

Fig.6.4 shows the effects of temperature change on bone remodelling rate at $r = b_0$ when $\phi_b - \phi_a = p(t) = P(t) = 0$. In general, low temperature induces more porous bone structures, while a warmer environment may improve the remodelling process with a less porous bone structure. After considering all other factors, it is expected that there is a preferred temperature under which an ideal remodelling rate may be achieved.

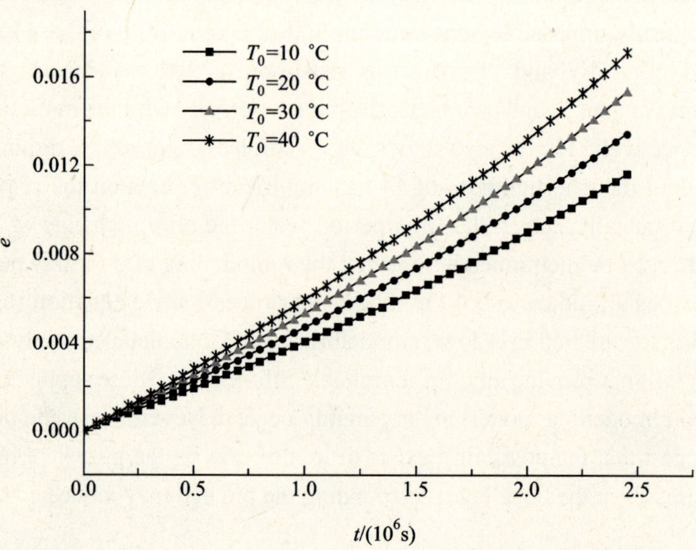

Fig. 6.4   Variation of e with time t for several temperatures ($\phi_b - \phi_a = p = P = 0$).

④ $\phi_b - \phi_a = -60$ V, -30 V, 30 V, and 60 V, $r = b_0$, and $T_0 = p = P = 0$.

Fig. 6.5 shows the variation of e with time t for various values of electric potential difference with $T_0 = p = P = 0$. It can be observed from Fig. 6.5 that there are no significant differences between the remodelling rates when the external electric potential difference $\phi_b - \phi_a$ changes from -60 V to 60 V, though it is observed that the remodelling rate increases as the electric potential difference deceases. However, the result does suggest that the remodelling process may be improved by exposing a bone to an electric field. Further theoretical and experimental studies are needed to investigate the implication of this in medical practice.

⑤ $\phi_b - \phi_a = -60$ V, -30 V, 30 V, and 60 V, $p(t) = 2$ MPa, $P(t) = 1500$ N, and $T_0 = 0$.

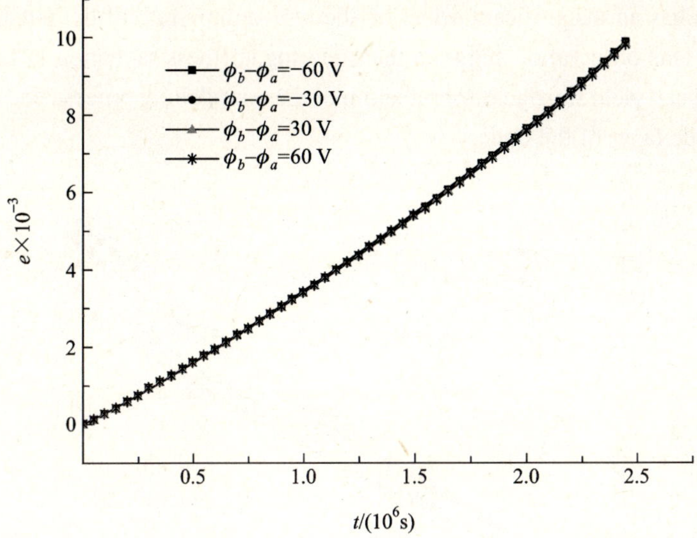

Fig. 6.5   Variation of e with time t for several potential differences ($T_0 = p = P = 0$)

This loading case is considered in order to study the coupling effect of electric and mechanical loads on bone remodelling rate. Fig.6.6 shows the numerical results of volume fraction change against different values of electric potential difference $\phi_b - \phi_a$ when T_0=0, $P(t)$=1500 N and $p(t)$=2 MPa. As observed in Fig. 6.5, it can again be seen from Fig. 6.6 that the bone remodelling rate increases along with the decrease of the potential difference $\phi_b-\phi_a$. The combination of electrical and mechanical loads results in significantly different values of the remodelling rate when different electrical fields are applied.

(2) A hollow, inhomogeneous circular cylindrical bone subjected to external loads.

The geometrical and material parameters of this problem are the same as those used in the above cases except that all material constants in Eq.(6.2.73) are now modified by a multiplier$[1-(1-\xi)(b-r)/(b-a)]$, where $0 \leqslant \xi \leqslant 1$ and represents a percentage reduction of stiffness at the inner surface of the bone. It is worth mentioning that by using the semi-analytical approach, the form of stiffness variation in the radial direction can be arbitrary. Fig.6.7 shows the results of e at the outside surface of the bone for ξ=1, 0.8 , 0.6 and 0.4 The external loads are p=4 MPa, P=1500 N, T=40, and $\phi_b - \phi_a$=30 V. In general, the remodelling rate declines as the initial stiffness of inner bone surface decreases. When time approaches infinity, it is observed that the stiffness reduction in the radial direction has an insignificant effect on the remodelling rate of the outside bone surface. This observation suggests that ignoring stiffness reduction in the radial direction can yield satisfactory prediction of the remodelling process occurring at the outside layer of the bone.

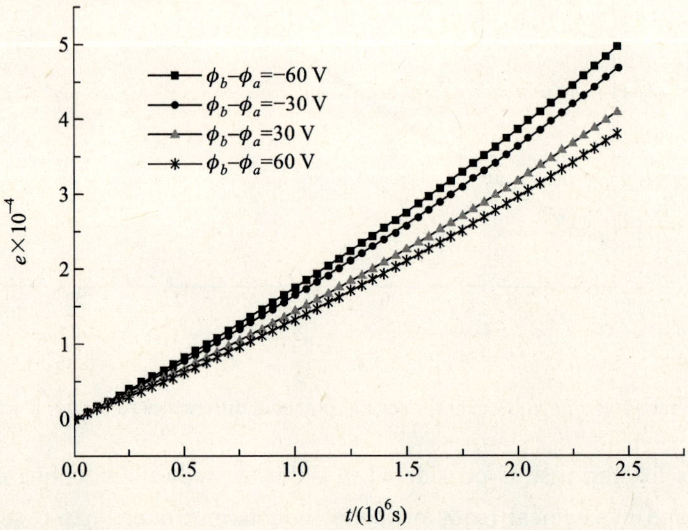

Fig. 6.6   Variation of e with time t for coupling loads (p=2 MPa, P=1500 N and T_0=0)

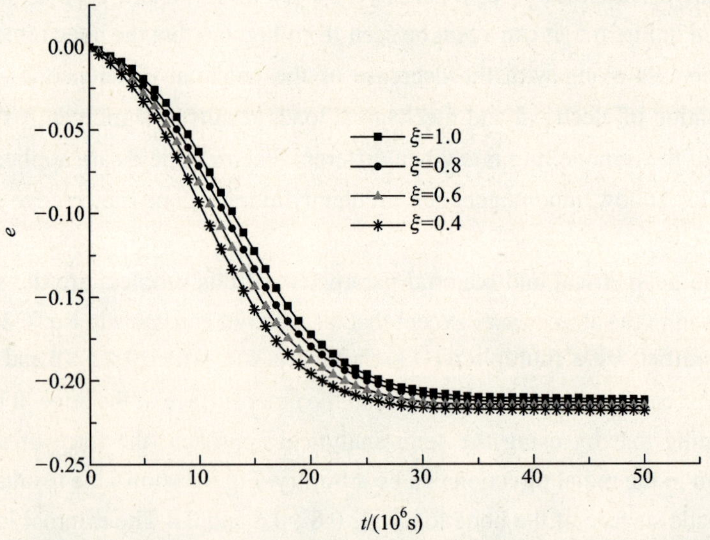

Fig. 6.7   Variation of e with time t for a inhomogeneous bone subjected to coupling loads (p=4 MPa, P=1500 N, T_0=40 and $\phi_b - \phi_a = 30$ V)

6.3 Thermo-electro-elastic surface bone remodelling

6.3.1 Equation for surface bone remodelling

The electroelastic model for surface remodelling described here is based on the work of Cowin and Buskirk [24]. They presented a hypothesis that the speed of the remodelling surface is linearly proportional to the strain tensor under the assumption of small strain

$$U(\boldsymbol{n},Q,t) = C_{ij}(\boldsymbol{n},Q)\left[\varepsilon_{ij}(Q,t) - \varepsilon_{ij}^{0}(Q,t)\right] \tag{6.3.1}$$

where $U(\boldsymbol{n}, Q, t)$ denote the speed of the remodelling surface normal to the surface at the surface point Q. It is assumed the velocity of the surface in any direction in the tangent plane is zero because the surface is not moving tangentially with respect to the body. n is the normal to the bone surface at the point Q, $\varepsilon_{ij}^{0}(Q,t)$ is a reference value of strain where no remodelling occurs, and $C_{ij}(\boldsymbol{n},Q)$ are surface remodelling rate coefficients which are, in general, dependent upon the point Q and the normal $\boldsymbol{n}$ to the surface at Q. Eq.(6.3.1) gives the normal velocity of the surface at the point Q as a function of the existing strain state at Q. If the strain state at Q. $\varepsilon_{ij}(Q,t)$, is equal to the reference strain state, $\varepsilon_{ij}^{0}(Q,t)$, then the velocity of the surface is zero and no remodelling occurs. If the right side of Eq.(6.3.1) is positive, the surface is growing by deposition of material. If, on the other hand, the right side of Eq.(6.3.1) is negative, the surface is resorbing.

Eq.(6.3.1) can be extended to include piezoelectric effects by adding some new terms as below [17,18]

$$U = C_{ij}(\boldsymbol{n},Q)\left[\varepsilon_{ij}(Q,t) - \varepsilon_{ij}^{0}(Q,t)\right] + C_{i}(\boldsymbol{n},Q)\left[E_{i}(Q,t) - E_{i}^{0}(Q,t)\right]$$

$$= C_{rr}\varepsilon_{rr} + C_{\theta\theta}\varepsilon_{\theta\theta} + C_{zz}\varepsilon_{zz} + C_{rz}\varepsilon_{rz} + C_{r}E_{r} + C_{z}E_{z} - C_{0} \tag{6.3.2}$$

where $C_{0} = C_{rr}\varepsilon_{rr}^{0} + C_{zz}\varepsilon_{zz}^{0} + C_{\theta\theta}\varepsilon_{\theta\theta}^{0} + C_{rz}\varepsilon_{rz}^{0} + C_{r}E_{r}^{0} + C_{z}E_{z}^{0}$, C_{i} are surface remodelling coefficients.

6.3.2　Differential field equation for surface remodelling rate

We now consider again the hollow circular cylinder of bone used in Section 6.2. The bone cylinder is subjected to the same external load and boundary conditions as those in Section 6.2.

　　Substituting Eqs. (6.2.35)~(6.2.39) into Eq.(6.3.2) yields

$$\begin{cases} U_e = N_1^e \dfrac{b^2}{b^2-a^2} + N_2^e \dfrac{1}{\ln\left(\dfrac{b}{a}\right)} + N_3^e \dfrac{1}{b^2-a^2} + N_4^e \dfrac{1}{a\ln\left(\dfrac{b}{a}\right)} - C_0^e \\[4em] U_p = N_1^p \dfrac{b^2}{b^2-a^2} + N_1^{p\,'} \dfrac{a^2}{b^2-a^2} + N_2^p \dfrac{1}{\ln\left(\dfrac{b}{a}\right)} + N_3^p \dfrac{1}{b^2-a^2} + \\[4em] N_4^p \dfrac{1}{b\ln\left(\dfrac{b}{a}\right)} + N_3' - C_0^p \end{cases} \tag{6.3.3}$$

where

$$N_1^e = \frac{1}{F_3^*}\left\{ c_{13}\left(\frac{c_{13}}{c_{11}}\lambda_{11} - \lambda_{33}\right)T_0 - c_{33}\left[\frac{\lambda_{11}}{2}(\frac{c_{12}}{c_{11}}-1)T_0 + p(t)\right]\right\}(C_{rr}^e + C_{\theta\theta}^e) +$$

$$\frac{1}{F_3^*}\left\{ 2c_{13}C_{zz}^e\left[\frac{\lambda_{11}}{2}(\frac{c_{12}}{c_{11}}-1)T_0 + p(t)\right] - (c_{11}+c_{12})C_{zz}^e\left(\frac{c_{13}}{c_{11}}\lambda_{11} - \lambda_{33}\right)T_0 \right\} +$$

$$\frac{(C_{rr}^e - C_{\theta\theta}^e)\left[\dfrac{\lambda_{11}}{2}(\dfrac{c_{12}}{c_{11}}-1)T_0 + p(t)\right]}{c_{11}-c_{12}} \tag{6.3.4}$$

$$N_2^e = \frac{1}{F_3^*}\left[\frac{c_{12}c_{33}}{2c_{11}}\lambda_{11}T_0 - \left(\frac{c_{13}}{c_{11}}\lambda_{11} - \frac{\lambda_{33}}{2}\right)c_{13}T_0\right](C_{rr}^e + C_{\theta\theta}^e) +$$

$$\frac{C_{zz}^e}{F_3^*}\left[(c_{11}+c_{12})\left(\frac{c_{13}}{c_{11}}\lambda_{11} - \frac{\lambda_{33}}{2}\right)T_0 - \frac{c_{12}c_{13}}{c_{11}}\lambda_{11}T_0\right] - \frac{C_{\theta\theta}^e\lambda_{11}T_0}{2c_{11}} \tag{6.3.5}$$

$$N_3^e = \frac{1}{F_3^*}\left[c_{13}\left(C_{rr}^e + C_{\theta\theta}^e\right) - (c_{11}+c_{12})C_{zz}^e\right]\frac{P(t)}{\pi} \tag{6.3.6}$$

$$N_4^e = -\left(\frac{e_{15}}{c_{44}}C_{zr}^e + C_r\right)(\phi_b - \phi_a) \qquad (6.3.7)$$

$$N_1^p = \frac{1}{F_3^*}\left\{c_{13}\left(\frac{c_{13}}{c_{11}}\lambda_{11} - \lambda_{33}\right)T_0 - c_{33}\left[\frac{\lambda_{11}}{2}(\frac{c_{12}}{c_{11}} - 1)T_0 + p(t)\right]\right\}(C_{rr}^p + C_{\theta\theta}^p) +$$

$$\frac{1}{F_3^*}\left\{2c_{13}C_{zz}^p\left[\frac{\lambda_{11}}{2}(\frac{c_{12}}{c_{11}} - 1)T_0 + p(t)\right] - (c_{11} + c_{12})C_{zz}^p\left(\frac{c_{13}}{c_{11}}\lambda_{11} - \lambda_{33}\right)T_0\right\} \qquad (6.3.8)$$

$$N_1^{p'} = -\frac{(C_{rr}^p - C_{\theta\theta}^p)\left[\dfrac{\lambda_{11}}{2}(\dfrac{c_{12}}{c_{11}} - 1)T_0 + p(t)\right]}{c_{11} - c_{12}} \qquad (6.3.9)$$

$$N_2^p = \frac{1}{F_3^*}\left[\frac{c_{12}c_{33}}{2c_{11}}\lambda_{11}T_0 - \left(\frac{c_{13}}{c_{11}}\lambda_{11} - \frac{\lambda_{33}}{2}\right)c_{13}T_0\right](C_{rr}^p + C_{\theta\theta}^p) +$$

$$\frac{C_{zz}^p}{F_3^*}\left[(c_{11} + c_{12})\left(\frac{c_{13}}{c_{11}}\lambda_{11} - \frac{\lambda_{33}}{2}\right)T_0 - \frac{c_{12}c_{13}}{c_{11}}\lambda_{11}T_0\right] - \frac{C_{\theta\theta}^p\lambda_{11}T_0}{2c_{11}} \qquad (6.3.10)$$

$$N_3^p = \frac{1}{F_3^*}\left[c_{13}\left(C_{rr}^p + C_{\theta\theta}^p\right) - \left(c_{11} + c_{12}\right)C_{zz}^p\right]\frac{P(t)}{\pi} \qquad (6.3.11)$$

$$N_4^p = -\left(\frac{e_{15}}{c_{44}}C_{zr}^p + C_r\right)(\phi_b - \phi_a) \qquad (6.3.12)$$

$$N_3' = \frac{\left(C_{\theta\theta}^p + C_{rr}^p\right)\lambda_{11}T_0}{2c_{11}} \qquad (6.3.13)$$

and the subscripts p and e refer to periosteal and endosteal, respectively. Since U_e and U_p are the velocities normal to the inner and outer surfaces of the cylinders, respectively, they are calculated as

$$U_e = -\frac{da}{dt}, \quad U_p = \frac{db}{dt} \qquad (6.3.14)$$

where the minus sign appearing in the expression for U_e denotes that the outward normal of the endosteal surface is in the negative coordinate direction. Substituting Eq.(6.3.14) into Eq.(6.3.3) yields

$$\left\{ \begin{aligned} -\frac{da}{dt} &= N_1^e \frac{b^2}{b^2-a^2} + N_2^e \frac{1}{\ln\left(\dfrac{b}{a}\right)} + N_3^e \frac{1}{b^2-a^2} + N_4^e \frac{1}{a\ln\left(\dfrac{b}{a}\right)} - C_0^e \\[3ex] \frac{db}{dt} &= N_1^p \frac{b^2}{b^2-a^2} + N_1^{p\,\prime} \frac{a^2}{b^2-a^2} + N_2^p \frac{1}{\ln\left(\dfrac{b}{a}\right)} + N_3^p \frac{1}{b^2-a^2} + \\[3ex] & \quad N_4^p \frac{1}{b\ln\left(\dfrac{b}{a}\right)} - C_0^{p\,\prime} \end{aligned} \right. \qquad (6.3.15)$$

where $C_0^{p\,\prime} = C_0^p - N_3'$.

6.3.3   Approximation for small changes in radii

It is apparent that Eq.(6.3.15) are non-linear and cannot, in general, be solved analytically. However, the equations can be approximately linearized when they are applied to solve problems with small changes in radii. In the bone surface remodelling process, we can assume that the radii of the inner and outer surface of the bone change very little compared to their original values. This means that the changes in $a(t)$ and $b(t)$ are small. This is a reasonable assumption from the viewpoint of physics of the problem. To introduce the approximation the non-dimensional parameters

$$\varepsilon = \frac{a}{a_0} - 1, \quad \eta = \frac{b}{b_0} - 1 \qquad (6.3.16)$$

are adopted in the following calculations. As a result, $a(t)$ and $b(t)$ can be written as

$$a(t) = \left[1 + \varepsilon(t)\right]a_0, \qquad b(t) = \left[1 + \eta(t)\right]b_0, \quad \varepsilon, \ \eta \ll 1 \qquad (6.3.17)$$

Since both ε and η are far smaller than one, their squares can be ignored from the equations. Consequently, we can have the following approximations:

$$\frac{b^2}{b^2-a^2} \cong L_0 + 2L_0^2 \frac{a_0^2}{b_0^2}(\varepsilon - \eta) \qquad (6.3.18)$$

$$\frac{a^2}{b^2-a^2} \cong L_0' + 2L_0'^2 \frac{b_0^2}{a_0^2}(\varepsilon - \eta) \qquad (6.3.19)$$

$$\frac{1}{b^2-a^2} \cong L_2 + 2L_2^2(a_0^2\varepsilon - b_0^2\eta) \qquad (6.3.20)$$

$$\frac{1}{\ln\left(\dfrac{b}{a}\right)} \cong L_1 + L_1^2(\varepsilon - \eta) \tag{6.3.21}$$

$$\frac{1}{a\ln\left(\dfrac{b}{a}\right)} \cong \frac{1}{a_0}L_1(1-\varepsilon) + \frac{1}{a_0}L_1^2(\varepsilon - \eta) \tag{6.3.22}$$

$$\frac{1}{b\ln\left(\dfrac{b}{a}\right)} \cong \frac{1}{b_0}L_1(1-\eta) + \frac{1}{b_0}L_1^2(\varepsilon - \eta) \tag{6.3.23}$$

where

$$L_0 = \frac{b_0^2}{b_0^2 - a_0^2} \tag{6.3.24}$$

$$L_0' = \frac{a_0^2}{b_0^2 - a_0^2} \tag{6.3.25}$$

$$L_1 = \frac{1}{\ln\left(\dfrac{b_0}{a_0}\right)} \tag{6.3.26}$$

$$L_2 = \frac{1}{b_0^2 - a_0^2} \tag{6.3.27}$$

Thus, Eq.(6.3.15) can be approximately represented in terms of ε and η, as follows

$$\begin{cases} \dfrac{d\varepsilon}{dt} = B_1\varepsilon + B_2\eta + B_3 \\[2mm] \dfrac{d\eta}{dt} = B_1'\varepsilon + B_2'\eta + B_3' \end{cases} \tag{6.3.28}$$

where

$$B_1 = -\frac{1}{a_0}\left(2L_0^2\frac{a_0^2}{b_0^2}N_1^e + L_1^2N_2^e - \frac{L_1N_4^e}{a_0} + \frac{L_1^2N_4^e}{a_0} + 2L_2^2a_0^2N_3^e\right) \tag{6.3.29}$$

$$B_2 = \frac{1}{a_0}\left(2L_0^2\frac{a_0^2}{b_0^2}N_1^e + L_1^2N_2^e + \frac{L_1^2N_4^e}{a_0} + 2L_2^2b_0^2N_3^e\right) \tag{6.3.30}$$

$$B_3 = -\frac{1}{a_0}\left(L_0N_1^e + L_1N_2^e + L_2N_3^e + \frac{L_1N_4^e}{a_0} - C_0^e\right) \tag{6.3.31}$$

$$B_1' = \frac{1}{b_0}\left(2L_0^2\frac{a_0^2}{b_0^2}N_1^p + L_1^2N_2^p + \frac{L_1^2N_4^p}{b_0} + 2L_2^2a_0^2N_3^p + 2L_0'^2\frac{b_0^2}{a_0^2}N_1^{p'}\right) \tag{6.3.32}$$

$$B_2' = -\frac{1}{b_0}\left(2L_0^2 \frac{a_0^2}{b_0^2} N_1^p + L_1^2 N_2^p - \frac{L_1 N_4^p}{b_0} + \frac{L_1^2 N_4^p}{b_0} + 2L_2^2 b_0^2 N_3^p + 2L_0'^2 \frac{b_0^2}{a_0^2} N_1^{p'} \right)$$

$$(6.3.33)$$

$$B_3' = \frac{1}{b_0}\left(L_0 N_1^p + L_1 N_2^p + L_2 N_3^p + \frac{L_1 N_4^p}{b_0} + L_0' N_1^{p'} - C_0^{p'} \right) \qquad (6.3.34)$$

6.3.4 Analytical solution of surface remodelling

An analytical solution of Eq.(6.3.28) can be obtained if smeared homogeneous property is assumed for bone material. In such a case, the inhomogeneous linear differential equations system (6.3.28) can be converted into the following homogeneous one:

$$\begin{cases} \dfrac{d\varepsilon'}{dt} = B_1 \varepsilon' + B_2 \eta' \\[2mm] \dfrac{d\eta'}{dt} = B_1' \varepsilon' + B_2' \eta' \end{cases} \qquad (6.3.35)$$

by introducing two new variables such that

$$\begin{cases} \varepsilon' = \varepsilon - \varepsilon_\infty \\ \eta' = \eta - \eta_\infty \end{cases} \qquad (6.3.36)$$

$$\varepsilon_\infty = \frac{1}{\det \boldsymbol{M}}(B_3' B_2 - B_3 B_2'), \qquad \eta_\infty = \frac{1}{\det \boldsymbol{M}}(B_3 B_1' - B_3' B_1) \qquad (6.3.37)$$

$$\boldsymbol{M} = \begin{bmatrix} B_1 & B_2 \\ B_1' & B_2' \end{bmatrix} \qquad (6.3.38)$$

$$\det \boldsymbol{M} = B_1 B_2' - B_1' B_2 \qquad (6.3.39)$$

The solution of Eq.(6.3.35), subject to the initial conditions that $\varepsilon(0) = 0$ and $\eta(0) = 0$, can be expressed in four possible forms that fulfil the physics of the problem, i.e., when $t \to \infty$, ε and η must be limited quantities, $a < b$ and the solution must be stable. The form of the solution depends on the roots of the following quadratic equation

$$s^2 - \operatorname{tr} \boldsymbol{M} s + \det \boldsymbol{M} = 0 \qquad (6.3.40)$$

where

$$\operatorname{tr} \boldsymbol{M} = B_1 + B_2' = s_1 + s_2 \qquad (6.3.41)$$

All the theoretically possible solutions are shown as follows:

Case 1   When $(B_1 - B_2')^2 + 4B_2 B_1' > 0$, $B_1 + B_2' < 0$ and $B_1 B_2' - B_2 B_1' > 0$,

Eq.(6.3.40) has two different roots, s_1 and s_2, both of which are real and distinct. Then the solutions of the equations are

$$\begin{cases} \varepsilon' = \dfrac{1}{s_1 - s_2}\left[\left(s_2\varepsilon_\infty - B_3\right)e^{-s_1 t} + \left(B_3 - s_1\varepsilon_\infty\right)e^{-s_2 t}\right] \\[2ex] \eta' = \dfrac{1}{s_1 - s_2}\left[\left(s_2\eta_\infty - B_3'\right)e^{-s_1 t} + \left(B_3' - s_1\eta_\infty\right)e^{-s_2 t}\right] \end{cases} \tag{6.3.42}$$

which can also be written as

$$\begin{cases} \varepsilon(t) = \varepsilon_\infty + \dfrac{1}{s_1 - s_2}\left[\left(s_2\varepsilon_\infty - B_3\right)e^{-s_1 t} + \left(B_3 - s_1\varepsilon_\infty\right)e^{-s_2 t}\right] \\[2ex] \eta(t) = \eta_\infty + \dfrac{1}{s_1 - s_2}\left[\left(s_2\eta_\infty - B_3'\right)e^{-s_1 t} + \left(B_3' - s_1\eta_\infty\right)e^{-s_2 t}\right] \end{cases} \tag{6.3.43}$$

The formulae for the variation of the radii, i.e., $a(t)$ and $b(t)$, with time can be obtained by substituting Eq.(6.3.43) into Eq.(6.3.16). Thus

$$\begin{cases} a(t) = a_0 + a_0\varepsilon_\infty + \dfrac{a_0}{s_1 - s_2}\left[\left(s_2\varepsilon_\infty - B_3\right)e^{-s_1 t} + \left(B_3 - s_1\varepsilon_\infty\right)e^{-s_2 t}\right] \\[2ex] b(t) = b_0 + b_0\eta_\infty + \dfrac{b_0}{s_1 - s_2}\left[\left(s_2\eta_\infty - B_3'\right)e^{-s_1 t} + \left(B_3' - s_1\eta_\infty\right)e^{-s_2 t}\right] \end{cases} \tag{6.3.44}$$

The final radii of the cylinder are then

$$\begin{cases} a_\infty = \lim_{t\to\infty} a(t) = a_0(1 + \varepsilon_\infty) \\[2ex] b_\infty = \lim_{t\to\infty} b(t) = b_0(1 + \eta_\infty) \end{cases} \tag{6.3.45}$$

Case 2 When $\left(B_1 - B_2'\right)^2 + 4B_2 B_1' = 0$, $B_1 \neq B_2'$ and $B_1 + B_2' < 0$, Eq.(6.3.40) has two equal roots, $B_2' + B_1$. The solutions of the equations are

$$\begin{cases} \varepsilon' = -\left\{\varepsilon_\infty + \left[\dfrac{B_1 - B_2'}{2}\varepsilon_\infty + B_2\eta_\infty\right]t\right\}e^{\frac{B_2' + B_1}{2}t} \\[2ex] \eta' = -\left\{\eta_\infty - \left[\dfrac{\left(B_2' - B_1\right)^2}{4B_2}\varepsilon_\infty - \dfrac{B_2' - B_1}{2}\eta_\infty\right]t\right\}e^{\frac{B_2' + B_1}{2}t} \end{cases} \tag{6.3.46}$$

which can also be written as

$$\begin{cases} \varepsilon(t) = \varepsilon_\infty - \left\{\varepsilon_\infty + \left[\dfrac{B_1 - B_2'}{2}\varepsilon_\infty + B_2\eta_\infty\right]t\right\}e^{\frac{B_2' + B_1}{2}t} \\[2ex] \eta(t) = \eta_\infty - \left\{\eta_\infty - \left[\dfrac{\left(B_2' - B_1\right)^2}{4B_2}\varepsilon_\infty - \dfrac{B_2' - B_1}{2}\eta_\infty\right]t\right\}e^{\frac{B_2' + B_1}{2}t} \end{cases} \tag{6.3.47}$$

The formulae for the variation of $a(t)$ and $b(t)$ with time can be obtained by substituting Eq.(6.3.47) into Eq.(6.3.16) as

$$\left\{ \begin{aligned} &a(t) = a_0 + a_0\varepsilon_\infty - a_0\left\{\varepsilon_\infty + \left[\frac{B_1 - B_2'}{2}\varepsilon_\infty + B_2\eta_\infty\right]t\right\}e^{\frac{B_2'+B_1}{2}t} \\ &b(t) = b_0 + b_0\eta_\infty - b_0\left\{\eta_\infty - \left[\frac{(B_2'-B_1)^2}{4B_2}\varepsilon_\infty - \frac{B_2'-B_1}{2}\eta_\infty\right]t\right\}e^{\frac{B_2'+B_1}{2}t} \end{aligned}\right. \tag{6.3.48}$$

The final radii of the cylinder are then

$$\left\{ \begin{aligned} a_\infty &= \lim_{t\to\infty} a(t) = a_0(1+\varepsilon_\infty) \\ b_\infty &= \lim_{t\to\infty} b(t) = b_0(1+\eta_\infty) \end{aligned}\right. \tag{6.3.49}$$

Case 3   When $B_1 = B_2' < 0$ and $B_2 = 0$, the solutions of the equations are

$$\left\{ \begin{aligned} \varepsilon' &= -\varepsilon_\infty e^{B_1 t} \\ \eta' &= -\left(B_1'\varepsilon_\infty t + \eta_\infty\right)e^{B_1 t} \end{aligned}\right. \tag{6.3.50}$$

which can also be written as

$$\left\{ \begin{aligned} \varepsilon(t) &= \varepsilon_\infty - \varepsilon_\infty e^{B_1 t} \\ \eta(t) &= \eta_\infty - \left(B_1'\varepsilon_\infty t + \eta_\infty\right)e^{B_1 t} \end{aligned}\right. \tag{6.3.51}$$

The formulae for the variation of $a(t)$ and $b(t)$ with time can be obtained by substituting Eq.(6.3.51) into Eq.(6.3.16) as follows:

$$\left\{ \begin{aligned} a(t) &= a_0 + a_0\varepsilon_\infty\left(1 - e^{B_1 t}\right) \\ b(t) &= b_0 + b_0\eta_\infty - b_0\left(B_1'\varepsilon_\infty t + \eta_\infty\right)e^{B_1 t} \end{aligned}\right. \tag{6.3.52}$$

The final radii of the cylinder are then

$$\left\{ \begin{aligned} a_\infty &= \lim_{t\to\infty} a(t) = a_0(1+\varepsilon_\infty) \\ b_\infty &= \lim_{t\to\infty} b(t) = b_0(1+\eta_\infty) \end{aligned}\right. \tag{6.3.53}$$

Case 4   When $B_1 = B_2' < 0$ and $B_1' = 0$, the solutions of the equations are

$$\left\{ \begin{aligned} \varepsilon' &= -\left(B_2\eta_\infty t + \varepsilon_\infty\right)e^{B_1 t} \\ \eta' &= -\eta_\infty e^{B_1 t} \end{aligned}\right. \tag{6.3.54}$$

which can also be written as

$$\left\{ \begin{aligned} \varepsilon(t) &= \varepsilon_\infty - \left(B_2\eta_\infty t + \varepsilon_\infty\right)e^{B_1 t} \\ \eta(t) &= \eta_\infty - \eta_\infty e^{B_1 t} \end{aligned}\right. \tag{6.3.55}$$

The formulae for the variation of the radii with time can be obtained by substituting Eq.(6.3.55) into Eq.(6.3.16). Thus

$$\begin{cases} a(t) = a_0 + a_0\varepsilon_\infty - a_0\left(B_2\eta_\infty t + \varepsilon_\infty\right)e^{B_1 t} \\ b(t) = b_0 + b_0\eta_\infty\left(1 - e^{B_1 t}\right) \end{cases} \qquad (6.3.56)$$

The final radii of the cylinder are then

$$\begin{cases} a_\infty = \lim_{t\to\infty} a(t) = a_0(1+\varepsilon_\infty) \\ b_\infty = \lim_{t\to\infty} b(t) = b_0(1+\eta_\infty) \end{cases} \qquad (6.3.57)$$

All the above solutions are theoretically valid. However, the first is the most likely solution to the problem, as it is physically possible when $t \to \infty$ [17, 18]. Therefore it can be used to calculate the bone surface remodelling.

6.3.5   Application of semi-analytical solution to surface remodelling of inhomogeneous bone

The semi-analytical solution presented in Section 6.2.4 can be used to calculate strains and stresses at any point on the bone surface. These results form the basis for surface bone remodelling analysis. This section presents applications of solution (6.2.61) to the analysis of surface remodelling behaviour in inhomogeneous bone.

It is noted that surface bone remodelling is a time-dependent process. The change in the radii (ε or η) can therefore be calculated by using the rectangular algorithm of integral (see Fig. 6.8). The procedure is described here. Firstly, let T_0 be the starting time and T be the length of time to be considered, and divide the time domain T into m equal interval $\Delta T = T/m$. At the time t, calculate the strain and electric field using Eqs. (6.2.35)$\sim$(6.2.39). The results are then substituted into Eq.(6.3.2) to determine the normal velocity of the surface bone remodelling. Assuming that ΔT is sufficiently small, we can replace U with its mean value $\bar{U}$ at each time interval $[t, t+\Delta T]$. The change in the radii (ε or η) at time t can thus be determined using the results of surface velocity. Accordingly, the strain and electric field are updated by considering the change in the radii. The updated strain and electric field are in turn used to calculate the normal surface velocity at the next time interval. This process is repeated up to the last time interval $[T_0 + (m-1)\Delta T,\ T_0 + T]$. Fig. 6.8 shows the rectangular-algorithm of integral when we replace U with its initial value U_t (rather than its mean value $\bar{U}$) at the time interval $[t, t+\Delta T]$.

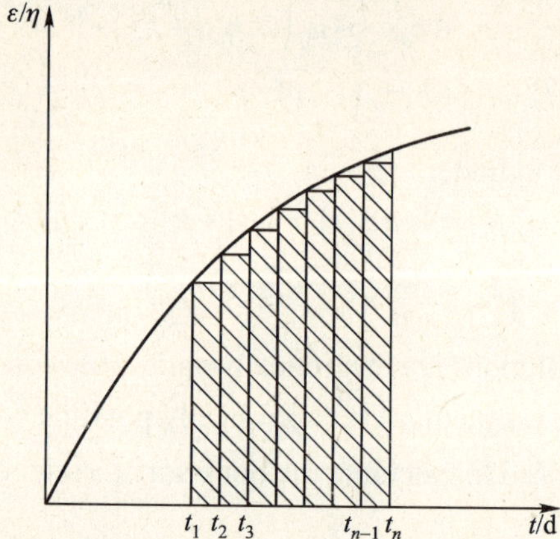

Fig. 6.8　Illustration of the rectangular algorithm

6.3.6　Surface remodelling equation modified by an inserting medullar pin

Substituting Eq.(6.2.68) into Eq.(6.3.15) yields

$$
\begin{cases}
-\dfrac{\mathrm{d}a}{\mathrm{d}t} = N_1^e \dfrac{b^2}{b^2-a^2} + N_2^e \dfrac{1}{\ln\left(\dfrac{b}{a}\right)} + N_3^e \dfrac{1}{b^2-a^2} + N_4^e \dfrac{1}{a\ln\left(\dfrac{b}{a}\right)} - \\[2ex]
\qquad M_1 p_1(t) \dfrac{b^2}{a^2-b^2} - C_0^e \\[3ex]
\dfrac{\mathrm{d}b}{\mathrm{d}t} = N_1^p \dfrac{b^2}{b^2-a^2} + N_1^{p'} \dfrac{a^2}{b^2-a^2} + N_2^p \dfrac{1}{\ln\left(\dfrac{b}{a}\right)} + N_3^p \dfrac{1}{b^2-a^2} - \\[2ex]
\qquad \left(M_2 \dfrac{b^2}{a^2-b^2} + M_3 \dfrac{a^2}{a^2-b^2} \right) p_1(t) + N_4^p \dfrac{1}{b\ln\left(\dfrac{b}{a}\right)} - C_0^{p'}
\end{cases}
\tag{6.3.58}
$$

where

$$
M_1 = \frac{(C_{rr}^e + C_{\theta\theta}^e)c_{33} - 2c_{13}C_{zz}^e}{F_3^*} - \frac{C_{rr}^e - C_{\theta\theta}^e}{c_{11} - c_{12}}
\tag{6.3.59}
$$

$$
M_2 = \frac{(C_{rr}^p + C_{\theta\theta}^p)c_{33} - 2c_{13}C_{zz}^p}{F_3^*}
\tag{6.3.60}
$$

$$M_3 = -\frac{C_{rr}^p - C_{\theta\theta}^p}{c_{11} - c_{12}} \tag{6.3.61}$$

It can be seen that Eq.(6.3.58) is similar to Eq.(6.3.15). It can also be simplified as

$$\begin{cases} \dfrac{d\varepsilon}{dt} = Y_1\varepsilon + Y_2\eta + Y_3 \\[2mm] \dfrac{d\eta}{dt} = Y_1'\varepsilon + Y_2'\eta + Y_3' \end{cases} \tag{6.3.62}$$

where

$$\begin{aligned} Y_1 = B_1 - M_1\Bigg[& H_5\left(\frac{\delta}{a_0} - H_1L_0 - H_2L_2 - H_3L_1\right) + \\ & H_4\left(2L_0^2H_1\frac{a_0^2}{b_0^2} + 2L_2^2H_2a_0^2 + H_3L_1^2 + \frac{\delta}{a_0}\right)\Bigg] \end{aligned} \tag{6.3.63}$$

$$\begin{aligned} Y_2 = B_2 + M_1\Bigg[& H_5\left(\frac{\delta}{a_0} - H_1L_0 - H_2L_2 - H_3L_1\right) + \\ & H_4\left(2L_0^2H_1\frac{a_0^2}{b_0^2} + 2L_2^2H_2b_0^2 + H_3L_1^2\right)\Bigg] \end{aligned} \tag{6.3.64}$$

$$Y_3 = B_3 + M_1H_4\left(\frac{\delta}{a_0} - H_1L_0 - H_2L_2 - H_3L_1\right) \tag{6.3.65}$$

$$\begin{aligned} Y_1' = B_1' - M_2\Bigg[& H_5\left(\frac{\delta}{a_0} - H_1L_0 - H_2L_2 - H_3L_1\right) + H_4 \times \\ & \left(2L_0^2H_1\frac{a_0^2}{b_0^2} + 2L_2^2H_2a_0^2 + H_3L_1^2 + \frac{\delta}{a_0}\right)\Bigg] + M_3\Bigg[H_7\left(\frac{\delta}{a_0} - H_1L_0 - H_2L_2 - H_3L_1\right) - \\ & H_6\left(2L_0^2H_1\frac{a_0^2}{b_0^2} + 2L_2^2H_2a_0^2 + H_3L_1^2 + \frac{\delta}{a_0}\right)\Bigg] \end{aligned} \tag{6.3.66}$$

$$\begin{aligned} Y_2' = B_2' + M_2\Bigg[& H_5\left(\frac{\delta}{a_0} - H_1L_0 - H_2L_2 - H_3L_1\right) + H_4 \times \\ & \left(2L_0^2H_1\frac{a_0^2}{b_0^2} + 2L_2^2H_2b_0^2 + H_3L_1^2\right)\Bigg] - M_3\Bigg[H_7\left(\frac{\delta}{a_0} - H_1L_0 - H_2L_2 - H_3L_1\right) - \\ & H_6\left(2L_0^2H_1\frac{a_0^2}{b_0^2} + 2L_2^2H_2b_0^2 + H_3L_1^2\right)\Bigg] \end{aligned} \tag{6.3.67}$$

$$Y_3' = B_3' + \left(M_2 H_4 + M_3 H_6 \right) \left(\frac{\delta}{a_0} - H_1 L_0 - H_2 L_2 - H_3 L_1 \right) \tag{6.3.68}$$

$$H_4 = \cfrac{1}{2\left(\cfrac{c_{33}}{F_3^*} + \cfrac{1}{c_{11} - c_{12}} \right) - \cfrac{2\mu + \lambda}{\mu(3\lambda + 2\mu)}\left(1 - \cfrac{a_0^2}{b_0^2} \right)} \tag{6.3.69}$$

$$H_5 = \cfrac{\cfrac{\left(4\mu + 2\lambda \right) a_0^2}{\mu(3\lambda + 2\mu) b_0^2}}{\left[2\left(\cfrac{c_{33}}{F_3^*} + \cfrac{1}{c_{11} - c_{12}} \right) - \cfrac{2\mu + \lambda}{\mu(3\lambda + 2\mu)}\left(1 - \cfrac{a_0^2}{b_0^2} \right) \right]^2} \tag{6.3.70}$$

$$H_6 = \cfrac{1}{\cfrac{2\mu + \lambda}{\mu(3\lambda + 2\mu)} + \left[2\left(\cfrac{c_{33}}{F_3^*} + \cfrac{1}{c_{11} - c_{12}} \right) - \cfrac{2\mu + \lambda}{\mu(3\lambda + 2\mu)} \right] \cfrac{b_0^2}{a_0^2}} \tag{6.3.71}$$

$$H_7 = \cfrac{2\left[2\left(\cfrac{c_{33}}{F_3^*} + \cfrac{1}{c_{11} - c_{12}} \right) - \cfrac{2\mu + \lambda}{\mu(3\lambda + 2\mu)} \right] \cfrac{b_0^2}{a_0^2}}{\left\{ \cfrac{2\mu + \lambda}{\mu(3\lambda + 2\mu)} + \left[2\left(\cfrac{c_{33}}{F_3^*} + \cfrac{1}{c_{11} - c_{12}} \right) - \cfrac{2\mu + \lambda}{\mu(3\lambda + 2\mu)} \right] \cfrac{b_0^2}{a_0^2} \right\}^2} \tag{6.3.72}$$

Eq.(6.3.62) is similar to Eq.(6.3.35) and can thus be solved by following the solution procedure described in Section 6.3.4.

6.3.7   Numerical examples

Consider again the femur used in Section 6.2.6. The geometrical and material coefficients of the femur are the same as those used in Section 6.2.6 except that the volume fraction change e is now taken to be zero here. In addition, the surface remodelling rate coefficients are assumed to be

$$C_{rr}^e = -9.6 \text{ m/d}, \quad C_{\theta\theta}^e = -7.2 \text{ m/d}, \quad C_{zz}^e = -5.4 \text{ m/d}$$

$$C_{zr}^e = -8.4 \text{ m/d}, \quad C_{rr}^p = -12.6 \text{ m/d}, \quad C_{\theta\theta}^e = -10.8 \text{ m/d}$$

$$C_{zz}^p = -9.6 \text{ m/d}, \quad C_{zr}^p = -12 \text{ m/d}$$

$$C_0^e = 0.000\,837\,3 \text{ m/d}, \quad C_0^p = 0.000\,158\,43 \text{ m/d}$$

and $\varepsilon_0 = 0$, $\eta_0 = 0$ are assumed.

In the following, numerical results are provided to show the effect of temperature and external electric load on the surface bone remodelling process.

While the results for the effects of mechanical loading, inserted pin, and material inhomogeneinity on the surface remodelling behaviour are omitted here, they can be found in [17,18].

(1) Effect of temperature change on surface bone remodelling. The temperature is assumed to change between $29.5\,^\circ\mathrm{C} \sim 30.5\,^\circ\mathrm{C}$, i.e. $T_0(t) =$ $29.5\,^\circ\mathrm{C}, 29.8\,^\circ\mathrm{C}$, $30\,^\circ\mathrm{C}, 30.2\,^\circ\mathrm{C}$, $30.5\,^\circ\mathrm{C}$, while the other external loadings are specified as: $\phi_b - \phi_a = 30$ V, $p(t) = 1\,\mathrm{MPa}$, $P(t){=}1500$ N. Fig. 6.9 and Fig. 6.10 show the effects of temperature change on bone surface remodelling. In general, the radii of the bone decrease when the temperature increases and they increase when the temperature decreases. It can also be seen from Figs. 6.9 and 6.10 that ε and η are almost the same. Since $a_0 < b_0$, the change of the outer surface radius is normally greater than that of the inner one. The area of the bone cross-section decreases as the temperature increases. This also suggests that a lower temperature is likely to induce thicker bone structures, while a warmer environment may improve the remodelling process with a less thick bone structure. This result seems to coincide with actual fact. Thicker and stronger bones maybe make a person living in Russia look stronger than one who lives in Vietnam. It should be mentioned here that how this change may affect the bone remodelling process is still an open question. As an initial investigation, the purpose of this study is to show how a bone may respond to thermal loads and to provide information for the possible use of imposed external temperature fields in medical treatment and in controlling the healing process of injured bones.

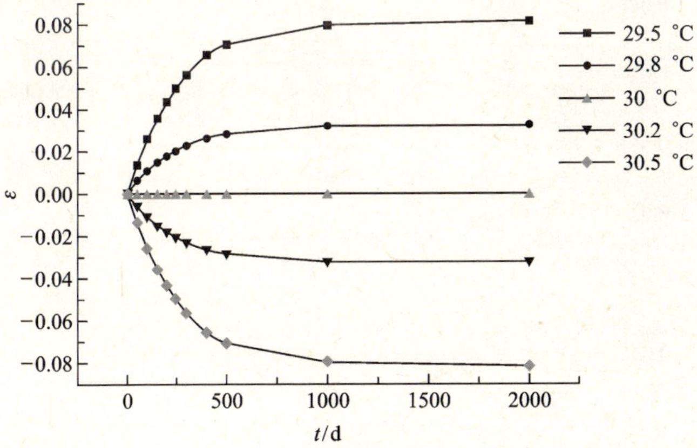

Fig. 6.9   Variation of ε with time t for several temperatures

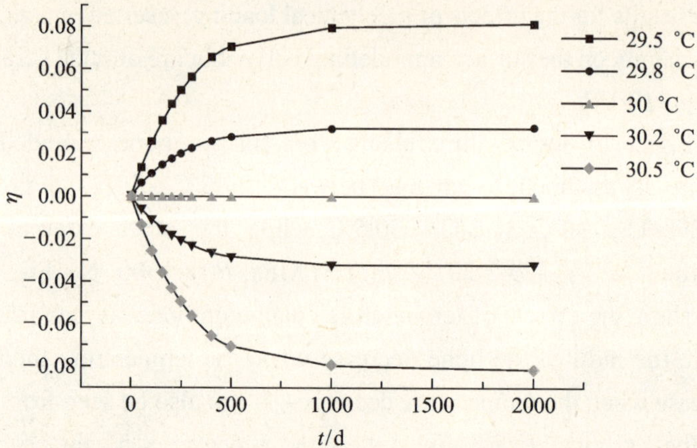

Fig. 6.10   Variation of η with time t for several temperatures

(2) Effect of external electrical potential on surface bone remodelling. In this case, the coupled loading is assumed as: $\phi_b - \phi_a = -60$ V, -30 V, 30 V, and 60 V, $p(t) = 1$ MPa, $P(t)=1500$ N, and $T_0=0$. Fig. 6.11 and Fig. 6.12 show the variation of ε and η with time t for various values of electric potential difference. It can be seen that the effect of the electric potential is just the opposite to that of temperature. A decrease in the intensity of electric field results in a decrease of the inner and outer surface radii of the bone by almost the same magnitude. Theoretically, the results suggest that the remodelling process may be improved by exposing a bone to an electric field. Clearly, further theoretical and

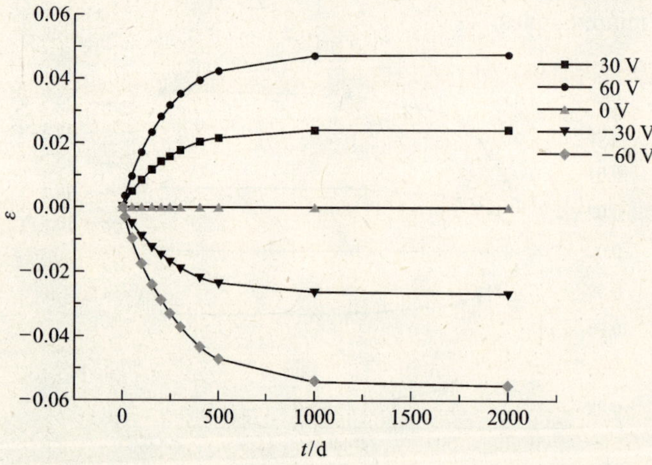

Fig. 6.11   Variation of ε with time t for several potential differences

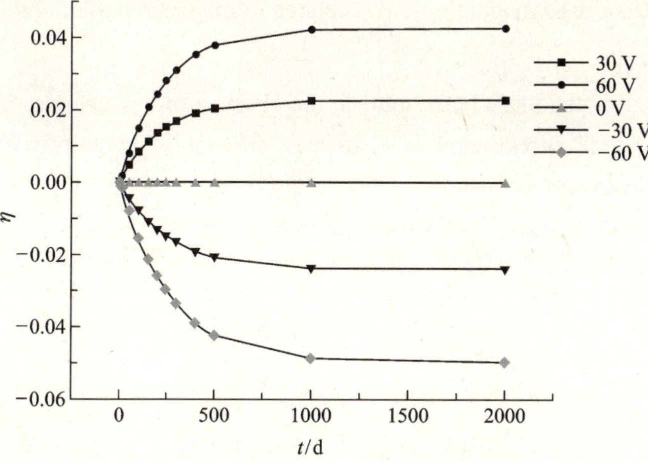

Fig. 6.12   Variation of η with time t for several potential differences

experimental studies are needed to investigate the implications of this in medical practice.

6.4   Extension to thermo-magneto-electro-elastic problem

6.4.1   Linear theory of thermo-magneto-electro-elastic solid

For a hollow circular cylinder composed linearly of a thermo-magneto-electro-elastic bone material subjected to axisymmetric loading, the field equations described in the previous two sections can still be used by adding the related magnetic terms as follows [25]:

$$\sigma_{rr} = c_{11}\varepsilon_{rr} + c_{12}\varepsilon_{\theta\theta} + c_{13}\varepsilon_{zz} - e_{31}E_z - \tilde{e}_{31}H_z - \lambda_{11}T$$

$$\sigma_{\theta\theta} = c_{12}\varepsilon_{rr} + c_{11}\varepsilon_{\theta\theta} + c_{13}\varepsilon_{zz} - e_{31}E_z - \tilde{e}_{31}H_z - \lambda_{11}T$$

$$\sigma_{zz} = c_{13}\varepsilon_{rr} + c_{13}\varepsilon_{\theta\theta} + c_{33}\varepsilon_{zz} - e_{33}E_z - \tilde{e}_{33}H_z - \lambda_{33}T$$

$$\sigma_{zr} = c_{44}\varepsilon_{zr} - e_{15}E_r - \tilde{e}_{15}H_r$$

$$D_r = e_{15}s_{zr} + \kappa_{11}E_r + \alpha_{11}H_r$$

$$D_z = e_{31}(s_{rr} + s_{\theta\theta}) + e_{33}s_{zz} + \kappa_{33}E_z + \alpha_{33}H_z - \rho_3 T$$

$$B_r = \tilde{e}_{15}s_{zr} + \alpha_{11}E_r + \mu_{11}H_r$$

$$B_z = \tilde{e}_{31}(s_{rr} + s_{\theta\theta}) + \tilde{e}_{33}s_{zz} + \alpha_{33}E_z + \mu_{33}H_z - \upsilon_3 T$$

$$h_r = k_r W_r, \qquad h_z = k_z W_z$$

$$(6.4.1)$$

The associated magnetic field is related to magnetic potential ψ, as

$$H_r = -\psi_{,r}, \quad H_z = -\psi_{,z} \tag{6.4.2}$$

For quasi-stationary behaviour, in the absence of heat source, free electric charge, electric current, and body forces, the set of equations for thermo-magneto-electro-elastic theory of bones is completed by adding Eqs. (6.2.2), (6.2.3) and following equation of equilibrium for magnetic induction to Eqs. (6.4.1) and (6.4.2)

$$\frac{\partial B_r}{\partial r} + \frac{\partial B_z}{\partial z} + \frac{B_r}{r} = 0 \tag{6.4.3}$$

6.4.2   Solution for internal bone remodelling

1. Equation for internal bone remodelling

The extended adaptive elastic theory presented in Section 6.2 is used and extended to include the piezomagnetic effect as follows [26]:

$$\dot{e} = A^*(e) + A_r^E(e)E_r + A_z^E(e)E_z + G_r^E(e)H_r + G_z^E(e)H_z +$$
$$A_{rr}^s(e)(\varepsilon_{rr} + \varepsilon_{\theta\theta}) + A_{zz}^s(e)\varepsilon_{zz} + A_{rz}^s(e)\varepsilon_{rz} \tag{6.4.4}$$

where $G_i^E(e)$ are newly introduced material coefficients dependent upon the volume fraction e.

2. Solution for a homogeneous hollow circular cylindrical bone

Consider again a hollow circular cylinder of bone, subjected to an external temperature change T_0, a quasi-static axial load P, an external pressure p, an electric potential load ϕ_a (or/and ϕ_b) and an magnetic potential load ψ_a (and/or ψ_b). The boundary conditions are

$$\begin{aligned}
T &= 0, & \sigma_{rr} &= \sigma_{r\theta} = \sigma_{rz} = 0, & \phi &= \phi_a, & \psi &= \psi_a, & r &= a \\
T &= T_0, & \sigma_{rr} &= -p, & \sigma_{r\theta} &= \sigma_{rz} = 0, & \phi &= \phi_b, & \psi &= \psi_b, & r &= b
\end{aligned} \tag{6.4.5}$$

and

$$\int_S \sigma_{zz}\,\mathrm{d}S = -P \tag{6.4.6}$$

where a and b denote, respectively, the inner and outer radii of the bone, and S is the cross-sectional area. For a long bone, it is assumed that, except for the axial displacement u_z, all displacements, temperature and electrical potential are independent of the z coordinate and that u_z may have linear dependence on z. Using Eqs. (6.2.2), (6.4.1), and (6.4.2), the differential equations (6.2.3) and

(6.4.3) can be written as

$$\left(\frac{\partial^2}{\partial r^2}+\frac{1}{r}\frac{\partial}{\partial r}\right)T=0,\qquad c_{11}\left(\frac{\partial^2}{\partial r^2}+\frac{1}{r}\frac{\partial}{\partial r}-\frac{1}{r^2}\right)u_r=\lambda_{11}\frac{\partial T}{\partial r}\tag{6.4.7}$$

$$c_{44}\left(\frac{\partial^2}{\partial r^2}+\frac{1}{r}\frac{\partial}{\partial r}\right)u_z+e_{15}\left(\frac{\partial^2}{\partial r^2}+\frac{1}{r}\frac{\partial}{\partial r}\right)\phi+\tilde{e}_{15}\left(\frac{\partial^2}{\partial r^2}+\frac{1}{r}\frac{\partial}{\partial r}\right)\psi=0\tag{6.4.8}$$

$$e_{15}\left(\frac{\partial^2}{\partial r^2}+\frac{1}{r}\frac{\partial}{\partial r}\right)u_z-\kappa_{11}\left(\frac{\partial^2}{\partial r^2}+\frac{1}{r}\frac{\partial}{\partial r}\right)\phi-\alpha_{11}\left(\frac{\partial^2}{\partial r^2}+\frac{1}{r}\frac{\partial}{\partial r}\right)\psi=0\tag{6.4.9}$$

$$\tilde{e}_{15}\left(\frac{\partial^2}{\partial r^2}+\frac{1}{r}\frac{\partial}{\partial r}\right)u_z-\alpha_{11}\left(\frac{\partial^2}{\partial r^2}+\frac{1}{r}\frac{\partial}{\partial r}\right)\phi-\mu_{11}\left(\frac{\partial^2}{\partial r^2}+\frac{1}{r}\frac{\partial}{\partial r}\right)\psi=0\tag{6.4.10}$$

The solution of displacements u_r, u_z, and electric potential ϕ to the problem above in the absence of magnetic field was presented in Section 6.2. This section extends the results in Section 6.2 to include the piezomagnetic effect. It is found that temperature T, displacement u_r, and electric potential ϕ are again given by Eqs. (6.2.13), (6.2.32), and (6.3.24), respectively, while u_z and ψ are as follows [26]:

$$u_z=\frac{z}{F_3^*}\left\{\left[F_1^*T_0-\frac{F_2^*T_0+P(t)}{\pi(b^2-a^2)}\right](c_{11}+c_{12})-2c_{13}\beta_1^*[\beta_2^*T_0+p(t)]-\right.$$

$$\left.\frac{2c_{13}c_{12}\bar{\omega}}{c_{11}}\right\}-\frac{e_{15}(\phi_b-\phi_a)\ln(r/a)}{c_{44}\ln(b/a)}-\frac{\tilde{e}_{15}(\psi_b-\psi_a)\ln(r/a)}{c_{44}\ln(b/a)}\tag{6.4.11}$$

$$\psi=\frac{\ln(r/a)}{\ln(b/a)}(\psi_b-\psi_a)+\psi_a\tag{6.4.12}$$

The strains, electric field, and magnetic field can be found by introducing the boundary conditions (6.4.5) and (6.4.6) into Eqs. (6.2.2) and (6.4.2). They are, respectively,

$$s_{rr}=\frac{1}{F_3^*}\left\{c_{33}\beta_1^*\left[\beta_2^*T_0+p(t)\right]+\bar{\omega}\frac{c_{33}c_{12}}{c_{11}}+\frac{F_2^*T_0+P(t)}{\pi(b^2-a^2)}c_{13}-F_1^*T_0c_{13}\right\}-$$

$$\frac{a^2\beta_1^*\left[\beta_2^*T_0+p(t)\right]}{r^2(c_{11}-c_{12})}+\frac{\bar{\omega}\ln(r/a)}{c_{11}}\tag{6.4.13}$$

$$s_{\theta\theta}=\frac{1}{F_3^*}\left\{c_{33}\beta_1^*\left[\beta_2^*T_0+p(t)\right]+\bar{\omega}\frac{c_{33}c_{12}}{c_{11}}+\frac{F_2^*T_0+P(t)}{\pi(b^2-a^2)}c_{13}-F_1^*T_0c_{13}\right\}+$$

$$\frac{a^2\beta_1^*\left[\beta_2^*T_0+p(t)\right]}{r^2(c_{11}-c_{12})}+\frac{\bar{\omega}\left[\ln(r/a)-1\right]}{c_{11}}\tag{6.4.14}$$

$$s_{zz} = \frac{1}{F_3^*}\left\{\left[F_1^*T_0 - \frac{F_2^*T_0 + P(t)}{\pi(b^2 - a^2)}\right](c_{11} + c_{12}) - 2c_{13}\beta_1^*\left[\beta_2^*T_0 + p(t)\right] - \frac{2c_{13}c_{12}\bar{\omega}}{c_{11}}\right\}$$

$$(6.4.15)$$

$$s_{rz} = -\frac{e_{15}(\phi_b - \phi_a)}{rc_{44}\ln(b/a)} - \frac{\alpha_{15}(\psi_b - \psi_a)}{rc_{44}\ln(b/a)} \tag{6.4.16}$$

$$E_r = -\frac{(\phi_b - \phi_a)}{r\ln(b/a)}, \qquad H_r = -\frac{(\psi_b - \psi_a)}{r\ln(b/a)} \tag{6.4.17}$$

Substituting Eqs. (6.4.13)$\sim$(6.4.17) into Eq.(6.4.4) yields [26]

$$\dot{e} = A^*(e) + \frac{2A_{rr}^s}{F_3^*}\left\{c_{33}\beta_1^*\left[\beta_2^*T_0 + p(t)\right] + \bar{\omega}\frac{c_{33}c_{12}}{c_{11}} + \frac{F_2^*T_0 + P(t)}{\pi(b^2 - a^2)}c_{13} - F_1^*T_0 c_{13}\right\} +$$

$$\frac{A_{rr}^s\bar{\omega}\left[2\ln(r/a) - 1\right]}{c_{11}} + \frac{A_{zz}^s}{F_3^*}\left\{\left[F_1^*T_0 - \frac{F_2^*T_0 + P(t)}{\pi(b^2 - a^2)}\right](c_{11} + c_{12}) -\right.$$

$$\left.2c_{13}\beta_1^*\left[\beta_2^*T_0 + p(t)\right] - \frac{2c_{13}c_{12}\bar{\omega}}{c_{11}}\right\} - \frac{\phi_b - \phi_a}{r\ln(b/a)}\left(A_r^E + \frac{e_{15}}{c_{44}}A_{zr}^s\right) -$$

$$\frac{\psi_b - \psi_a}{r\ln(b/a)}\left(G_r^E + \frac{\alpha_{15}}{c_{44}}A_{zr}^s\right) \tag{6.4.18}$$

Eq.(6.4.18) can be solved in a way similar to that described in Section 6.2.

3. Numerical assessment

As a numerical illustration of the analytical solution presented above, we consider again the femur used in Subsection 6.2.6. The material parameters used here are the same as those given in Subsection 6.2.6. The additional material constants for magnetic field are

$$\tilde{e}_{15} = 550(1 + e)\ \text{N/Am}, \qquad G_r^{E0} = G_r^{E1} = 1.5 \times 10^{-8}\ \text{m/(A} \cdot \text{d)}$$

We investigate the change of the volume fraction of bone matrix material from its reference value, which is denoted by e, in the transverse direction at several specific times. We also distinguish two loading cases to investigate the influence of magnetic and coupling loads on the bone structure.

(1) $p(t) = 0$, $P = 1500\ \text{N}$, $T_0(t) = 0\ ^\circ\text{C}$, $\phi_b - \phi_a = 0$, $\psi_b - \psi_a = 1\ \text{A}$.

Fig. 6.13 shows the variation of e with time t along the radii of bone when the loading case is $p(t) = 0$, $P = 1500\ \text{N}$, $T_0(t) = 0\ ^\circ\text{C}$, $\phi_b - \phi_a = 0$, $\psi_b - \psi_a = 1\text{A}$.

It can be seen from Fig. 6.13 that a magnetic load has a similar influence on bone structure to an electric load. A magnetic load can also inhomogenize an initially homogeneous bone structure through the bone remodeling process. But essentially further experimental and theoretical investigations need to be developed to ascertain the exact remodeling rate coefficients and to discover the importance of the role played by magnetic stimuli.

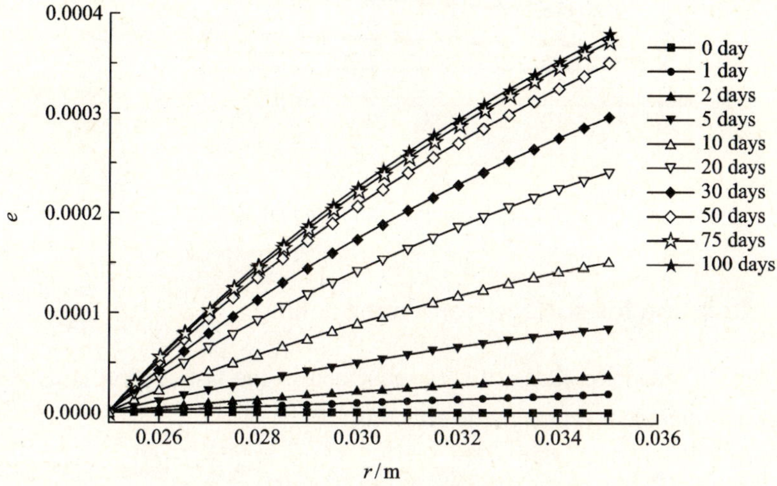

Fig. 6.13   Variation of e with time t along the radii for magnetic load

(2) $p(t) = 1\,\text{MPa}$, $P = 1500\,\text{N}$, $T_0(t) = 0.1\,^\circ\text{C}$, $\phi_b - \phi_a = 30\,\text{V}$, $\psi_b - \psi_a = 1\,\text{A}$.

Fig. 6.14 shows the variation of e with time t in the transverse direction when subjected to coupling loads. The above loading case is considered to study the coupling effect of magnetoelectric and mechanical loads on bone structure. It can be seen from Fig. 6.14 that the function of coupled loads is the superimposition of the single loads. However, they are not simply linearly superposed. Further, the properties of bone tissue change more sharply under coupled loads than when it is subjected to only one load. The combination of the magnetic, electric, thermal and mechanical loads results in significant change in bone structure and properties of bone tissues. This indicates that loading coupled fields is more effective in modifying bone structure than loading only one kind of field.

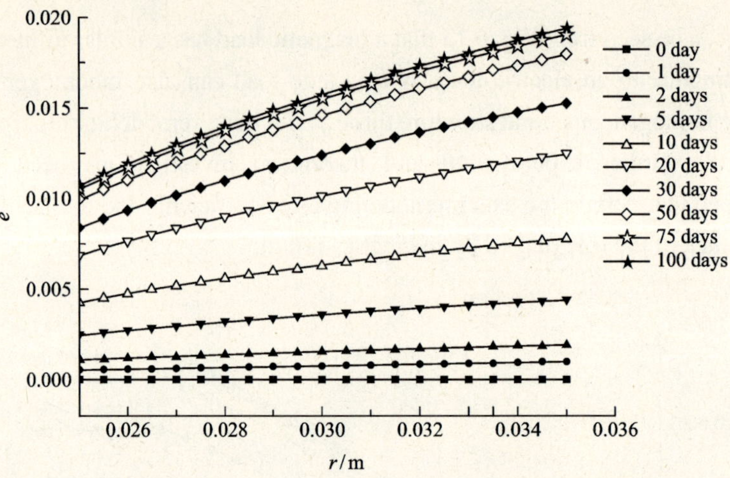

Fig. 6.14   Variation of e with time t along the radii for coupling loads

6.4.3   Solution for surface bone remodelling

The extended adaptive elastic theory presented in the last section is used and extended to include piezomagnetic effect as [27]

$$U = C_{ij}(\boldsymbol{n},Q)\left[s_{ij}(Q) - s_{ij}^0(Q)\right] + C_i\left[E_i(Q) - E_i^0(Q)\right] + G_i\left[H_i(Q) - H_i^0(Q)\right]$$

$$= C_{rr}s_{rr} + C_{\theta\theta}s_{\theta\theta} + C_{zz}s_{zz} + C_{rz}s_{rz} + C_r E_r + C_z E_z + G_r H_r + G_z H_z - C_0$$

$$\tag{6.4.19}$$

where $C_0 = C_{rr}s_{rr}^0 + C_{zz}s_{zz}^0 + C_{\theta\theta}s_{\theta\theta}^0 + C_{rz}s_{rz}^0 + C_r E_r^0 + C_z E_z^0 + G_r H_r^0 + G_z H_z^0$, G_i is a surface remodeling coefficient.

Substituting Eqs. (6.4.13)$\sim$(6.4.17) into Eq.(6.3.19) yields [27]

$$\begin{cases} U_e = N_1^e \dfrac{b^2}{b^2 - a^2} + N_2^e \dfrac{1}{\ln\left(\dfrac{b}{a}\right)} + N_3^e \dfrac{1}{b^2 - a^2} + N_4^e \dfrac{1}{a\ln\left(\dfrac{b}{a}\right)} - C_0^e \\[4em] U_p = N_1^p \dfrac{b^2}{b^2 - a^2} + N_1^{p'} \dfrac{a^2}{b^2 - a^2} + N_2^p \dfrac{1}{\ln\left(\dfrac{b}{a}\right)} + N_3^p \dfrac{1}{b^2 - a^2} + \\[4em] \qquad N_4^p \dfrac{1}{b\ln\left(\dfrac{b}{a}\right)} + N_2^{p'} - C_0^p \end{cases} \tag{6.4.20}$$

where

$$N_1^e = \frac{1}{F_3^*}\left\{ c_{13}\left(\frac{c_{13}}{c_{11}}\beta_1 - \beta_3\right)T_0 - c_{33}\left[\frac{\beta_1}{2}\left(\frac{c_{12}}{c_{11}}-1\right)T_0 + p(t)\right]\right\}\left(C_{rr}^e + C_{\theta\theta}^e\right) +$$

$$\frac{1}{F_3^*}\left\{2c_{13}C_{zz}^e\left[\frac{\beta_1}{2}\left(\frac{c_{12}}{c_{11}}-1\right)T_0 + p(t)\right] - (c_{11}+c_{12})C_{zz}^e\left(\frac{c_{13}}{c_{11}}\beta_1 - \beta_3\right)T_0\right\} +$$

$$\frac{(C_{rr}^e - C_{\theta\theta}^e)\left[\dfrac{\beta_1}{2}\left(\dfrac{c_{12}}{c_{11}}-1\right)T_0 + p(t)\right]}{c_{11}-c_{12}}$$

$$(6.4.21)$$

$$N_2^e = \frac{1}{F_3^*}\left[\frac{c_{12}c_{33}}{2c_{11}}\beta_1 T_0 - \left(\frac{c_{13}}{c_{11}}\beta_1 - \frac{\beta_3}{2}\right)c_{13}T_0\right]\left(C_{rr}^e + C_{\theta\theta}^e\right) +$$

$$\frac{C_{zz}^e}{F_3^*}\left[(c_{11}+c_{12})\left(\frac{c_{13}}{c_{11}}\beta_1 - \frac{\beta_3}{2}\right)T_0 - \frac{c_{12}c_{13}}{c_{11}}\beta_1 T_0\right] - \frac{C_{\theta\theta}^e \beta_1 T_0}{2c_{11}} \tag{6.4.22}$$

$$N_3^e = \frac{1}{F_3^*}\left[c_{13}\left(C_{rr}^e + C_{\theta\theta}^e\right) - (c_{11}+c_{12})C_{zz}^e\right]\frac{P(t)}{\pi} \tag{6.4.23}$$

$$N_4^e = -\left(\frac{e_{15}}{c_{44}}C_{zr}^e + C_r\right)(\phi_b - \phi_a) - \left(\frac{\alpha_{15}}{c_{44}}C_{zr}^e + G_r\right)(\psi_b - \psi_a) \tag{6.4.24}$$

$$N_1^p = \frac{1}{F_3^*}\left\{ c_{13}\left(\frac{c_{13}}{c_{11}}\beta_1 - \beta_3\right)T_0 - c_{33}\left[\frac{\beta_1}{2}\left(\frac{c_{12}}{c_{11}}-1\right)T_0 + p(t)\right]\right\}\left(C_{rr}^p + C_{\theta\theta}^p\right) +$$

$$\frac{1}{F_3^*}\left\{2c_{13}C_{zz}^p\left[\frac{\beta_1}{2}\left(\frac{c_{12}}{c_{11}}-1\right)T_0 + p(t)\right] - (c_{11}+c_{12})C_{zz}^p\left(\frac{c_{13}}{c_{11}}\beta_1 - \beta_3\right)T_0\right\}$$

$$(6.4.25)$$

$$N_1^{p'} = -\frac{(C_{rr}^p - C_{\theta\theta}^p)\left[\dfrac{\beta_1}{2}\left(\dfrac{c_{12}}{c_{11}}-1\right)T_0 + p(t)\right]}{c_{11}-c_{12}}, \quad N_2^{p'} = \frac{(C_{\theta\theta}^p + C_{rr}^p)\beta_1 T_0}{2c_{11}} \tag{6.4.26}$$

$$N_2^p = \frac{1}{F_3^*}\left[\frac{c_{12}c_{33}}{2c_{11}}\beta_1 T_0 - \left(\frac{c_{13}}{c_{11}}\beta_1 - \frac{\beta_3}{2}\right)c_{13}T_0\right]\left(C_{rr}^p + C_{\theta\theta}^p\right) +$$

$$\frac{C_{zz}^p}{F_3^*}\left[(c_{11}+c_{12})\left(\frac{c_{13}}{c_{11}}\beta_1 - \frac{\beta_3}{2}\right)T_0 - \frac{c_{12}c_{13}}{c_{11}}\beta_1 T_0\right] - \frac{C_{\theta\theta}^p \beta_1 T_0}{2c_{11}} \tag{6.4.27}$$

$$N_3^p = \frac{1}{F_3^*}\left[c_{13}\left(C_{rr}^p + C_{\theta\theta}^p\right) - (c_{11}+c_{12})C_{zz}^p\right]\frac{P(t)}{\pi} \tag{6.4.28}$$

$$N_4^p = -\left(\frac{e_{15}}{c_{44}}C_{zr}^p + C_r\right)(\phi_b - \phi_a) - \left(\frac{\alpha_{15}}{c_{44}}C_{zr}^p + G_r\right)(\psi_b - \psi_a) \tag{6.4.29}$$

and the subscripts p and e refer to periosteal and endosteal surfaces, respectively. Since U_e and U_p are the velocities normal to the inner and outer surfaces of the cylinders, respectively, they are calculated as

$$U_e = -\frac{da}{dt}, \qquad U_p = \frac{db}{dt} \tag{6.4.30}$$

where the minus sign appearing in the expression for U_e denotes that the outward normal of the endosteal surface is in the negative coordinate direction. Thus, equations (6.4.20) can be written as [27]

$$\left\{ \begin{aligned}
-\frac{da}{dt} &= N_1^e \frac{b^2}{b^2 - a^2} + N_2^e \frac{1}{\ln\left(\dfrac{b}{a}\right)} + N_3^e \frac{1}{b^2 - a^2} + N_4^e \frac{1}{a\ln\left(\dfrac{b}{a}\right)} - C_0^e \\
\frac{db}{dt} &= N_1^p \frac{b^2}{b^2 - a^2} + N_1^{p'} \frac{a^2}{b^2 - a^2} + N_2^p \frac{1}{\ln\left(\dfrac{b}{a}\right)} + N_3^p \frac{1}{b^2 - a^2} + \\
&\qquad N_4^p \frac{1}{b\ln\left(\dfrac{b}{a}\right)} - C_0^{p'}
\end{aligned} \right. \tag{6.4.31}$$

where $C_0^{p'} = C_0^p - N_3'$.

These equations are quite similar to those presented in Section 6.3 except for the additional terms related to magnetic field. Their solution procedure is similar to that in Section 6.3 and we omit it here for conciseness.

As a numerical illustration of the analytical solutions above, we consider the femur used in Subsection 6.4.2. The material constants are assumed to be the same as those in Section 6.3. The additional surface remodelling constant for magnetic field is

$$G_r = 10^{-10} \text{ m/d}$$

We distinguish the following three loading cases:

(1) $T_0(t) = -0.5 \,^{\circ}\text{C}$, $-0.2 \,^{\circ}\text{C}$, $0 \,^{\circ}\text{C}$, $0.2 \,^{\circ}\text{C}$, $0.5 \,^{\circ}\text{C}$, $\phi_b - \phi_a = 30$ V, $\psi_b - \psi_a = 1$ A, $p(t) = 1$ MPa, $P(t) = 1500$ N.

Fig. 6.15 shows the effects of temperature change on bone surface remodelling. In general, the radii of the bone decrease when the temperature increases and they increase when the temperature decreases. It can also be seen from Fig. 6.15 that ε and η are almost the same. Since $a_0 < b_0$, the change of the outer surface radius is normally greater than that of the inner surface radius.

The area of the bone cross section decreases as the temperature increases. This also suggests that a lower temperature is likely to induce thicker bone structures, while a warmer environment may improve the remodeling process with a less thick bone structure. It should be mentioned here that how this change may affect the bone remodeling process is still an open question. As an initial investigation, the purpose of this section to show how a bone may response to thermal loads and to provide information for possible use of imposed external temperature fields in medical treatment and controlling the healing process of injured bone.

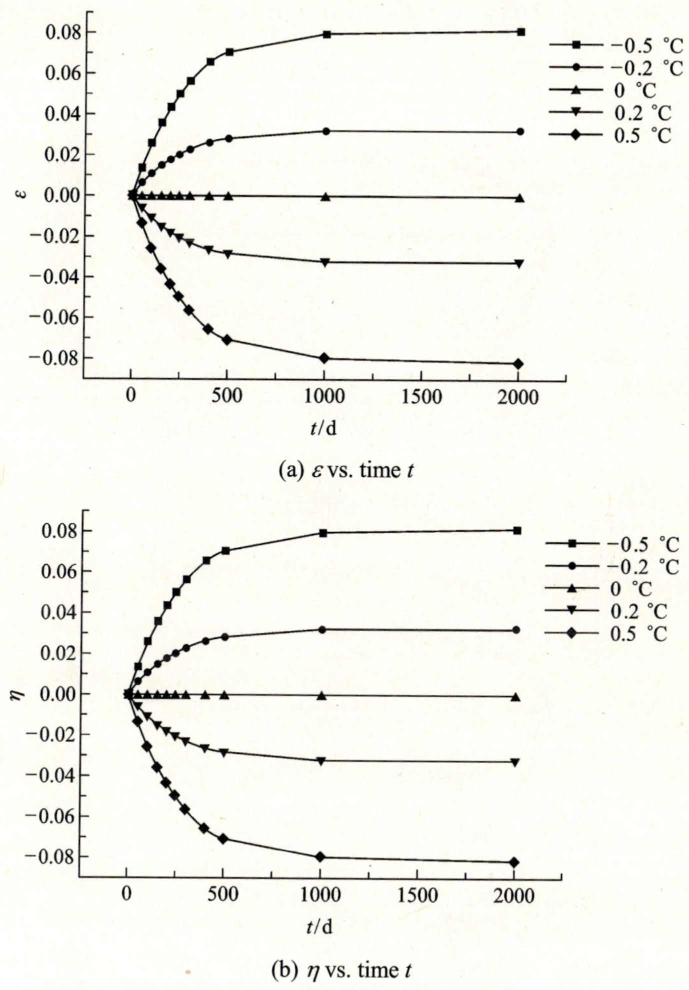

(a) ε vs. time t

(b) η vs. time t

Fig. 6.15   Variation of ε and η with time t for several temperature changes

(2) $\phi_b - \phi_a = -60$ V, -30 V, 30 V, and 60 V, $p(t) = 1$ Mpa, $P(t) = 1500$ N, $\psi_b - \psi_a = 1$ A and $T_0 = 0$.

Fig. 6.16 shows the variation of ε and η with time t for various values of electric potential difference. It can be seen that the effect of the electric potential is the opposite to that of temperature. A decrease of the intensity of electric field results in a decrease of the inner and outer surface radii of the bone by almost the same magnitude. Theoretically, the results suggest that the remodelling process may be improved by exposing a bone to an electric field. It is evident that further theoretical and experimental studies are needed to investigate the implication of this for medical practice.

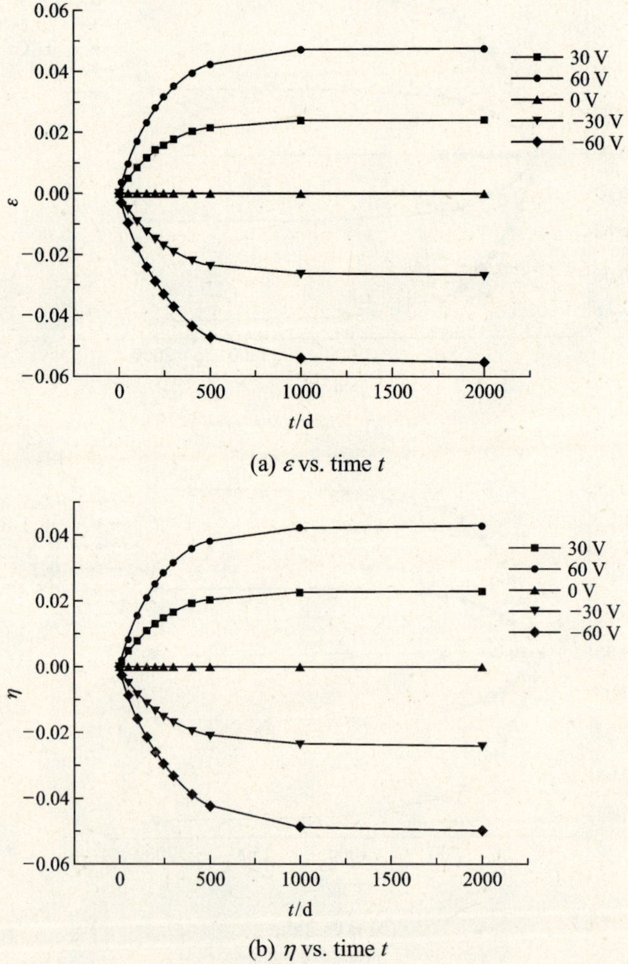

(a) ε vs. time t

(b) η vs. time t

Fig.6.16   Variation of ε and η with time t for several electric potential differences

(3) $\psi_b - \psi_a = -2$ A, -1 A, 1 A, and 2 A, $p(t) = 1$ MPa $P(t) = 1500$ N, $\phi_b - \phi_a = 30$ V and $T_0 = 0$.

Fig. 6.17 shows the variation of ε and η with time t for various values of magnetic potential difference. The changes in the outer and inner surfaces of the bone due to magnetic influence are similar to those for electric field as shown in Section 6.2.

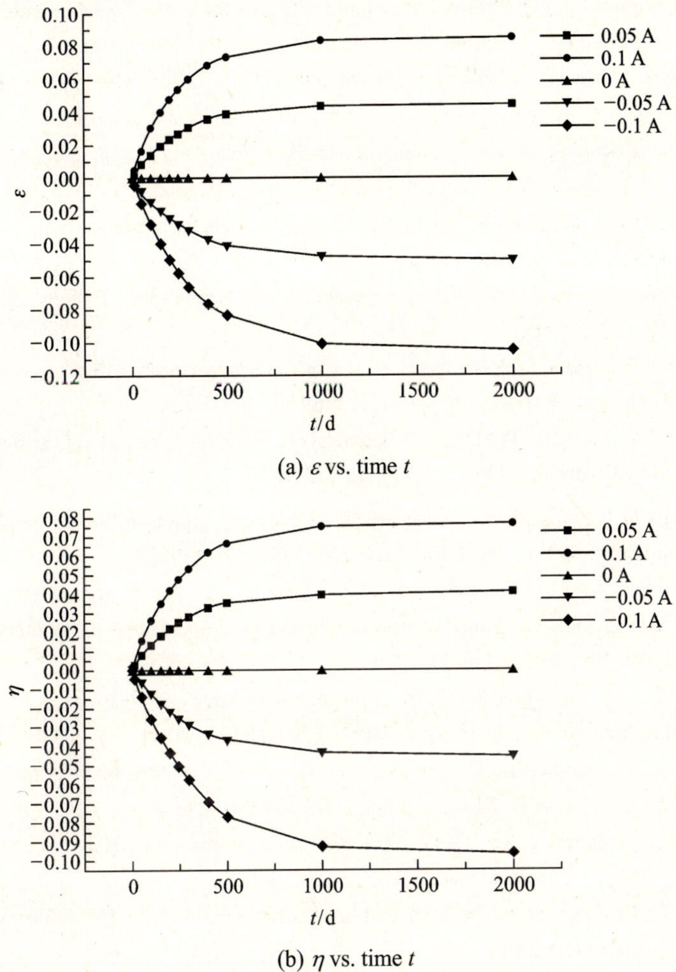

(a) ε vs. time t

(b) η vs. time t

Fig. 6.17　Variation of ε and η with time t for several magnetic potential differences

References

[1]　Fukada E, Yasuda I. On the piezoelectric effect of bone. J. phys. Soc. Japan, 1957, 12:

1158-1162.

[2] Fukada E, Yasuda I. piezoelectric effects in collagen. Jap. J. Appl. Phys.,1964, 3: 117-121.

[3] Kryszewski M. Fifty years of study of the piezoelectric properties of macromolecular structured biological materials. Acta Physica Polonica A, 2004, 105: 389-408.

[4] Robiony M, Polini F, Costa F, et al. Piezoelectric bone cutting in multioiece maxillary osteotomies. J. Oral Maxillofac Surg., 2004, 62:759-761.

[5] Qin Qinghua, Ye J Q. Thermolectroelastic solutions for internal bone remodeling under axial and transverse loads. Int. J. Solids Struct., 2004, 41: 2447-2460.

[6] Gjelsvik A. Bone remodelling and piezoelectricity—I. J. Biomech., 1973, 6:69-77.

[7] Williams W S, Breger L. Analysis of stress distribution and piezoelectric response in cantilever bending of bone and tendon. Ann. N.Y. Acad. Sci., 1974, 238: 121-130.

[8] Guzelsu N. A piezoelectric model for dry bone tissue. J. Biomech., 1978, 11: 257-267.

[9] Johnson M W. Williams W S, Gross D. Ceramic models for piezoelectricity in dry bone. J. Biomech., 1980, 13: 565-573.

[10] Demiray H. Electro-mechanical remodelling of bones. Int. J. Eng. Sci., 1983, 21:1117-1126.

[11] Aschero G, Gizdulich P, Mango F, et al. Converse piezoelectric effect detected in fresh cow femur bone. J. Biomech., 1996, 29: 1169-1174.

[12] Aschero G, Gizdulich P, Mango F. Statistical characterization of piezoelectric coefficient d_{23} in cow bone. J. Biomech., 1999, 32:573-577.

[13] Fotiadis D I, Foutsitzi G, Massalas C V. Wave propagation in human long bones of arbitrary cross-section. Int. J. Eng. Sci., 2000, 38: 1553-1591.

[14] El-Naggar A M, Abd-Alla A M, Mahmoud S R. Analytical solution of electro-mechanical wave propagation in long bones. Appl. Mathe. Computation, 2001, 119: 77-98.

[15] Ahmed S M, Abd-Alla A M. Electromechanical wave propagation in a cylindrical poroelastic bone with cavity. Appl. Mathe. Computation, 2002, 133: 257-286.

[16] Silva C C, Thomazini D, Pinheriro A G, et al. Collagen-hydroxyapatite films: piezoelectric properties. Mat. Sci. Eng., 2001, B86:210-218.

[17] Qin Qinghua, Qu C Y, Ye J Q. Thermolectroelastic solutions for surface bone remodeling under axial and transverse loads. Biomaterials, 2005, 26: 6798-6810.

[18] Qu C Y. Research on the behaviour of bone surface remodelling (in Chinese). Tianjin: Tianjin University, 2005.

[19] Mindlin R D. Equations of high frequency vibrations of thermopiezoelectric crystal plates. Int. J. Solids Struct., 1974, 10: 625-637.

[20] Cowin S C, Hegedus D H. Bone remodelling I: theory of adaptive elasticity. J. Elasticity, 1976, 6: 313-326.

[21] Cowin S C, Van Buskirk W C. Internal bone remodeling induced by a medullary pin. J. Biomech., 1978, 11: 269-275.

[22] Hegedus D H, Cowin S C. Bone remodelling II: small strain adaptive elasticity. J. Elasticity, 1976, 6: 337-352.

[23] Ye J Q. Laminated composite plates and shells: 3D modeling. London: Springer-Verlag, 2002.

[24] Cowin S C, Buskirk W C V. Surface bone remodelling induced by a medullary pin. J. Biomech., 1979, 12: 269-276.

[25] Gao C F, Naotake N. Thermal-induced interfacial cracking of magneto-electroelastic materials. Int. J. Eng. Sci., 2004, 42: 1347-1360.

[26] Qu C Y, Qin Q H. Evolution of bone structure under axial and transverse loads. Struct. Eng. Mech., 2006, 24: 19-29.

[27] Qu C Y, Qin Q H, Kang Y L. Thermomagnetoelectroelastic prediction of the bone surface remodeling under axial and transverse loads//Proc. of 9th International Conference on Inspection, Appraisal, Repairs & Maintenance of Structures, October 20-21, Fuzhou, China, 2005: 373-380.

Chapter 7 Effective coupling properties of heterogeneous materials

Intelligent materials have found increasing applications in engineering structures, especially in adaptive structure systems, as sensing and actuating devices and components. Composite materials have been developed to create smart properties through coupling of the mechanical and non-mechanical properties. A micromechanics analysis of intelligent composite materials is very helpful for studying their property-structure relationships and guiding the design and optimization of the new materials.

As an example, composite materials consisting of a piezoelectric phase and a piezomagnetic phase have attracted significant interest in recent years. Such materials exhibit considerable multi-field coupling properties, i.e. electro-magneto-mechanical coupling. They evidence a remarkably large magnetoelectric coefficient, the coupling coefficient between static electric and magnetic fields, which is absent in either constituent. Magnetoelectric coupling, a new product property in the composite, is created through the interaction between the piezoelectric phase and the piezomagnetic phase. The product properties of such composite offer great opportunities to engineers for the design of new materials.

In 1974, Van Run [1] et al. reported the fabrication and magneto-electric effect of a composite consisting of $BaTiO_3$ (piezoelectric phase) and $CoFe_2O_4$ (piezomagnetic phase). The magnetoelectric coefficient is two orders larger than that of Cr_2O_3, which had the highest magnetoelectric coefficient among single-phase materials known at that time. Bracke and Van Vliet [2] reported a broad band magnetoelectric transducer with a flat frequency response using composite materials. Since then, much theoretical and experimental work for investigation of the magneto-electric coupling effect has been carried out. A summary of this topic was given in Chapter 4.

In fact, the coupling effect exists in many materials. A simple example is piezoelectric material which exhibits electro-mechanical coupling properties [3,4]. In general, a composite consisting of constituents with coupling properties can exhibit effective coupling properties and create, in some situations, the product properties. Table 7.1 summaries the coupling effect of several composites.

Table 7.1　Coupling properties of composites

Properties of matrix	Properties of inclusion	Effective properties	Product properties
Mechanical	Thermo-mechanical	Thermo-mechanical	
Electro-mechanical	Defect	Electro-mechanical	
Mechanical	Electro-mechanical	Electro-mechanical	
Thermo-mechanical	Electro-mechanical	Electro-mechanical	Thermo-electric
Magneto-mechanical	Defect	Magneto-mechanical	
Electro-mechanical	Magneto-mechanical	Magneto-electro-mechanical	Magneto-electric
Mechanical	Electro-mechanical/ magneto-mechanical	Magneto-electro-mechanical	Magneto-electric
Magneto-electro-mechanical	Defect	Magneto-electro-mechanical	
Thermo-magneto-electro-mechanical	Defect	Thermo-magneto-electro-mechanical	
Thermo-mechanical	Magneto-electro-mechanical	Thermo-magneto-electro-mechanical	Thermo-mechanical, thermo-magnetic

In this chapter, the effective coupling properties of heterogeneous materials are emphasized. The basic methods for homogenization of heterogeneous materials are described and several numerical results are presented.

7.1　Basic equations for multifield coupling

Consider a thermo-magneto-electro-mechanical coupling problem. For a thermo-magneto-electro-elastic material, the basic equations can be summarized as follows. The governing equations are

$$\sigma_{ij,j} + b_i = 0, \quad D_{i,i} + b_e = 0, \quad B_{i,i} + b_m = 0 \tag{7.1.1}$$

The constitutive equations are

$$\left.\begin{array}{l}\sigma_{ij} = c_{ijkl}\varepsilon_{kl} - e_{lij}E_l - \tilde{e}_{lij}H_l - \lambda_{ij}T \\[2mm] D_i = e_{ikl}\varepsilon_{kl} + \kappa_{il}E_l + \alpha_{il}H_l - \rho_iT \\[2mm] B_i = \tilde{e}_{ikl}\varepsilon_{kl} + \alpha_{il}E_l + \mu_{il}H_l - \nu_iT \end{array}\right\} \qquad (7.1.2)$$

The gradient equations are

$$\varepsilon_{ij} = \frac{1}{2}(u_{i,j} + u_{j,i}), \quad E_i = -\phi_{,i}, \quad H_i = -\psi_{,i} \qquad (7.1.3)$$

$$q_{i,i} = 0, \quad q_i = -k_{ij}T_{,j} \qquad (7.1.4)$$

For convenience of writing, we use the same notations here as in previous chapters. Definite vectors $\boldsymbol{Z}$ and $\boldsymbol{\Pi}$ as

$$\boldsymbol{Z} = \begin{bmatrix} \varepsilon_{11} & \varepsilon_{22} & \varepsilon_{33} & 2\varepsilon_{23} & 2\varepsilon_{13} & 2\varepsilon_{12} \\ -E_1 & -E_2 & -E_3 & -H_1 & -H_2 & -H_3 \end{bmatrix} \qquad (7.1.5)$$

$$\boldsymbol{\Pi} = \begin{bmatrix} \sigma_{11} & \sigma_{22} & \sigma_{33} & \sigma_{23} & \sigma_{13} & \sigma_{12} \\ D_1 & D_2 & D_3 & B_1 & B_2 & B_3 \end{bmatrix} \qquad (7.1.6)$$

Thus the constitutive equations (7.1.2) can be rewritten in compact form

$$\boldsymbol{\Pi} = \boldsymbol{EZ} - \boldsymbol{\Gamma}T \qquad (7.1.7)$$

Where coefficient $\boldsymbol{E}$ is 12th order symmetric matrix in the form

$$\boldsymbol{E} = \begin{bmatrix} \boldsymbol{c} & \boldsymbol{e}^{\mathrm{T}} & \tilde{\boldsymbol{e}}^{\mathrm{T}} \\ \boldsymbol{e} & -\boldsymbol{\kappa} & -\boldsymbol{\alpha} \\ \tilde{\boldsymbol{e}} & -\boldsymbol{\alpha} & -\boldsymbol{\mu} \end{bmatrix} \qquad (7.1.8)$$

and

$$\boldsymbol{\Gamma} = \begin{bmatrix} -\lambda \\ -\rho \\ -\nu \end{bmatrix} \qquad (7.1.9)$$

In Eq.(7.1.8), 6×6 matrix $\boldsymbol{c}$ is a 4th order stiffness tensor; 3×6 matrix $\boldsymbol{e}$ is a 3rd order piezoelectric tensor; 6×3 matrix $\boldsymbol{e}^{\mathrm{T}}$ is the transpose of $\boldsymbol{e}$; 3×6 matrix $\tilde{\boldsymbol{e}}$ is a 3rd piezomagnetic tensor; $\tilde{\boldsymbol{e}}^{\mathrm{T}}$ is the transpose of $\tilde{\boldsymbol{e}}$; 3×3 matrix $\boldsymbol{\kappa}$ denotes a dielectric tensor; 3×3 matrix $\boldsymbol{\alpha}$ is a magneto-electric coefficient matrix; 3×3 $\boldsymbol{\mu}$ is a magnetic permeability tensor; 6th dimensional vector λ is a thermo-mechanical coefficient; 3rd dimensional vector ρ is a thermo-electric coefficient; 3rd dimensional vector ν is a thermo-magnetic coefficient.

Here we use a new notation system here. A repeated subscript represents summation from 1 to 3. A repeat capital letter stands for the summation from 1 to 5. For instance, $T_JU_J = T_jU_j + T_4U_4 + T_5U_5$. Using this notation, we have

$$U_I = \begin{cases} u_i, & I = 1,2,3 \\ \phi, & I = 4 \\ \varphi, & I = 5 \end{cases}, \quad \Pi_{iJ} = \begin{cases} \sigma_{ij}, & J = 1,2,3 \\ D_i, & J = 4 \\ B_i, & J = 5 \end{cases}, \quad Z_{Mn} = \begin{cases} \varepsilon_{mn}, & M = 1,2,3 \\ -E_n, & M = 4 \\ -H_n, & M = 5 \end{cases}$$

$$(7.1.10)$$

$$E_{iJMn} = \begin{cases} c_{ijmn}, & J,M = 1,2,3 \\ e_{ijn}, & M = 4, J = 1,2,3 \\ \tilde{e}_{ijn}, & M = 5, J = 1,2,3 \\ e_{ijn}, & J = 4, M = 1,2,3 \\ -\kappa_{in}, & J = 4, M = 4 \\ -\alpha_{in}, & J = 4, M = 5 \\ \tilde{e}_{ijn}, & J = 5, M = 1,2,3 \\ -\alpha_{in}, & J = 5, M = 4 \\ -\mu_{in}, & J = 5, M = 5 \end{cases}, \quad \Gamma_{iJ} = \begin{cases} \lambda_{ij} & J = 1,2,3 \\ \rho_i & J = 4 \\ \nu_i & J = 5 \end{cases} \qquad (7.1.11)$$

These are called, for convenience, the generalized displacements, stresses, strains and generalized stiffness and thermal coefficients, respectively. Eq.(7.1.7) can be rewritten as

$$\Pi_{iJ} = E_{iJKl}Z_{Kl} - \Gamma_{iJ}T \qquad (7.1.12)$$

In this chapter, we focus on the effective coupling properties of a composite with a hierarchic microstructure. As with the situation of an elastic composite, we summarize the homogenization methods to predict the effective coupling properties of the composite. Here the direct average method, indirect average method and mathematical homogenization method are briefly reviewed.

(1) The direct method is based on the average of local fields to calculate effective coupling properties. For instance, a uniform strain is applied to a piezoelectric composite and the average of the electric displacement is calculated. Then it is possible to calculate the effective piezoelectric coefficients of the composite. In general, local fields can be calculated by a numerical method, such as the boundary element method (BEM) [5] or finite element method (FEM) [6].

(2) The indirect method extends the elastic inclusion theory into multi-field coupling problems. There are many methods along this line to predict the effective piezoelectric [7-10], thermo-electric [11,12], magneto-electric [13-16] and piezoelectric-piezomagnetic [17-19] properties of composites.

(3) Mathematical homogenization is applied to multi-field composites

with periodic microstructure. The displacement, electric field and magnetic field are asymptotically expanded on two scales. An averaging procedure over a volume is used to calculate the effective coupling properties. Aboudi [20] has presented a micromechanical model to evaluate the effective thermo-magneto-electro-mechanical properties of composites.

7.2 Direct method

Neglecting the thermal effect, the constitutive equation for a coupling problem is

$$\boldsymbol{\Pi} = \boldsymbol{E}\boldsymbol{Z} \tag{7.2.1}$$

In the direct method, the average fields of generalized strain $\boldsymbol{Z}$ and generalized stress $\boldsymbol{\Pi}$ are first evaluated, then the effective coupling properties can be obtained. In detail, for a heterogeneous medium, consider a RVE subjected to a specific boundary condition and calculate the local field, $\boldsymbol{Z}$ and $\boldsymbol{\Pi}$, by a numerical method, such as BEM and FEM. Then a volume averaging is carried out and the homogenized fields, $\bar{\boldsymbol{Z}}$ and $\bar{\boldsymbol{\Pi}}$, can be obtained by

$$\bar{\boldsymbol{Z}} = \frac{1}{V}\int_{\Omega} \boldsymbol{Z}\,\mathrm{d}\Omega, \quad \bar{\boldsymbol{\Pi}} = \frac{1}{V}\int_{\Omega} \boldsymbol{\Pi}\,\mathrm{d}\Omega \tag{7.2.2}$$

The effective stiffness $\bar{\boldsymbol{E}}$ can be determined by

$$\bar{\boldsymbol{\Pi}} = \bar{\boldsymbol{E}}\bar{\boldsymbol{Z}} \tag{7.2.3}$$

Expanding Eq.(7.2.3), we can obtain

$$\bar{\Pi}_i = \bar{E}_{ij}\bar{Z}_j \tag{7.2.4}$$

Applying uniaxial $\bar{Z}_1 = 1$ and other $\bar{Z}_i = 0$ on right hand side vector $\bar{\boldsymbol{Z}}$ and calculating the all $\bar{\Pi}_i$ of the left hand side vector $\bar{\boldsymbol{\Pi}}$, we can obtain the effective coupling stiffness coefficients $\bar{E}_{i1}$. In the same method, $\bar{E}_{i2}$ can be obtained. Ultimately, all components of $\bar{E}_{ij}$ are found.

This procedure is generally operated within a RVE. To reflect the periodicity of microstructure of a composite, a periodic boundary condition should be applied to a RVE. The following is a brief description of periodic boundary conditions for displacement, electric field and magnetic field.

Denote x_i as a point on the boundary of a RVE, d_i the periodicity of the RVE in the corresponding direction. The point on the opposite boundary of the RVE is $x_i + d_i$. Without loss of generality, the periodic boundary condition can read

$$u_i(x_j + d_j) = u_i(x_j) + \left\langle \frac{\partial u_i}{\partial x_k} \right\rangle d_k \tag{7.2.5a}$$

$$\phi(x_j + d_j) = \phi(x_j) + \left\langle \frac{\partial \phi}{\partial x_k} \right\rangle d_k \tag{7.2.5b}$$

$$\bar{\omega}(x_j + d_j) = \bar{\omega}(x_j) + \left\langle \frac{\partial \bar{\omega}}{\partial x_k} \right\rangle d_k \tag{7.2.5c}$$

where $\langle \ \rangle$ is the average of the quantity. u_i, ϕ and $\bar{\omega}$ are displacements, electric potential and magnetic potential, respectively.

7.3 Indirect method

Consider a heterogeneous magneto-electro-elastic medium subjected to uniform Z_{Ji}^0 and uniform temperature T. The macroscopic constitutive equation of the composite can be written as

$$\bar{\Pi}_{iJ} = \bar{E}_{iJKl}\bar{Z}_{Kl} - \bar{\Gamma}_{iJ}T \tag{7.3.1}$$

A homogeneous boundary condition leads to

$$\bar{Z}_{Kl} = \bar{Z}_{Kl}^0 \tag{7.3.2}$$

We decompose the strain and stress induced by external Z_{Ji}^0 and temperature T as

$$Z_{Ji} = Z_{Ji}^{I} + Z_{Ji}^{II}, \quad \Pi_{Ji} = \Pi_{Ji}^{I} + \Pi_{Ji}^{II} \tag{7.3.3}$$

where superscripts I and II stand for the external load and temperature, respectively. Obviously, we have

$$\bar{Z}_{Ji}^{I} = Z_{Ji}^0, \quad \bar{Z}_{Ji}^{II} = 0 \tag{7.3.4}$$

Then Eq.(7.3.1) can be written as

$$\Pi_{iJ}^{I} = E_{iJMn}Z_{Mn}^{I} \tag{7.3.5a}$$

$$\Pi_{iJ}^{II} = E_{iJMn}Z_{Mn}^{II} - \Gamma_{iJ}T \tag{7.3.5b}$$

Averaging of the equations yields

$$\bar{\Pi}_{iJ}^{I} = \bar{E}_{iJMn}Z_{Mn}^0 \tag{7.3.6a}$$

$$\bar{\Pi}_{iJ}^{II} = -\bar{\Gamma}_{iJ}T \tag{7.3.6b}$$

where $\bar{E}_{iJMn}$ and $\bar{\Gamma}_{iJ}$ are the effective coupling properties to be found.

For a binary composite, the constitutive equations of the phases are

$$\boldsymbol{\Pi}_1^{I} = \boldsymbol{E}_1\boldsymbol{Z}_1, \quad \boldsymbol{\Pi}_2^{I} = \boldsymbol{E}_2\boldsymbol{Z}_2 \tag{7.3.7}$$

where the subscript denotes the phase. A simple mixture law gives

$$c_1 Z_1^{\mathrm{I}} + c_2 Z_2^{\mathrm{I}} = \bar{Z} \qquad (7.3.8a)$$

$$c_1 \Pi_1^{\mathrm{I}} + c_2 \Pi_2^{\mathrm{I}} = \bar{\Pi}^{\mathrm{I}} = \bar{E}\bar{Z} \qquad (7.3.8b)$$

where c_1 and c_2 denote the phase volume fractions. Substituting Eq.(7.3.7) into Eq.(7.3.8b) and employing Eq.(7.3.8a), we can obtain

$$\begin{aligned}
\bar{E}\bar{Z} &= c_1 E_1 Z_1^{\mathrm{I}} + c_2 E_2 Z_2^{\mathrm{I}} \\
&= E_1(\bar{Z} - c_2 Z_2^{\mathrm{I}}) + c_2 E_2 Z_2^{\mathrm{I}} \qquad (7.3.9) \\
&= E_1 \bar{Z} + c_2(E_2 - E_1)Z_2^{\mathrm{I}}
\end{aligned}$$

Introducing the concentration factor A, which connects the quantities in the inclusion and the effective quantities, we have

$$Z_2^{\mathrm{I}} = A\bar{Z} \qquad (7.3.10)$$

Thus, Eq.(7.3.9) can be expressed by the concentration factor as

$$\bar{E} = E_1 + c_2(E_2 - E_1)A \qquad (7.3.11)$$

The concentration factor A can be calculated by single inclusion theory where an inclusion is embedded in an infinite matrix. That is

$$A = [I + SE_1^{-1}(E_2 - E_1)]^{-1} \qquad (7.3.12)$$

where S is the Eshelby tensor depending on the shape of the inclusion. Some publications have already presented calculation of Eshelby tensor [15, 21, 22].

Note that the concentration factor determined by Eq.(7.3.12) is based on single inclusion theory and can be used in the situation of dilute composite. For the finite volume fraction of an inclusion, a modification is carried out by the Mori-Tanaka method (see reference [23] for an elastic problem)

$$A^{\mathrm{MT}} = A(c_1 I + c_2 A)^{-1} \qquad (7.3.13)$$

Replacing A in Eq.(7.3.11) by A^{MT}, we can find the effective coupling properties. Similarly, a derivation of thermal properties by using Eq.(7.3.6b) yields

$$\bar{\Gamma} = \Gamma_2 + (\bar{E} - E_2)(E_1 - E_2)^{-1}(\Gamma_1 - \Gamma_2) \qquad (7.3.14)$$

It is evident from the above derivation that the calculation of Eshelby tensor is a key step. As in the elastic problem, the Eshelby tensor can be calculated only for an ellipsoidal-shaped inclusion. This implies that the indirect method is available for an ellipsoidal-shaped inclusion. Here a discussion of the Eshelby tensor in the multi-field framework is given.

Consider first an elastic inclusion problem. In a small local domain Ω of an infinite isotropic matrix, there is a local strain ε_{ij}^* which is a non-elastic deformation, such as thermal expansion, transformation, pre-strain or plastic deformation. Such a strain was named eigenstrain by Mura [24]. A self-balanced stress induced by the eigenstrain is called eigenstress. An inclusion theory has been developed to investigate the elastic solution for an inclusion embedded in an infinite matrix.

For uniform eigenstrain ε_{ij}^* in an inclusion Ω, the perturbing strain induced by eigenstrain in the matrix is derived by Eshelby [25]. It is

$$\varepsilon_{ij} = S_{ijkl}\varepsilon_{kl}^* \tag{7.3.15}$$

where ε_{ij} is the perturbing strain in the matrix. S_{ijkl} is the Eshelby tensor depending on the shape of the inclusion and properties of the matrix. Under the situation of ellipsoidal inclusion, the Eshelby tensor is a constant. It is concluded that the perturbing strain induced by a uniform eigenstrain is a constant if the shape of the inclusion is ellipsoidal. The corresponding stress in the matrix is expressed by Hooke's law

$$\sigma_{ij}^0 = c_{ijkl}\varepsilon_{kl} \tag{7.3.16}$$

and the stress in the inclusion is

$$\sigma_{ij}^{\mathrm{I}} = c_{ijkl}(\varepsilon_{kl} - \varepsilon_{kl}^*) \tag{7.3.17}$$

Now we consider situation of multifield coupling. Denote the generalized eigenstrain in an inclusion by Z_{Mn}^*. It has the form

$$Z_{Mn}^* = \begin{cases} \varepsilon_{mn}^*, & M \leqslant 3 \\ -E_n^*, & M = 4 \\ -H_n^*, & M = 5 \end{cases} \tag{7.3.18}$$

where ε_{mn}^* is the stress-free eigenstrain, E_n^* is the eigen-electric field which is electric displacement-free. H_n^* is the magnetic field that is magnetic flux-free. The strain, electric field and magnetic field induced by the generalized eigenstrain in an inclusion can be expressed in terms of the Eshelby tensor

$$Z_{Mn} = S_{MnAb}Z_{Ab}^* = \begin{cases} \varepsilon_{mn} = S_{mnAb}Z_{Ab}^*, & M \leqslant 3 \\ -E_n = S_{4nAb}Z_{Ab}^*, & M = 4 \\ -H_n = S_{5nAb}Z_{Ab}^*, & M = 5 \end{cases} \tag{7.3.19}$$

The stress, electric displacement and magnetic flux in the matrix can be written as

$$\Pi_{iJ}^{0} = E_{iJMn} Z_{Mn} \tag{7.3.20}$$

The corresponding quantities in the inclusion are

$$\Pi_{iJ}^{1} = E_{iJMn}(Z_{Mn} - Z_{Mn}^{*}) \tag{7.3.21}$$

It is obvious that the Eshelby tensor is an ensemble of nine tensors, because the eigenstrain can induce the deformation, electric displacement and magnetic flux for magneto-electro-mechanical coupling. For a transversely isotropic medium, the nine tensors have the following forms:

$$S_{mnab} = \frac{1}{8\pi}\left[c_{iJAb}(M_{mjin} + M_{njim}) + \tilde{e}_{iab}(M_{m5in} + M_{n5im}) \right] \tag{7.3.22a}$$

$$S_{mn4b} = \frac{-1}{8\pi}\kappa_{ib}(M_{m4in} + M_{n4im}) \tag{7.3.22b}$$

$$S_{mn5b} = \frac{1}{8\pi}\left[\tilde{e}_{bij}(M_{mjin} + M_{njim}) - \Gamma_{ib}(M_{m5in} + M_{n5im}) \right] \tag{7.3.22c}$$

$$S_{4nab} = \frac{1}{4\pi}(c_{iJAb}M_{4jin} + \tilde{e}_{iab}M_{45in}) \tag{7.3.22d}$$

$$S_{4n4b} = -\frac{1}{4\pi}\kappa_{ib}M_{44in} \tag{7.3.22e}$$

$$S_{4n5b} = \frac{1}{4\pi}(\tilde{e}_{bij}M_{4jin} - \Gamma_{ib}M_{45in}) \tag{7.3.22f}$$

$$S_{5nab} = \frac{1}{4\pi}(c_{iJAb}M_{5jin} + \tilde{e}_{iab}M_{55in}) \tag{7.3.22g}$$

$$S_{5n4b} = -\frac{1}{4\pi}\kappa_{ib}M_{54in} \tag{7.3.22h}$$

$$S_{5n5b} = \frac{1}{4\pi}(\tilde{e}_{bij}M_{5jin} - \Gamma_{ib}M_{55in}) \tag{7.3.22i}$$

where M_{MJin} are functions depending on the properties of the matrix and the shape of the inclusion. Detailed formulations have been published in the literature [14,23]. It can be verified that the Eshelby tensor is symmetric.

$$S_{MnAb} = \begin{cases} S_{mnab} = S_{nmab} = S_{nmba} = S_{mnba}, & M \leqslant 3, A \leqslant 3 \\ S_{mn4b} = S_{nm4b}, & M \leqslant 3, A = 4 \\ S_{mn5b} = S_{nm5b}, & M \leqslant 3, A = 5 \\ S_{4nab} = S_{4nba}, & M = 4, A \leqslant 3 \\ S_{5nab} = S_{5nba}, & M = 5, A \leqslant 3 \end{cases} \tag{7.3.23}$$

For a transversely isotropic magneto-electro-elastic medium, the Eshelby tensor has only 28 independent non-zero components. These components have been given in reference [14].

7.4   Two-scale expansion method

The method under discussion is a generalized form of the two-scale expansion method as applied in the elastic problem. This method can be used in the multifield coupling materials with periodic microstructure.

7.4.1   Asymptotic expansion of fields

Consider a composite with periodic microstructure. There are two coordinate systems. $x = (x_1, x_2, x_3)$ represents the global coordinate and $y = (y_1, y_2, y_3)$ stands for the local coordinate. They define a periodically repeatable RVE of the composite. The characteristic length of a RVE (e.g. the average size of the domain) is much smaller than the characteristic length of the whole structure. The global and local coordinates are related by

$$y_i = \frac{x_i}{\delta} \tag{7.4.1}$$

where δ is a small scaling parameter characterizing the size of a RVE. It means that the RVE can be viewed as a point in the scale of the whole structure.

Through the two coordinate systems, the displacements u_i, electric potential ξ and magnetic potential η can be asymptotically expanded in terms of the small parameter. For the displacements u_i, we have

$$u_i(x, y) = u_{0i}(x, y) + \delta u_{1i}(x, y) + \cdots \tag{7.4.2}$$

For a periodicity Y of the microstructure, we have

$$u_i(x, y) = u_i(x, y + Y) \tag{7.4.3}$$

Due to the change of coordinates from the global to the local systems the following relation must be employed in evaluating the derivative of a field quantity

$$\frac{\partial}{\partial x_i} \rightarrow \frac{\partial}{\partial x_i} + \frac{1}{\delta} \frac{\partial}{\partial y_i} \tag{7.4.4}$$

The displacement u_{0i} in Eq.(7.4.2) is the mean value of u_i and hence is independent of y_i. Let

$$u_{0i} = u_{0i}(x) \equiv \bar{u}_i \tag{7.4.5}$$

Eq.(7.4.2) can be rewritten as

$$u_i(x, y) = \bar{u}_i(x) + \delta \tilde{u}_i(x, y) + \cdots \tag{7.4.6}$$

As in the elastic situation, $\bar{u}_i(x)$ is the mean displacement, $\tilde{u}_i(x, y)$ is the perturbing displacement which is an unknown periodic function. Thus the physical interpretation of Eq.(7.4.6) is that the real displacement u_i is oscillating rapidly around the mean displacement $\bar{u}_i$ due to the inhomogeneity from the microscopic point of view. In conjunction with Eq.(7.4.4), the strain is determined by the displacement expansion (7.4.2), that is

$$\varepsilon_{ij} = \bar{\varepsilon}_{ij} + \tilde{\varepsilon}_{ij}(x, y) + O(\delta) \tag{7.4.7}$$

where

$$\bar{\varepsilon}_{ij} = \frac{1}{2}\left(\frac{\partial \bar{u}_i}{\partial x_j} + \frac{\partial \bar{u}_j}{\partial x_i} \right) \tag{7.4.8a}$$

$$\tilde{\varepsilon}_{ij} = \frac{1}{2}\left(\frac{\partial \tilde{u}_i}{\partial x_j} + \frac{\partial \tilde{u}_j}{\partial x_i} \right) \tag{7.4.8b}$$

This shows that the strain components can be represented as a sum of the average strain $\bar{\varepsilon}_{ij}$ and the perturbing strain $\tilde{\varepsilon}_{ij}$. It can easily be shown that

$$\frac{1}{V}\int_V \varepsilon_{ij} \mathrm{d}V = \frac{1}{V}\int_V (\bar{\varepsilon}_{ij} + \tilde{\varepsilon}_{ij})\mathrm{d}V = \bar{\varepsilon}_{ij} \tag{7.4.9}$$

where V is the volume of the RVE. This stems directly from the periodicity of the perturbing strain, implying that the average of the perturbing strain taken over the RVE vanishes. For a homogeneous material it is obvious that the perturbing displacements and strains vanish identically. Using Eq.(7.4.7), we can readily represent the displacements in the form

$$u_i(x, y) = \bar{\varepsilon}_{ij} x_j + \delta \tilde{u}_i + O(\delta^2) \tag{7.4.10a}$$

Similarly, an asymptotic expansion of the electric and magnetic potentials can be utilized to yield

$$E_i = \bar{E}_i(x) + \tilde{E}_i(x, y) + O(\delta) \tag{7.4.10b}$$

$$H_i = \bar{H}_i(x) + \tilde{H}_i(x, y) + O(\delta) \tag{7.4.10c}$$

where the average electric and magnetic fields are given by

$$\bar{E}_i(x) = -\frac{\partial \bar{\phi}}{\partial x_i} \tag{7.4.11a}$$

$$\bar{H}_i(\boldsymbol{x}) = -\frac{\partial \bar{\eta}}{\partial x_i} \qquad (7.4.11b)$$

whereas the corresponding perturbing fields are determined from

$$\tilde{E}_i(\boldsymbol{x},\boldsymbol{y}) = -\frac{\partial \tilde{\phi}}{\partial x_i} \qquad (7.4.12a)$$

$$\tilde{H}_i(\boldsymbol{x},\boldsymbol{y}) = -\frac{\partial \tilde{\eta}}{\partial x_i} \qquad (7.4.12b)$$

Consequently, just as in Eq.(7.4.10a) for the displacements, the electric and magnetic potentials take the form

$$\phi(\boldsymbol{x},\boldsymbol{y}) = -\bar{E}_j x_j + \tilde{\phi} + O(\delta^2) \qquad (7.4.13a)$$

$$\eta(\boldsymbol{x},\boldsymbol{y}) = -\bar{H}_j x_j + \tilde{\eta} + O(\delta^2) \qquad (7.4.13b)$$

The coefficient tensor E_{iJKl} is a periodic function defined in a RVE in terms of the local coordinates

$$E_{iJKl}(\boldsymbol{x}) = E_{iJKl}(\boldsymbol{x},\boldsymbol{y}) \qquad (7.4.14)$$

Substituting Eq.(7.4.10) into the constitutive equations (7.1.1), (7.1.2) and (7.1.3), respectively, and differentiating with respect to the local coordinate y_j leads, respectively, to the following three equations (assuming isothermal conditions):

$$\frac{\partial}{\partial y_j}\left[c_{ijkl}(\bar{\varepsilon}_{kl} + \tilde{\varepsilon}_{kl}) - e_{kij}(\bar{E}_k + \tilde{E}_k) - \tilde{e}_{kij}(\bar{H}_k + \tilde{H}_k) \right] = 0 \qquad (7.4.15a)$$

$$\frac{\partial}{\partial y_j}\left[e_{jkl}(\bar{\varepsilon}_{kl} + \tilde{\varepsilon}_{kl}) + \kappa_{jk}(\bar{E}_k + \tilde{E}_k) + \alpha_{jk}(\bar{H}_k + \tilde{H}_k) \right] = 0 \qquad (7.4.15b)$$

$$\frac{\partial}{\partial y_j}\left[\tilde{e}_{jkl}(\bar{\varepsilon}_{kl} + \tilde{\varepsilon}_{kl}) + \alpha_{jk}(\bar{E}_k + \tilde{E}_k) + \mu_{jk}(\bar{H}_k + \tilde{H}_k) \right] = 0 \qquad (7.4.15c)$$

Define the stress as follows:

$$\sigma_{ij}^0 = c_{ijkl}\bar{\varepsilon}_{kl} - e_{kij}\bar{E}_k - \tilde{e}_{kij}\bar{H}_k \qquad (7.4.16a)$$

$$\tilde{\sigma}_{ij} = c_{ijkl}\tilde{\varepsilon}_{kl} - e_{kij}\tilde{E}_k - \tilde{e}_{kij}\tilde{H}_k \qquad (7.4.16b)$$

where Eq.(7.4.16b) represents the perturbing stress. Similarly we define

$$D_i^0 = e_{ikl}\bar{\varepsilon}_{kl} + \kappa_k\bar{E}_k + \alpha_{ik}\bar{H}_k \qquad (7.4.17a)$$

$$\tilde{D}_i = e_{ikl}\tilde{\varepsilon}_{kl} + \kappa_k\tilde{E}_k + \alpha_{ik}\tilde{H}_k \qquad (7.4.17b)$$

$$B_i^0 = \tilde{e}_{ikl}\bar{\varepsilon}_{kl} + \alpha_k\bar{E}_k + \mu_{ik}\bar{H}_k \qquad (7.4.18a)$$

$$\tilde{B}_i = \tilde{e}_{ikl}\tilde{\varepsilon}_{kl} + \alpha_k\tilde{E}_k + \mu_{ik}\tilde{H}_k \qquad (7.4.18b)$$

Thus the differential equations can be formed

$$\frac{\partial \tilde{\sigma}_{ij}}{\partial y_j} + \frac{\partial \sigma_{ij}^0}{\partial y_j} = 0 \tag{7.4.19a}$$

$$\frac{\partial \tilde{D}_i}{\partial y_i} + \frac{\partial D_i^0}{\partial y_i} = 0 \tag{7.4.19b}$$

$$\frac{\partial \tilde{B}_i}{\partial y_i} + \frac{\partial B_i^0}{\partial y_i} = 0 \tag{7.4.19c}$$

The coupling governing equation (7.4.19) form the strong form of the equilibrium and Maxwell's equations. It is readily seen that the first terms in Eq.(7.4.19) involve the unknown perturbing periodic displacements $\tilde{u}_i$, electric potential $\tilde{\phi}$ and magnetic potential $\tilde{\eta}$, while the second terms in these equations produce pseudo-body forces.

For given values of the average strain $\bar{\varepsilon}_{kl}$, average electric field $\bar{E}_i$ and average magnetic field $\bar{H}_i$ the perturbing displacements $\tilde{u}_i$, electric field $\tilde{\phi}$ and magnetic field $\tilde{\eta}$ can be determined by Eq.(7.4.19). The periodic boundary conditions must be prescribed at the boundaries of the RVE.

7.4.2   Effective coupling properties

To connect the perturbing strains, electric and magnetic fields with the average strains, electric and magnetic fields, we define a matrix $\tilde{A}$ as follows:

$$\tilde{Z} = \tilde{A}(y)\bar{Z} \tag{7.4.20}$$

With Eq.(7.4.10), we can rewrite the strains, electric and magnetic fields as follows:

$$Z = \bar{Z} + \tilde{A}(y)\bar{Z} = \left[I + \tilde{A}(y) \right]\bar{Z} = A(y)\bar{Z} \tag{7.4.21}$$

where I is the unit matrix. $A(y)$ was named the electro-magneto-elastic concentration matrix by Aboudi [20].

Aboudi [20] presented a procedure to obtain the electro-magneto-elastic concentration matrix $A(y)$. To find $A(y)$, a series of problems must be solved as follows. Solve Eq.(7.4.19) in conjunction with the periodic boundary conditions with $\varepsilon_{11} = 1$, and all other components of $\bar{Z}$ being equal to zero. The solution of these coupled differential equations readily provides A_{i1} $(i = 1, 2, \cdots, 12)$. This procedure is repeated with $\varepsilon_{22} = 1$ and all other components of $\bar{Z}$ being equal to zero which provides A_{i2} and so on.

Once the matrix $A(y)$ has been determined, it is possible to compute the effective matrix of the coefficients $\overline{E}$ of the composite. Substitution of Z given by Eq.(7.4.21) into Eq.(7.2.1) (assuming isothermal conditions) yields

$$\overline{\Pi} = E(y)A(y)\overline{Z} \tag{7.4.22}$$

Taking the average of both sides of Eq.(7.4.22) over a RVE yields the average stresses, electric displacements and magnetic flux densities in the composite in terms of the average strains, electric and magnetic fields, namely

$$\overline{\Pi} = \overline{E}\overline{Z} \tag{7.4.23}$$

where

$$\overline{E} = \frac{1}{V}\int E(y)A(y)\mathrm{d}V \tag{7.4.24}$$

The structure of the square 12th order symmetric matrix is of the form

$$\overline{E} = \begin{bmatrix} c^* & e^* & \tilde{e}^{*\mathrm{T}} \\ e^* & -\kappa^* & -\alpha^* \\ \tilde{e}^* & -\alpha^* & -\mu^* \end{bmatrix} \tag{7.4.25}$$

Where $c^*, e^*, \tilde{e}^*, \kappa^*, \alpha^*, \mu^*$ are the effective elastic stiffness, piezoelectric, piezomagnetic, dielectric, magnetic permeability and electromagnetic coefficients, respectively.

In order to incorporate the thermal effects in the composite, we utilize Levin's [26] result to establish the effective thermal stress λ_{ij}^* tensor, pyroelectric ρ_i^* and pyromagnetic ν_i^* coefficients. This approach was also followed by Dunn [27] to establish the effective thermal moduli of piezoelectric composites. To this end we define the following vector of thermal stresses, pyroelectric and pyromagnetric coefficients:

$$\Gamma = \begin{bmatrix} \lambda_1 & \lambda_2 & \lambda_3 & \lambda_4 & \lambda_5 & \lambda_6 & -\rho_1 & -\rho_2 & -\rho_3 & -\nu_1 & -\nu_2 & -\nu_3 \end{bmatrix} \tag{7.4.26}$$

The corresponding global or effective vector is defined by

$$\overline{\Gamma} = \begin{bmatrix} \lambda_1^* & \lambda_2^* & \lambda_3^* & \lambda_4^* & \lambda_5^* & \lambda_6^* & -\rho_1^* & -\rho_2^* & -\rho_3^* & -\nu_1^* & -\nu_2^* & -\nu_3^* \end{bmatrix} \tag{7.4.27}$$

According to Levin's result, the relation between Γ and $\overline{\Gamma}$ is given in terms of the matrix A.

$$\overline{\Gamma} = \frac{1}{V}\int A^{\mathrm{T}}\Gamma \mathrm{d}V \tag{7.4.28}$$

where A^{T} is the transpose of A. The above relation provides the effective thermal stress vector λ^*, pyroelectric vector ρ^* and pyromagnetic vector

v^* of the composite.

Finally, we can collect the relative equations to form a coupled constitutive equation of the electro-magneto-thermo-elastic composite as follows:

$$\bar{\Pi} = \bar{E}\bar{Z} - \bar{\Gamma}T \qquad (7.4.29)$$

The coefficients of thermal expansion α_i and the associated pyroelectric constants P_i and the pyromagnetic M_i of the constituent material can be assembled to form the vector

$$\Omega = \begin{bmatrix} \alpha_1 & \alpha_2 & \alpha_3 & \alpha_4 & \alpha_5 & \alpha_6 \\ P_1 & P_2 & P_3 & M_1 & M_2 & M_3 \end{bmatrix} \qquad (7.4.30)$$

This vector can be given by

$$\Omega = E^{-1}\Gamma \qquad (7.4.31)$$

where E and Γ are given by Eqs. (7.1.8) and (7.1.9), respectively.

In the same manner, the effective coefficients of thermal expansion α_i^* and the associated pyroelectric constants P_i^* and the pyromagnetic constants M_i^* of the composite can be assembled into the vector

$$\bar{\Omega} = \begin{bmatrix} \alpha_1^* & \alpha_2^* & \alpha_3^* & \alpha_4^* & \alpha_5^* & \alpha_6^* \\ P_1^* & P_2^* & P_3^* & M_1^* & M_2^* & M_3^* \end{bmatrix} \qquad (7.4.32)$$

Once $\bar{E}$ and $\bar{\Gamma}$ have been established, this vector can be readily determined from

$$\bar{\Omega} = \bar{E}^{-1}\bar{\Gamma} \qquad (7.4.33)$$

Aboudi [20] has given numerical results for the electro-magneto-mechanical properties of an electro-magneto-elastic medium.

7.5 FE computation of effective coupling properties

As an alternative to the concept of a coupling concentration matrix of Aboudi [20] detailed in the previous section, Yang [29] and Wang [30] presented a procedure to calculate the effective coupling properties in the frame of a two-scale expansion method in conjunction with the FEM. The present discussion focuses on the coupling properties of piezoelectric materials.

Using Eqs. (7.4.7) and (7.4.10), we can write the generalized stress as

$$\Pi = \Pi^0 + \tilde{\Pi} \qquad (7.5.1)$$

Thus the governing equation (7.4.19) becomes

$$\nabla \Pi = \nabla(\Pi^0 + \tilde{\Pi}) = 0 \tag{7.5.2}$$

where

$$\nabla \Pi = \Pi_{iJ,J} = \begin{cases} \sigma_{ij,j}, & J = 1,2,3 \\ D_{i,i}, & J = 4 \\ B_{i,i}, & J = 5 \end{cases} \tag{7.5.3}$$

In order to solve the strain concentration factor by means of the FEM, the variational form of the equilibrium equation must be given prior consideration. For an arbitrary virtual generalized displacement δU, the integration of equilibrium equation on a periodicity Y can be written as

$$\int_Y (\delta U)^{\mathrm{T}} \nabla \tilde{\Pi} \mathrm{d}V + \int_Y (\delta U)^{\mathrm{T}} \nabla \Pi^0 \mathrm{d}V = 0 \tag{7.5.4}$$

Integrating by parts leads

$$\int_Y (\delta Z)^{\mathrm{T}} \tilde{\Pi} \mathrm{d}V + \int_Y (\delta Z)^{\mathrm{T}} \Pi^0 \mathrm{d}V + b.c. = 0 \tag{7.5.5}$$

where δZ is the virtual strain corresponding to δU. $b.c.$ stands for the boundary terms. Note that the arbitrary virtual generalized displacement is equal to zero on the boundary, so that the boundary term in Eq.(7.5.5) vanishes. Substituting Eq.(7.4.20) and Eq.(7.4.23) into Eq.(7.5.5), we obtain

$$\int_Y (\delta Z)^{\mathrm{T}} E(\tilde{A}\bar{Z}) \mathrm{d}V + \int_Y (\delta Z)^{\mathrm{T}} E\bar{Z} \mathrm{d}V = 0 \tag{7.5.6}$$

Take $\bar{Z}$ as follows, respectively

$$\begin{bmatrix} 1 \\ 0 \\ 0 \\ 0 \\ 0 \end{bmatrix}, \quad \begin{bmatrix} 0 \\ 1 \\ 0 \\ 0 \\ 0 \end{bmatrix}, \quad \begin{bmatrix} 0 \\ 0 \\ 1 \\ 0 \\ 0 \end{bmatrix}, \quad \begin{bmatrix} 0 \\ 0 \\ 0 \\ 1 \\ 0 \end{bmatrix}, \quad \begin{bmatrix} 0 \\ 0 \\ 0 \\ 0 \\ 1 \end{bmatrix} \tag{7.5.7}$$

Then the above integration can be expressed as

$$\int_Y (\delta Z)^{\mathrm{T}} E\tilde{A}^k \mathrm{d}V = -\int_Y (\delta Z)^{\mathrm{T}} E^k \mathrm{d}V \tag{7.5.8}$$

where k is the independently variable index. For plane problems each index changes from 1 to 5, corresponding to the strain modes in Eq.(7.5.7). $\tilde{A}^k$ is a vector consisting of the k th column of the matrix $\tilde{A}$. E^k is a vector consisting of the k th column of the matrix E.

To compute the vector $\tilde{A}^k$, as in the elastic case, we introduce a function ψ^k to satisfy

$$\tilde{A}^k = L\psi^k \tag{7.5.9}$$

where L is an operator matrix which can be determined by the gradient equations. The FE interpolation of the function ψ^k is

$$\psi^k = N\overline{\psi}^k \tag{7.5.10}$$

where N is the shape function, $\overline{\psi}^k$ is the value of function ψ^k at nodes. We can obtain

$$\tilde{A}^k = L\psi^k = B\overline{\psi}^k \tag{7.5.11}$$

where $B = LN$ is the generalized strain matrix. The discrete form of the virtual strain δZ is taken as $\delta Z = B\delta\overline{\psi}^k$. By using Eq.(7.5.8), the standard FE form can be expressed as

$$\left(\int_Y B^{\mathrm{T}} EB\mathrm{d}V\right)\overline{\psi}^k = \int_Y B^{\mathrm{T}} E^k \mathrm{d}V \tag{7.5.12}$$

This equation provides a procedure for FE calculation for ψ^k ($k = 1,2,3,4,5$). Thus the effective stiffness matrix can be calculated by

$$\overline{E} = \frac{1}{Y}\int_Y E(I + B\overline{\psi})\mathrm{d}V \tag{7.5.13}$$

where

$$\overline{\psi} = \begin{bmatrix} \overline{\psi}^1 & \overline{\psi}^2 & \overline{\psi}^3 & \overline{\psi}^4 & \overline{\psi}^5 \end{bmatrix}$$

The integrations in Eqs. (7.5.12) and (7.5.13) can be calculated by the numerical integrations in each element.

7.6   Numerical examples

Consider a plane strain problem of a piezoelectric solid with elastic inclusions or voids. It is assumed that the materials are transversely isotropic in 1-3 plane and the polarization is along the 3rd-direction. The coupling constitutive equation is described in a matrix form

$$\begin{bmatrix} \sigma_1 \\ \sigma_3 \\ \sigma_5 \\ D_1 \\ D_3 \end{bmatrix} = \begin{bmatrix} C_{11} & C_{13} & 0 & 0 & e_{31} \\ C_{13} & C_{33} & 0 & 0 & e_{33} \\ 0 & 0 & C_{55} & e_{15} & 0 \\ 0 & 0 & e_{15} & -\kappa_{11} & 0 \\ e_{31} & e_{33} & 0 & 0 & -\kappa_{33} \end{bmatrix}\begin{bmatrix} \varepsilon_1 \\ \varepsilon_3 \\ \varepsilon_5 \\ -E_1 \\ -E_3 \end{bmatrix} \tag{7.6.1}$$

or in a compact form

$$\Pi = EZ \tag{7.6.2}$$

The piezoelectricity ceramic material $BaTiO_3$ is studied here. The material properties are listed in Table 7.2.

Table 7.2   BaTiO$_3$ material properties

Elastic constants/GPa				Piezoelectric constants /(C/m^2)			Dielectric constants /(10^{-9}C^2/Nm^2)	
C_{11}	C_{13}	C_{33}	C_{55}	e_{51}	e_{31}	e_{33}	κ_{11}	κ_{33}
150	66	150	44	11.4	−4.35	17.5	9.86	11.15

7.6.1   Piezoelectric solid with voids

The effective properties of the piezoelectric material containing circular voids are calculated for different volume fractions of the voids. Fig.7.1 shows the curves of the effective properties versus the volume fractions of the voids. Here the numerical result for $\overline{C}_{11}/C_{11}$ is compared with that of the direct method by a BEM [5]. Good agreement between the two methods is demonstrated in this figure.

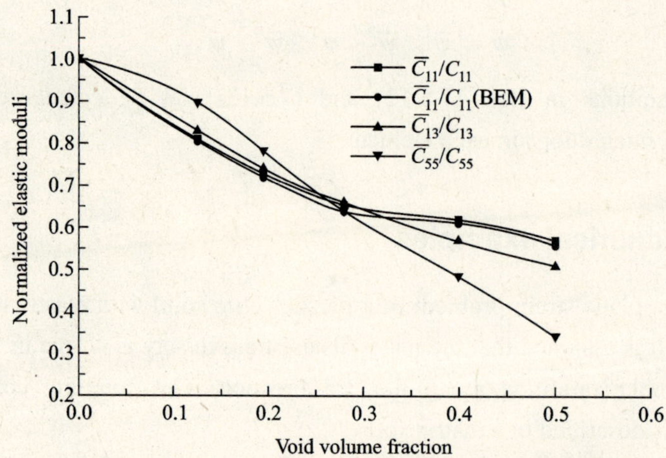

Fig.7.1   Effective elastic moduli of piezoelectric material with voids

Fig.7.2 shows the variation curves of the effective piezoelectric moduli with the void volume fractions for a piezoelectric solid containing the voids. It can be seen that the value of $\overline{e}_{31}/e_{31}$ increases as the volume fraction of voids increases, whereas the opposite conclusion can be obtained for $\overline{e}_{33}/e_{33}$ and $\overline{e}_{51}/e_{51}$.

Fig.7.3 shows the curves illustrating the variation of effective dielectric moduli with the void volume fractions. It can be concluded that an increase in the void volume fraction leads to enhanced $\bar{\kappa}_{33}/\kappa_{33}$ and a decrease of $\bar{\kappa}_{11}/\kappa_{11}$.

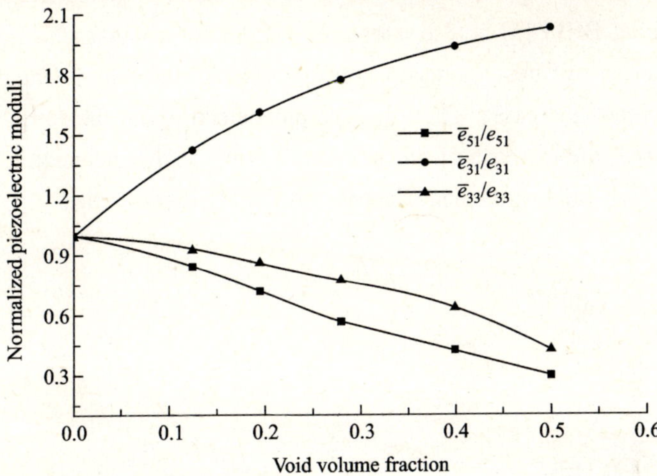

Fig. 7.2   Effective piezoelectric moduli of piezoelectric solid with voids

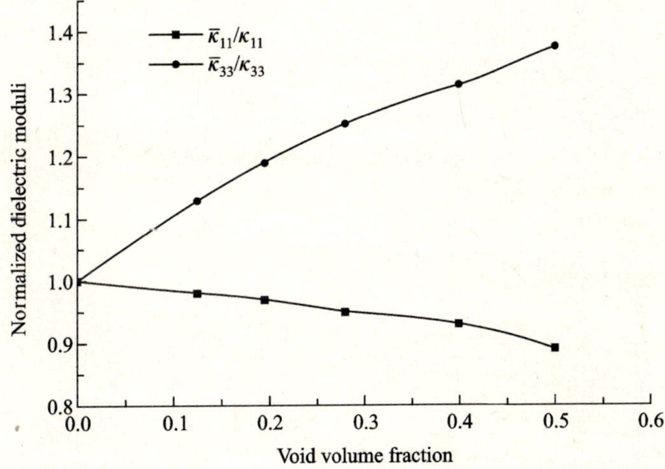

Fig.7.3   Effective dielectric moduli of piezoelectric solid containing voids

7.6.2 Rigid inclusions

Consider the effective properties for a piezoelectric solid containing rigid inclusions. The rigid inclusion is modeled by a much greater stiffness than that of the matrix. The piezoelectric coefficients and the dielectric coefficients of the rigid inclusions vanish. The numerical results for the effective properties of the composite are illustrated in Figs.7.4~7.6. A result obtained from the direct method using BEM [5] is also shown in Fig.7.4 for comparison. It is evident that the elastic stiffness increases considerably as the volume fraction of the rigid inclusions increases. The effective piezoelectric and dielectric properties decrease with the increase of the volume fraction of the inclusion because the rigid inclusion has lost its piezoelectric and dielectric properties.

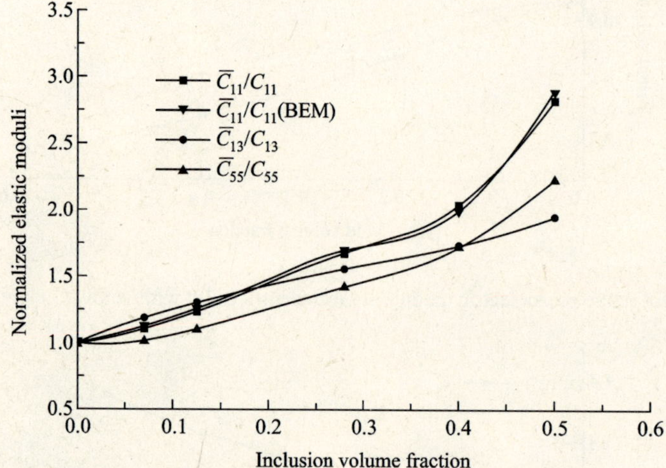

Fig. 7.4   Effective elastic stiffness of rigid inclusion composite

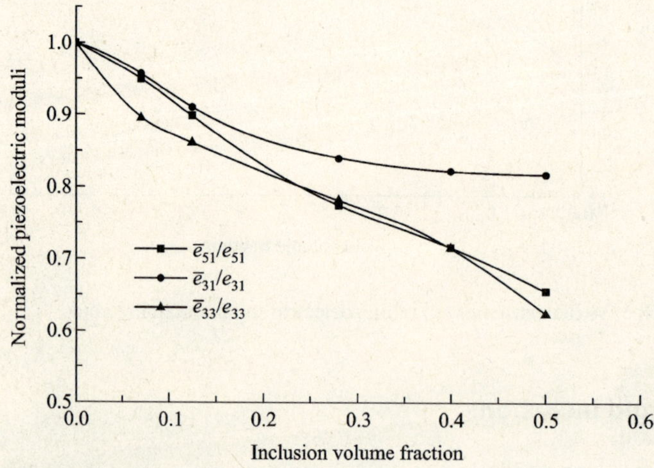

Fig.7.5   Effective piezoelectric properties of rigid inclusion composite

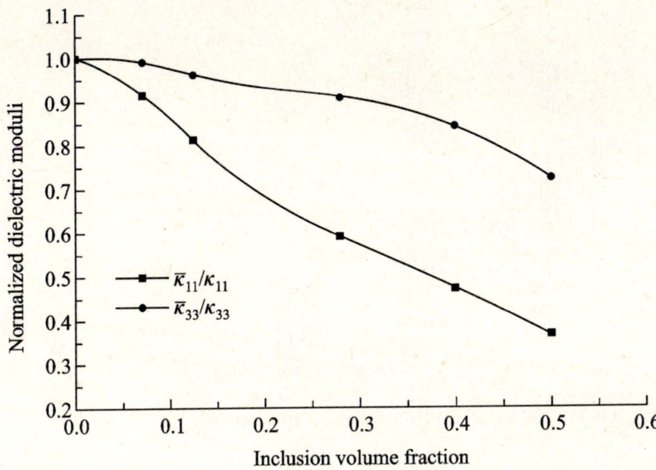

Fig.7.6   Effective dielectric properties of rigid inclusion composite

7.6.3 Piezoelectric composite

A composite of a piezoelectric matrix reinforced by another piezoelectric material is taken into account in this section. The material constants of phases are listed in Table 7.3.

Table 7.3   Material constants of piezoelectric composite

	Elastic constants/GPa				Piezoelectric constants/(C/m^2)			Dielectric constants /$(10^{-9}$C^2/Nm$^2)$	
	C_{11}	C_{13}	C_{33}	C_{55}	e_{51}	e_{31}	e_{33}	κ_{11}	κ_{33}
Matrix	50	14.5	50	3.4	2.2	-3	3.2	12	11.5
Inclusion	139	78	139	25.6	12.7	-5.2	15.1	6.5	5.6

The effective properties calculated are shown in Figs. 7.7~7.9, where the effective moduli are normalized by the properties of the matrix denoted by superscripts 0. The curves are given through variations of the effective properties versus the inclusion volume fractions. The elastic reinforcement effect and the enhanced piezoelectric properties can be determined in the present investigation. However the reduction of the dielectric properties is seen in Fig. 7.9.

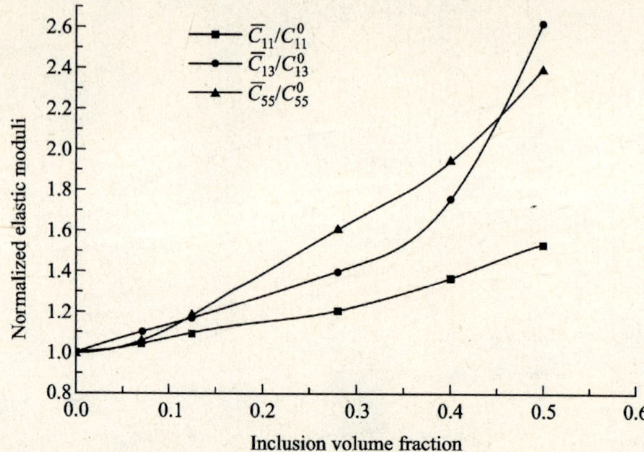

Fig.7.7   Effective elastic moduli of piezoelectric composite

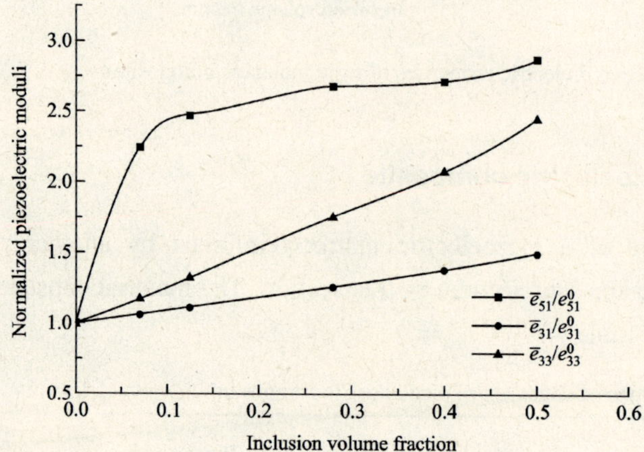

Fig.7.8   Effective piezoelectric moduli of composite

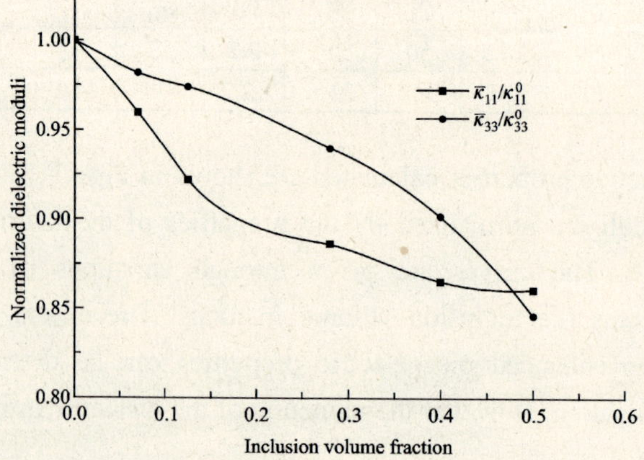

Fig.7.9   Effective dielectric moduli of composite

References

[1] Van Run A M J G, Terrell D R, Scholing J H. An in situ grown eutectic magneto-electric composite material.J Mater Sci, 1974, 9: 1710-1714.

[2] Bracke L P M, Van Vliet R G, A broadband magneto-electric transducer using a composite material. Int J Electron, 1981, 51: 255–262.

[3] Uchino K. Piezoelectric actuators and ultrasonic motors. Boston: Kluwer, 1997.

[4] Qin Qinghua. Fracture mechanics of piezoelectric materials. Southampton: WIT press, 2001.

[5] Qin Qinghua. Material properties of piezoelectric composites by BEM and homogenization method. Composites Structure, 2004, 66: 295-299.

[6] Lee J, Boyd J G, Lagoudas D C. Effective properties of three-phase electro-magneto-elastic composites. Int J Engineering Science, 2005, 43: 790-825.

[7] Wang B. Three dimensional analysis of an ellipsoidal inclusion in a piezoelectric material. Int J Solids Struct, 1992, 29: 293-308.

[8] Benveniste Y. Exact results concerning the local fields and effective properties in piezoelectric composites. ASME Journal of Engineering Materials and technology, 1994, 26: 260-267.

[9] Dunn M L, Taya M. An analysis of composite materials containing ellipsoidal piezoelectric inhomogeneities. Proc Roy Soc London A, 1993, 443: 265-287.

[10] Chen T. Piezoelectric properties of multiphase fibrous composites: some theoretical results. J Mech. Phys. Solids, 1993, 41, 1781-1794.

[11] Huang J H. Micromechanics determinations of thermoelectroelastic fields and effective thermoelectroelastic moduli of piezoelectric composites. Materals Science and Engineer B, 1996, 39:163-172.

[12] Levin V M, The effective thermoelectroelastic properties of microinhomogeneous materials. Int J Solids and Structures, 1999, 36: 2683-2705.

[13] Harshe G, Dougherty J P, Newnham R E. Theoretical modelling of 3±0/0±3 magnetoelectric composites. Int J Appl Electromagn Mater, 1993, 4: 161-171.

[14] Huang J H and Kuo W S. The analysis of piezoelectric/piezomagnetic composite materials containing ellipsoidal inclusions. Journal of Applied Physics, 1997, 81: 4889-4898.

[15] Wu T L, Huang J H. Closed-form solutions for the magnetoelectric coupling coefficients in composites with piezoelectric and piezomagnetic phases. Int J Solids Strtuct, 2000, 37: 2981-3009.

[16] Benveniste Y. Magnetoelectric effect in fibbrous composites with piezoelectric and magnetostrictive phases. Physical Review B, 1995, 51: 16424-16427.

[17] Nan C W. Magnetoelectric effect in composites of piezoelectric and piezomagnetic phases. Physical Review B, 1994, 50: 6082-6088.

[18] Li J Y, Dun M L. Micromechanics of magnetoelectroelastric composite materials:

average field and effective behavior. Inte J Mat Syst Struct, 1998, 9: 404-416.

[19] Li J Y. Magnetoellectroelastic multi-inclusion and inhomoogeneity problems and their applications in composite materials. Int J Engineering Science, 2000, 38: 1993-2011.

[20] Aboudi J. Micromechanical analysis of fully coupled electro-magneto-thermo-elastic multiphase composites. Smart Materials and structures, 2001, 10:867-877.

[21] Li J Y, Dun M L. Anisotropic coupled-field inclusion and inhomogeneity problems. Philosophical Magazine A, 1998, 77: 1341-1350.

[22] Huang J H, Chiu Y H, Liu H K. Magneto-electro-elastic Eshelby tensors for a piezoelectric-piezomagnetic composite reinforced by ellipsoidal inclusions. J Appl Phys, 1998, 83: 5364-5370.

[23] Yang Qingsheng. Microstructural mechanics and design of composite materials(in Chinese). Beijing: Chinese Tiedao Publishing House, 2000.

[24] Mura T. Micromechanics of defeats in solids. Dordrecht: Martinus Nijhoff, 1987.

[25] Eshelby J D. The determination of the elastic field of an ellipsoidal inclusion and related problems. Proc R Soc London, Ser. A, 1957, 241:376-396.

[26] Levin V M. On the coefficients of thermal expansion of heterogeneous materials. Mechanics of Solids, 1967, 2: 58-61.

[27] Dun M L. Micromechanics of coupled electroelastic composites: effective thermal expansion and pyroelectric coefficients. J Appl Phys, 1993, 73: 5131-5140.

[28] Aboudi J. Micromechanical prediction of the effective behavior of fully coupled electro-magneto-thermo-elastic multiphase composites. NASA Report, CR-2000-209787, 2000.

[29] Yang Qingsheng. Multifield coupling behavior of smart composites (in Chinese)//Y M Guo. Proceedings of NCCM-14, Yichang, China, 2006: 706-709.

[30] Wang H M. Analysis of the effective properties of piezoelectric composites(in Chinese). Beijing: Beijing University of Technology, 2006.

Chapter 8 Effective properties of thermo-piezoelectricity

8.1 Introduction

In Chapters 3 and 4, we presented a linear theory of multifield materials and its solutions for some special problems such as analytical expressions of a 2D thermo-piezoelectric plate with a crack of finite length. Based on the theoretical results presented in those two chapters, micromechanics models including the generalized self-consistent method, differential approach, and Mori-Tanaka method are presented in this chapter to predict effective material properties of defective multifield materials and heterogeneous materials.

It is well known that piezoelectric ceramics are brittle materials. Thus, they may develop various microdefects such as microcracks, delamination, and microvoids during the production process and service period. This drawback has driven the development of composites of piezoelectric ceramics combined with piezoelectrically inactive polymers or other ductile materials that exhibit higher toughness than the piezoelectric ceramic alone [1]. To accurately predict the effects of microdefects on material performance and to assist mechanical engineers in developing piezoelectric composites for electromechanical transducers and engineering smart material applications, it is valuable to develop reliable theories to predict the changes in material performance due to the presence of these microdefects and the effects of the material properties and microstructural geometry of the constituents on the effective electroelastic behaviour of the composite. Early in 1978, Newnham [1] et al. presented a connectivity theory based on the combination of mechanics of material type parallel and series models to predict effective pyroelectric behaviour. Banno [2] generalized the connectivity theory to include the effects of a discontinuous reinforcement phase of particle reinforced piezocomposites. Grekov [3] et al. further presented a concentric cylinder model for evaluating effective electroelastic properties of piezocomposites reinforced by long fibres. Dunn and Taya [4] studied the overall properties of piezoelectric composites containing interacting inhomogeneities using dilute method, self-consistent model, differential approach, and Mori-Tanaka method and obtained an explicit expression in a surface integral form for coupled electroelastic Eshelby tensors. With regard to the determination of effective thermal expansion and pyroelectric properties, Dunn [5] evaluated the effective pyroelectric properties of two-phase composites, again using the four micromechanics models mentioned above. Benveniste [6] showed that the effective thermal-stress constants and pyroelectric coefficients are related to the corresponding isothermal electroelastic moduli in two-phase media. For multiphase media, Benveniste [7] further indicated that the effective thermal-stress constants and pyroelectric coefficients follow from knowledge of the influence functions related to an electromechanical loading of the composite aggregate. Chen [8] obtained some formulae for the prediction of overall thermo-electro-elastic moduli of multiphase fibrous composites, using the self-consistent and Mori-Tanaka methods. Qin and Yu [9] and Mai [10] et al. presented effective thermo-electro-elastic properties of cracked piezoelectric solids using the self-consistent and Mori-Tanaka methods. Benvensite and Dvorak [11] showed that for a two-phase system, exact connections can be obtained not only between the effective moduli, but also among the local pointwise fields induced by a uniform electromechanical loading. Later, the connections were generalized to study piezoelectric fibrous composites of three or four phases [12,13]. The phase boundaries are cylindrical but otherwise the microgeometry is totally arbitrary. Qin [14-16] et al. developed a family of micromechanics models for evaluating defective thermo-piezoelectric materials. Levin [17,18] et al. developed self-consistent formulations for estimating the effective properties of piezocomposites with ellipsoidal inclusions. Using a generalized eigenstrain approach, Huang [19] obtained a unified explicit expression for the coupled electroelastic Eshelby tensors for piezoelectric ellipsoidal inclusions in a transversely isotropic medium. Based on the equivalent inclusion method and the Mori-Tanaka approach, Huang and Kuo [20], and Kuo and Huang [21] investigated the effective material behaviour of piezocomposite containing short fibres. They found that the longitudinal and in-plane shear moduli increased with fibre length, while the other moduli, piezoelectric and dielectric constants decreased. The method in [20,21] was later used to analyse effective material behaviour affected by microvoids [22] and to establish a statistical micromechanics model [23]. Using a method of 2-scale asymptotic expansions, Wojnar [24] analyzed the homogenization process of a piezoelectric periodic composite in which thermal effects are taken into account. Eduardo [25] et al. used the 2-scale method to investigate anti-plane problems of thermo-piezoelectric fibrous composites. Hori and Nemat-Nasser [26] generalized the Hashin–Shtrikman variational principle to the coupled problem of piezoelectricity and presented the upper and lower bounds for the effective moduli of heterogeneous piezoelectric materials. Based on the concept of a cell model, Poizat and Sester [27] studied 1-3 and 0-3 composites made of piezoceramic fibres embedded in a soft non-piezoelectric matrix; Beckert [28] et al. estimated the relevant effective electromechanical parameters of composites continuously reinforced with coated piezoelectric fibres; Li [29] et al. examined the influence of void volume fraction, void distribution, void shape and configuration on the effective properties of voided piezoelectric ceramics; Berger [30] et al. presented an asymptotic homogenization method and its numerical model for 1-3 periodic composites made of piezoceramic fibres embedded in a soft non-piezoelectric matrix. Using the self-consistent approach, the orientation distribution function, and traditional Voigt-Reuss averages, Li [31] evaluated the effective electroelastic moduli of textured piezoelectric polycrystalline aggregates. Jiang [32] et al. presented a generalized self-consistent method for analysing the effective electroelastic behaviour of anti-plane fibrous piezocomposites by means of a three-phase confocal elliptical cylinder model. Recently, Wang [33] et al. combined a micromechanics approach with a boundary node element to evaluate the effective electroelastic properties of transversely isotropic piezoelectric materials containing randomly distributed voids. Qin [34] developed a micromechanics-boundary element algorithm for predicting defective piezoelectric materials. Most of the developments in this field can also be found in [35-42]. This chapter, however, focuses on the results presented in [4,5,9,10,14-16,19,34].

8.2   Micromechanics model of thermo-piezoelectricity with microcracks

8.2.1   Basic formulation of two-phase thermo-piezoelectricity

It can be seen from the discussion in Section 4.2.4 that the resulting multifield theory is concerned with the piezoelectric analog of the uncoupled theory of thermoelasticity where the magnetic, electric, and elastic fields are fully coupled, but temperature enters the problem only through the constitutive equations. As a result of this, the effective conductivity and the effective electroelastic or magnetoelectroelatic constants can be determined independently, while evaluation of the effective thermal expansion and pyroelectric coefficients requires information about both of them. Accordingly, our derivation is divided into three major steps: First, develop formulations for effective conductivity; then find expressions for effective electroelastic (or magneto-electro-elastic) constants; and finally, derive effective thermal expansion and pyroelectric coefficients based on the results obtained from the first two steps. To illustrate this process, we consider a two-dimensional piezoelectric plate weakened by microcracks. If this is a generalized plane stress problem, its thermo-electro-elastic constitutive relationship can be obtained by extending Eq.(3.2.31) adding thermal terms. The addition rule is based on Eq.(3.6.6):

$$
\begin{bmatrix} \sigma_1 \\ \sigma_3 \\ \sigma_5 \\ D_1 \\ D_3 \end{bmatrix} = \begin{bmatrix} c_{11} & c_{13} & 0 & 0 & e_{31} \\ c_{13} & c_{33} & 0 & 0 & e_{33} \\ 0 & 0 & c_{55} & e_{15} & 0 \\ 0 & 0 & e_{15} & -\kappa_{11} & 0 \\ e_{31} & e_{33} & 0 & 0 & -\kappa_{33} \end{bmatrix} \begin{bmatrix} \varepsilon_1 \\ \varepsilon_3 \\ \varepsilon_5 \\ -E_1 \\ -E_3 \end{bmatrix} - \begin{bmatrix} \lambda_{11} \\ \lambda_{33} \\ 0 \\ 0 \\ \rho_3 \end{bmatrix} T
\tag{8.2.1}
$$

$$
\begin{bmatrix} h_1 \\ h_3 \end{bmatrix} = \begin{bmatrix} k_{11} & k_{13} \\ k_{13} & k_{33} \end{bmatrix} \begin{bmatrix} W_1 \\ W_3 \end{bmatrix}
\tag{8.2.2}
$$

where heat intensity is defined by

$$
W_i = -\frac{\partial T}{\partial x_i}
\tag{8.2.3}
$$

If we choose heat flow h_i, stress σ_i, and electric displacement D_i as independent variables, the constitutive equations (8.2.1) and (8.2.2) become

$$
\begin{bmatrix} \varepsilon_1 \\ \varepsilon_3 \\ \varepsilon_5 \\ -E_1 \\ -E_3 \end{bmatrix} = \begin{bmatrix} f_{11} & f_{13} & 0 & 0 & p_{31} \\ f_{13} & f_{33} & 0 & 0 & p_{33} \\ 0 & 0 & f_{55} & p_{15} & 0 \\ 0 & 0 & p_{15} & -\beta_{11} & 0 \\ p_{31} & p_{33} & 0 & 0 & -\beta_{33} \end{bmatrix} \begin{bmatrix} \sigma_1 \\ \sigma_3 \\ \sigma_5 \\ D_1 \\ D_3 \end{bmatrix} + \begin{bmatrix} \alpha_{11} \\ \alpha_{33} \\ 0 \\ 0 \\ \gamma_3 \end{bmatrix} T
\tag{8.2.4}
$$

$$\begin{bmatrix} H_1 \\ H_3 \end{bmatrix} = \begin{bmatrix} \rho_{11} & \rho_{13} \\ \rho_{13} & \rho_{33} \end{bmatrix} \begin{bmatrix} h_1 \\ h_3 \end{bmatrix} \tag{8.2.5}$$

where ρ_{ij} = heat resistivity. These equations can also be written in matrix form as

$$\boldsymbol{h} = \boldsymbol{kH}, \quad \boldsymbol{H} = \boldsymbol{\rho h} \tag{8.2.6}$$

$$\boldsymbol{\Pi} = \boldsymbol{EZ} - \boldsymbol{\Gamma T}, \quad \boldsymbol{Z} = \boldsymbol{F\Pi} + \boldsymbol{\alpha T} \tag{8.2.7}$$

with

$$\boldsymbol{h} = \begin{bmatrix} h_1 & h_2 \end{bmatrix}^{\mathrm{T}}, \quad \boldsymbol{H} = \begin{bmatrix} H_1 & H_2 \end{bmatrix}^{\mathrm{T}} \tag{8.2.8}$$

$$\boldsymbol{\Pi} = \begin{bmatrix} \sigma_1 & \sigma_3 & \sigma_5 & D_1 & D_3 \end{bmatrix}^{\mathrm{T}} \tag{8.2.9}$$

$$\boldsymbol{Z} = \begin{bmatrix} Z_{11} & Z_{22} & Z_{12} & Z_{31} & Z_{32} \end{bmatrix}^{\mathrm{T}} = \begin{bmatrix} \varepsilon_1 & \varepsilon_3 & \varepsilon_5 & -E_1 & -E_3 \end{bmatrix}^{\mathrm{T}} \tag{8.2.10}$$

Generally, a crack may be viewed as an inclusion with zero mechanical stiffness. Thus, micromechanics theories of a cracked piezoelectric solid can be established based on some fundamental results in the theory of two-phase media. In the case of two-phase materials, the volume average of a physical variable F is defined by

$$\overline{F} = v_1 \overline{F_1} + v_2 \overline{F_2} \tag{8.2.11}$$

where subscripts "1" and "2" denote the matrix and inclusion phases, respectively, v_1 and v_2 their volume (or area) fractions, and overbar denotes the volume (or area for 2D analysis) average of a quantity over a representative volume element Ω, i.e.,

$$(\overline{\bullet}) = \frac{1}{\Omega} \int_{\Omega} (\bullet) \mathrm{d}\Omega \tag{8.2.12}$$

The effective properties represented by the effective heat conductivity k_{ij}^* (or effective heat resistivity ρ_{ij}^*), the effective generalized stiffness E_{ij}^* (or generalized compliancy F_{ij}^*), and the effective generalized stress-thermal coefficients Γ_{ij}^* (or generalized thermal expansion α_{ij}^*) of the cracked piezoelectric solid can be defined by the concept of the volume average (8.2.12) as

$$\overline{\boldsymbol{q}} = \boldsymbol{k}^* \overline{\boldsymbol{H}}, \quad \overline{\boldsymbol{H}} = \boldsymbol{\rho}^* \overline{\boldsymbol{q}} \tag{8.2.13}$$

$$\overline{\boldsymbol{\Pi}} = \boldsymbol{E}^* \overline{\boldsymbol{Z}} - \boldsymbol{\Gamma}^* \overline{T}, \quad \overline{\boldsymbol{Z}} = \boldsymbol{F}^* \overline{\boldsymbol{\Pi}} + \boldsymbol{\alpha}^* \overline{T} \tag{8.2.14}$$

Since the effective conductivity and the effective electroelastic constants can be determined independently, the applied remote temperature change T_∞

is set to be zero when we study effective electroelastic constants. Following the average energy theorem [43], we have

$$\overline{T} = T_\infty = 0 \tag{8.2.15}$$

In this case, Eq.(8.2.14) becomes

$$\overline{\boldsymbol{\Pi}} = \boldsymbol{E}^* \overline{\boldsymbol{Z}}, \quad \overline{\boldsymbol{Z}} = \boldsymbol{F}^* \overline{\boldsymbol{\Pi}} \tag{8.2.16}$$

Making use of Eq.(8.2.11), the average generalized stress and strain can be written as

$$\overline{\boldsymbol{\Pi}} = v_1 \overline{\boldsymbol{\Pi}}_1 + v_2 \overline{\boldsymbol{\Pi}}_2, \quad \overline{\boldsymbol{Z}} = v_1 \overline{\boldsymbol{Z}}_1 + v_2 \overline{\boldsymbol{Z}}_2 \tag{8.2.17}$$

Substituting Eq.(8.2.17) into Eq.(8.2.16) and noting that $\boldsymbol{\Pi}_i = \boldsymbol{E}_i \boldsymbol{Z}_i$, we obtain

$$\boldsymbol{E}^* = \boldsymbol{E}_1 + (\boldsymbol{E}_2 - \boldsymbol{E}_1)\boldsymbol{A}_2 v_2 \tag{8.2.18}$$

$$\boldsymbol{F}^* = \boldsymbol{F}_1 + (\boldsymbol{F}_2 - \boldsymbol{F}_1)\boldsymbol{B}_2 v_2 \tag{8.2.19}$$

where the symmetric tensors $\boldsymbol{A}_2$ and $\boldsymbol{B}_2$ are defined by the linear relations [15]

$$\overline{\boldsymbol{Z}}_2 = \boldsymbol{A}_2 \boldsymbol{Z}_\infty, \quad \overline{\boldsymbol{\Pi}}_2 = \boldsymbol{B}_2 \boldsymbol{\Pi}_\infty \tag{8.2.20}$$

with $\boldsymbol{Z}_\infty$ and $\boldsymbol{\Pi}_\infty$ being the remote generalized stress and strain fields applied the effective medium.

Eq.(8.2.20) cannot be used directly to analyze problems with voids or cracks due to the difficulty in evaluating $\overline{\boldsymbol{Z}}_2$. To bypass this problem, we consider first the case when inclusions become voids. This implies that $\boldsymbol{E}_2 \to 0$, $\boldsymbol{F}_2 \to \infty$. In this case, the voids under consideration can be thought of as being filled with air, which has a dielectric constant approximately three orders of magnitude smaller than the dielectric constants of piezoelectric materials. The consequence of this fact is that the boundary conditions on the hole boundary are given by $\boldsymbol{\Pi} \cdot \boldsymbol{n} = 0$, where $\boldsymbol{n}$ is outward normal to the hole boundary. This is also equivalent to setting $\boldsymbol{E}_2 = 0$, where $\boldsymbol{E}_2$ stands for the material constants of the hole-phase. Then, Eqs.(8.2.18) and (8.2.19) become

$$\boldsymbol{E}^* = \boldsymbol{E}_1(\boldsymbol{I} - \boldsymbol{A}_0 v_2) \tag{8.2.21}$$

$$\boldsymbol{F}^* = \boldsymbol{F}_1(\boldsymbol{I} + \boldsymbol{B}_0 v_2) \tag{8.2.22}$$

where $\boldsymbol{I}$ is the uniat tensor, $\boldsymbol{A}_0$ is $\boldsymbol{A}_2$ of Eq.(8.2.18) for voids, and $\boldsymbol{B}_0$ is defined by [15]

$$\overline{\boldsymbol{Z}}_2 = \boldsymbol{F}_1 \boldsymbol{B}_0 \boldsymbol{\Pi}_\infty \tag{8.2.23}$$

The interpretation of $\overline{\boldsymbol{Z}}_2$ in Eq.(8.2.23) follows the average strain theorem [43]

$$(\bar{Z}_{ij})_2 = \frac{1}{\Omega_2}\int_{\Omega_2} Z_{ij}\,d\Omega_2 = \frac{1}{2\Omega_2}\int_{\partial\Omega_2}\left\{[1+H(i-3)]U_i n_j + U_j n_i\right\}d\Omega_2 \qquad (8.2.24)$$

where Ω_2 and $\partial\Omega_2$ are the total area and boundary of the voids, $n = \begin{bmatrix} n_1 & n_2 & 0\end{bmatrix}^T$ is the normal local to the void surface, $U = \begin{bmatrix} U_1 & U_2 & U_3\end{bmatrix}^T = \begin{bmatrix} u_1 & u_3 & \phi\end{bmatrix}^T$, and $H(i)$ is the Heaviside step function.

Cracks are defined as very flat voids of vanishing height and thus also of vanishing area. Multiplying both sides of Eq.(8.2.24) by v_2 and considering the limit of flattening out in cracks, i.e., $v_2 \to 0$, one has

$$\lim_{v_2 \to 0}[(\bar{Z}_{ij})_2 v_2] = \frac{1}{2A}\int_L \left\{[1+H(i-3)]\Delta U_i n_j + \Delta U_j n_i\right\}dl = X_{ij} \qquad (8.2.25)$$

where $L = l_1 \cup l_2 \cup \cdots \cup l_N$, l_i is the length of the ith crack, N the number of cracks within the representative area element, $\Delta(\bullet)$ stands for the jump of a quantity across the crack faces. For convenience, we define [44]

$$P = \lim_{v_2 \to 0}(A_0 v_2), \qquad Q = \lim_{v_2 \to 0}(B_0 v_2) \qquad (8.2.26)$$

Hence Eqs.(8.2.21) and (8.2.22) can be rewritten as

$$E^* = E_1(I - P) \qquad (8.2.27)$$
$$F^* = F_1(I + Q) \qquad (8.2.28)$$

with the relation

$$X = PZ_\infty = F_1 Q\Pi_\infty \qquad (8.2.29)$$

Thus, the estimation of the integral (8.2.25) and thus P (or Q) is the key to predicting the effective electroelastic moduli E^* and F^*. The approximation of the integral (8.2.25) through use of various micromechanics models is the subject of the subsequent sections.

8.2.2   Effective conductivity

It can be seen from the discussion in Subsection 8.2.1 that the key point for evaluating the effective properties of a cracked piezoelectric solid is to determine the concentration factors P and Q, and thus to calculate the integral (8.2.25). For a cracked piezoelectric sheet subjected to a set of far fields $W_{i\infty}$ or $h_{i\infty}$, i.e.,

$$T(s) = -W_{i\infty}x_i \qquad (8.2.30)$$

or

$$h_n(s) = h_{i\infty}n_i \tag{8.2.31}$$

where s stands for arc coordinate on boundary, subscript "∞" represents far field. Using the boundary conditions (8.2.30) and (8.2.31), the effective conductivity can be determined in the following way. For a cracked piezoelectric sheet, we have from the definition of average field and Eq.(8.2.25) [9]

$$\overline{W}_i = (\overline{W}_i)_M + \frac{1}{A}\sum_{k=1}^{N}\int_{l_k}\Delta T n_i \mathrm{d}l \tag{8.2.32}$$

$$\overline{h}_i = (\overline{h}_i)_M \tag{8.2.33}$$

where "M" represents the quantity associated with the matrix, and ΔT stands for the jump of temperature field across crack faces

$$\Delta T(x) = T_{(U)}(x) - T_{(L)}(x) \tag{8.2.34}$$

with subscripts "(U)" and "(L)" denoting the quantity associated with the upper and lower faces of the crack, respectively. If all cracks are assumed to have the same length and orientation, Eqs.(8.2.32) and (8.2.33) can be further written as

$$k_{ij}^* W_{j\infty} = k_{ijM}W_{j\infty} - \frac{k_{ijM}}{A}\sum_{k=1}^{N}\int_{l_k}\Delta T n_j \mathrm{d}l = k_{ijM}(W_{j\infty} - \overline{W}_{jc}) \tag{8.2.35}$$

$$\rho_{ij}^* h_{j\infty} = \rho_{ijM}h_{j\infty} - \frac{1}{A}\sum_{k=1}^{N}\int_{l_k}\Delta T n_j \mathrm{d}l = \rho_{ijM}h_{j\infty} + \overline{W}_{jc} \tag{8.2.36}$$

where subscript "c" denotes the quantity associated with crack. By comparing Eq.(8.2.36) with Eq.(8.2.25), we see that the concentration factors P and Q can be expressed as follows

$$\lim_{v_2 \to 0}(\overline{W}_c v_2) = \frac{1}{A}\int_L \Delta T n \mathrm{d}l = \rho_M Q h_\infty = P W_\infty = X \tag{8.2.37}$$

It can be seen from Eq.(8.2.36) that the solution of ΔT along the crack line is required for calculating effective heat resistivity ρ_{ij}^*. For a piezoelectric sheet with a number of cracks, it is very difficult to obtain an analytical solution of ΔT when the interactions among cracks are taken into account. In the following, we show how to use micromechanics algorithms to evaluate ΔT, and then determine ρ_{ij}^* and k_{ij}^*.

(1) Dilute method.

In the dilute assumption we assume that the interaction among cracks in an infinite plate can be ignored. The concentration factor P is then obtained from the solution of the auxiliary problem of a single crack embedded in an infinite intact plate (see Fig.8.1a). For an infinite plate with a horizontal crack and subjected to the far field $W_{2\infty}$, the temperature jump across the crack faces was obtained by Atkinson and Clements [45]

$$\Delta T(x) = \frac{4kW_{2\infty}}{k_{11}}(a^2 - x^2)^{1/2} \qquad (8.2.38)$$

where $k = (k_{11}k_{22} - k_{12}^2)^{1/2}$, and a is the length of the crack. Since with the dilute method we assume that there is no interaction among cracks, the constants k and k_{11} in Eq.(8.2.38) can be taken as k_M and k_{11M}. Thus, the concentration factor P^{DIL} can be expressed as

$$P_{11}^{\text{DIL}} = P_{12}^{\text{DIL}} = P_{21}^{\text{DIL}} = 0, \quad P_{22}^{\text{DIL}} = \frac{2\pi k_M \varepsilon}{k_{11M}} \qquad (8.2.39)$$

where superscript "DIL" stands for the quantity associated with dilute method, and $\varepsilon = Na^2/\Omega$ is the so-called crack density parameter. Substituting Eq.(8.2.39) into Eqs.(8.2.27) and (8.2.28) yields

$$k_{ij}^* = \begin{cases} k_{22M} - k_{22M}\dfrac{2\pi\varepsilon k_M}{k_{11M}}, & \text{for } i = j = 2 \\[3mm] k_{ijM}, & \text{otherwise} \end{cases} \qquad (8.2.40)$$

When $\varepsilon \ll 1$, we have

$$\frac{k_{22}^*}{k_{22M}} = 1 - 2\pi\varepsilon\frac{k_M}{k_{11M}} \approx \frac{1}{1 + 2\pi\varepsilon\dfrac{k_M}{k_{11M}}} \qquad (8.2.41)$$

It should be pointed out that the formulation obtained above applies to problems in which all cracks have the same length and are in the horizontal direction.

(2) Self-consistent method.

In the self-consistent method [16], for each crack, the effect of crack interaction is taken into account approximately by embedding each crack directly in the effective medium (see Fig.8.1b), i.e., the medium having the as yet unknown material properties of the cracked matrix. Obviously, with this method the same form is used in Eq.(8.2.39), except that the subscript "M" is replaced by "$*$", i.e.

$$P_{11}^{\text{SC}} = P_{12}^{\text{SC}} = P_{21}^{\text{SC}} = 0, \quad P_{22}^{\text{SC}} = \frac{2\pi k^* \varepsilon}{k_{11}^*} \qquad (8.2.42)$$

where superscript "SC" stands for the quantity associated with self-consistent method.

(3) Mori-Tanaka method.

It can be seen from the discussion above that the dilute method is based on the solution of a single crack embedded in infinite matrix subjected to the far field W_∞ (see Fig.8.1a). In this case, Eq.(8.2.37) becomes

$$X = P^{\mathrm{DIL}}W_\infty \tag{8.2.43}$$

Alternatively, the self-consistent method is based on the solution of a single crack embedded in an infinite unknown effective material subjected also to the far field W_∞ (see Fig.8.1b). With this method. Eq.(8.2.37) becomes

$$X = P^{\mathrm{SC}}W_\infty \tag{8.2.44}$$

In contrast, the Mori-Tanaka method [46] is based on the solution for a single crack embedded in an intact matrix subjected to an applied heat intensity equal to the as yet unknown average field $\bar{W}_1$ in the matrix (see Fig.8.1c), which means that the introduction of cracks in the matrix results in a value of X given by [16]

$$X = P^{\mathrm{MT}}W_\infty = P^{\mathrm{DIL}}\bar{W}_1 \tag{8.2.45}$$

Thus, the key point is calculation of $\bar{W}_1$. One method is to use Eq.(8.2.17)$_2$ and rewrite it in the form

$$\bar{W}_1 v_1 = W_\infty - \bar{W}_2 v_2 \tag{8.2.46}$$

For a cracked plate, noting that $v_1 \to 1$ and $v_2 \to 0$, we have

$$\lim_{v_1 \to 1}(\bar{W}_1 v_1) = \bar{W}_1 = W_\infty - \lim_{v_2 \to 0}(\bar{W}_2 v_2) = (I - P^{\mathrm{MT}})W_\infty \tag{8.2.47}$$

where superscript "MT" denotes the quantity associated with the Mori-Tanaka method. For example, P^{MT} stands for the concentration factor associated with the Mori-Tanaka method. Substituting Eq.(8.2.47) into Eq.(8.2.45) yields

$$P^{\mathrm{MT}} = P^{\mathrm{DIL}}(I + P^{\mathrm{DIL}})^{-1} \tag{8.2.48}$$

Making use of Eq.(8.2.39), we obtain

$$P_{11}^{\mathrm{MT}} = P_{12}^{\mathrm{MT}} = P_{21}^{\mathrm{MT}} = 0, \quad P_{22}^{\mathrm{MT}} = \frac{P_{22}^{\mathrm{DIL}}}{1 + P_{22}^{\mathrm{DIL}}} \tag{8.2.49}$$

It can be seen from Eq.(8.2.49) that when $\varepsilon \ll 1$, $P_{22}^{\mathrm{MT}} \approx P_{22}^{\mathrm{DIL}}$.

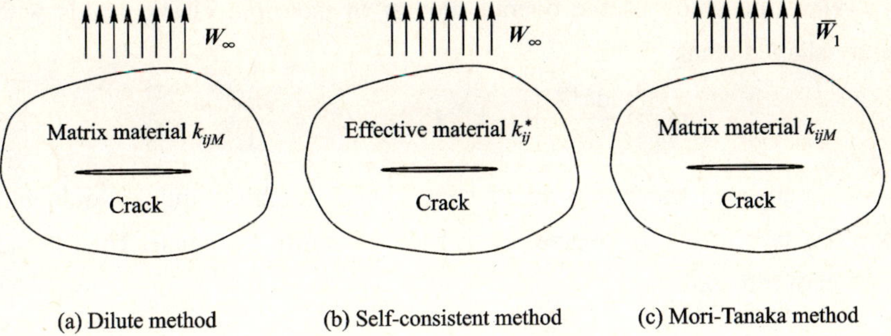

Fig.8.1 Three typical micromechanics models

Another way to calculate $\overline{W}_1$ is to use the results from the dilute method [Eq.(8.2.41)]. To this end, assume the average heat intensity $\overline{W}_{2M}$ in the matrix as

$$\overline{W}_{2M} = W_{2\infty} + \tilde{W}_{2p} \tag{8.2.50}$$

where $W_{2\infty}$ and $\tilde{W}_{2p}$ are, respectively, remote heat intensity perpendicular to the crack line and perturbed heat intensity due to the presence of the crack. With the assumption of the Mori-Tanaka method, $\overline{W}_{2c}$ in Eq.(8.2.35) can be written as [46]

$$\overline{W}_{2c} = \lim_{v_2 \to 0} (\overline{W}_2 v_2) = 2\pi\varepsilon(W_{2\infty} + \tilde{W}_{2p})\frac{k_M}{k_{11M}} \tag{8.2.51}$$

Substituting Eq.(8.2.51) into Eq.(8.2.47) yields

$$W_{2\infty} = \left(1 + 2\pi\varepsilon\frac{k_M}{k_{11M}}\right)(W_{2\infty} + \tilde{W}_{2p}) \tag{8.2.52}$$

Hence

$$P_{22}^{MT} = \frac{2\pi\varepsilon k_M / k_{11M}}{1 + 2\pi\varepsilon k_M / k_{11M}} \tag{8.2.53}$$

It can be seen from Eq.(8.2.53) that this procedure obtains the same results as Eq.(8.2.49).

(4) Differential method.

As was pointed out in Chapter 2, the essence of the differential scheme is the construction of the final cracked medium from the intact material through successive replacement of an incremental area of the current cracked material with that of the cracks [47]. The result obtained below is along the lines given in [44] in the study of the overall moduli of isotropic elastic solids with a penny-shaped crack

$$\lim_{v_2 \to 0} \left(v_2 \frac{\mathrm{d}\boldsymbol{k}^{\mathrm{DS}}}{\mathrm{d}v_2} \right) = -\boldsymbol{k}^{\mathrm{DS}} \boldsymbol{P}^{\mathrm{DS}}, \quad P_{22}^{\mathrm{DS}} = \frac{2\pi k^{\mathrm{DS}} \varepsilon}{k_{11}^{\mathrm{DS}}} \tag{8.2.54}$$

Assume that the cracks are obtained by flattening elliptical voids which have the axes a and ap, where p can be made infinitely small. Then the area fraction of the voids is

$$v_2 = \frac{\pi p \sum a^2}{A} = \pi p \varepsilon \tag{8.2.55}$$

Inserting Eq.(8.2.55) into Eq.(8.2.54) and noting that $\mathrm{d}v_2 = \pi p \mathrm{d}\varepsilon$, we have

$$\varepsilon \frac{\mathrm{d}\boldsymbol{k}^{\mathrm{DS}}}{\mathrm{d}\varepsilon} = -\boldsymbol{k}^{\mathrm{DS}} \boldsymbol{P}^{\mathrm{DS}} \tag{8.2.56}$$

with initial condition

$$\boldsymbol{k}^{\mathrm{DS}} \big|_{\varepsilon=0} = \boldsymbol{k}_1 \tag{8.2.57}$$

where superscript "DS" stands for the quantity associated with the differential scheme. Eq.(8.2.56) represents a set of 2×2 coupled nonlinear ordinary differential equations, which can be solved using certain numerical methods, such as the well-known 4th order Runge-Kutta integration scheme.

(5) Generalized self-consistent method.

The generalized self-consistent method considered here is based on the effective cracked medium model shown in Fig.8.2 [48], a crack of length $2a$ embedded in an elliptical matrix material, which in turn is embedded in a material with the as yet unknown effective property of a microcracked solid. The major axis of the elliptical matrix is chosen to be aligned along the crack line, and the area of the surrounding matrix is chosen so as to preserve the corresponding crack density in the matrix. Based on this understanding, the major and minor axes of the ellipse in Fig.8.2 are assumed to be [48]

$$a^* = a + \delta, \quad b^* = \delta \tag{8.2.58}$$

where δ is determined by

$$\varepsilon = \frac{a^2}{\pi(a+\delta)\delta} = \frac{Na^2}{A} \tag{8.2.59}$$

Since it is impossible to find the analytical temperature field for the effective cracked medium model, the approach presented in [48] is used to calculate ΔT. The method is based on the minimum potential principle of the following functional:

$$J(T) = \int_{S^M} k_{ijM} T_{,i} T_{,j}\, ds + \int_{S^{EM}} k_{ij}^* T_{,i} T_{,j}\, ds - \int_{-a}^{a} \Delta T h_{i\infty} n_i\, dc \qquad (8.2.60)$$

where S^M and S^{EM} are, respectively, regions occupied by the matrix and effective medium, and T is a kinematic admissible temperature field. Among all the kinematic admissible temperature fields, the exact temperature field gives the minimum potential energy.

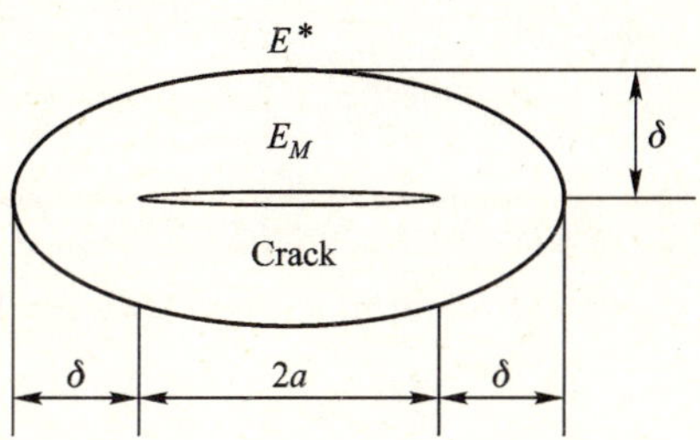

Fig.8.2   Effective cracked medium model for generalized self-consistent method

Let T^M be the temperature field for an infinite matrix medium containing a crack of length $2a$ and subjected to the far field $h_{2\infty}$, and let T^{EM} be the temperature field for an infinite effective medium having the as yet unknown material properties of a cracked matrix, where inside the medium, there is a crack of length $2a$ and it is subjected to the far field $h_{2\infty}$. These temperature fields have been given in $[16, 45]$ as

$$\Delta T^I = \frac{4h_{2\infty}}{k^I}(a^2 - x_1^2)^{1/2}, \quad I{=}M,\ EM \qquad (8.2.61)$$

$$T_{(U)}^I = \frac{2h_{2\infty}}{k^I}\operatorname{Re}\left[(a^2 - z_t^2)^{1/2} + iz_t\right] = t_{(U)}^I h_{2\infty}, \quad x_2 > 0,\ I{=}M,\ EM \qquad (8.2.62)$$

$$T_{(L)}^I = -\frac{2h_{2\infty}}{k^I}\operatorname{Re}\left[(a^2 - \bar{z}_t^2)^{1/2} + i\bar{z}_t\right] = t_{(L)}^I h_{2\infty}, \quad x_2 < 0,\ I{=}M,\ EM \qquad (8.2.63)$$

where superscripts "M" and "EM" represent the quantity associated with the solution in an infinite matrix medium and an infinite effective medium, respectively.

The approximate temperature field T is assumed to be the linear superimposition of the above two solutions

$$T = q_{2\infty}(\xi^M T^M + \xi^{EM} T^{EM}) \qquad (8.2.64)$$

where $T^I = t_{(U)}^I$, when $x_2 > 0$, otherwise $T^I = t_{(L)}^I$, and ξ^M and ξ^{EM} are the constants to be determined by the principle of minimum potential energy. To determine ξ^M and ξ^{EM}, substituting Eq.(8.2.64) into Eq.(8.2.60) and the vanishing variation of Eq.(8.2.60) with respect to ξ^M and ξ^{EM} yields

$$\begin{bmatrix} I_{11} & I_{12} \\ I_{12} & I_{22} \end{bmatrix}\begin{bmatrix} \xi^M \\ \xi^{EM} \end{bmatrix} = a^2\pi\begin{bmatrix} 1/k^M \\ 1/k^* \end{bmatrix} \tag{8.2.65}$$

where

$$I_{11} = \int_{S^M} k_{ijM} T_{,i}^M T_{,j}^M \,\mathrm{d}s + \int_{S^{EM}} k_{ij}^* T_{,i}^M T_{,j}^M \,\mathrm{d}s \tag{8.2.66}$$

$$I_{12} = \int_{S^M} k_{ijM} T_{,i}^{EM} T_{,j}^M \,\mathrm{d}s + \int_{S^{EM}} k_{ij}^* T_{,i}^{EM} T_{,j}^M \,\mathrm{d}s \tag{8.2.67}$$

$$I_{22} = \int_{S^M} k_{ijM} T_{,i}^{EM} T_{,j}^{EM} \,\mathrm{d}s + \int_{S^{EM}} k_{ij}^* T_{,i}^{EM} T_{,j}^{EM} \,\mathrm{d}s \tag{8.2.68}$$

Thus, ξ^M and ξ^{EM} can be determined by solving Eq.(8.2.65), and then substituting the solution of ξ^M and ξ^{EM} into Eq.(8.2.63) and subsequently into Eq.(8.2.34) for determining ΔT. It can be seen from Eqs.(8.2.37) and (8.2.39) that

$$\boldsymbol{P}^{\mathrm{GSC}} = \xi^M \boldsymbol{P}^{\mathrm{DIL}} + \xi^{EM} \boldsymbol{P}^{\mathrm{SC}} \tag{8.2.69}$$

where superscript "GSC" denotes the quantity associated with generalized self-consistent method.

8.2.3   Effective electroelastic constants

To obtain the relations between the effective electroelastic moduli of a cracked medium, the following auxiliary problem is considered:

$$\boldsymbol{t}(s) = \boldsymbol{\Pi}_\infty \boldsymbol{n} \tag{8.2.70}$$

or

$$\boldsymbol{U}(s) = \boldsymbol{Z}_\infty \boldsymbol{x} \tag{8.2.71}$$

and

$$T(s) = 0 \tag{8.2.72}$$

where $\boldsymbol{x} = \begin{bmatrix} x_1 & x_2 \end{bmatrix}^{\mathrm{T}} = \begin{bmatrix} x & y \end{bmatrix}^{\mathrm{T}}$ is a position vector. When the boundary conditions (8.2.71)~(8.2.72) are applied, it follows from the energy theorem [43]

$$\bar{\boldsymbol{\Pi}} = \boldsymbol{\Pi}_\infty \quad \text{or} \quad \bar{\boldsymbol{Z}} = \boldsymbol{Z}_\infty \tag{8.2.73}$$

and

$$\bar{T} = 0 \tag{8.2.74}$$

In the case of a cracked body, the average stress $\bar{\boldsymbol{\Pi}}$ and strain $\bar{\boldsymbol{Z}}$ defined on the basis of the integral average are [16]

$$\bar{\boldsymbol{\Pi}} = \bar{\boldsymbol{\Pi}}_M, \qquad \bar{\boldsymbol{Z}} = \bar{\boldsymbol{Z}}_M + \bar{\boldsymbol{Z}}_c \tag{8.2.75}$$

where $\bar{\boldsymbol{Z}}_c$ can be calculated through use of Eq.(8.2.25), i.e.,

$$\bar{Z}_{ijc} = \lim_{v_2 \to 0}\left[(\bar{Z}_{ij})_2 v_2\right] = \frac{1}{2\Omega}\int_L \left\{[1 + H(i-3)]\Delta U_i n_j + \Delta U_j n_i\right\} dl = X_{ij} \tag{8.2.76}$$

with ΔU_i being the jump of generalized displacement field across the crack

faces. Thus, Eq.(8.2.75) can be further written as

$$\boldsymbol{E}^* \bar{\boldsymbol{Z}}_\infty = \boldsymbol{E}_M \boldsymbol{Z}_\infty - \boldsymbol{E}_M \bar{\boldsymbol{Z}}_c \tag{8.2.77}$$

$$\boldsymbol{F}^* \boldsymbol{\Pi}_\infty = \boldsymbol{F}_M \boldsymbol{\Pi}_\infty + \bar{\boldsymbol{Z}}_c \tag{8.2.78}$$

It can be seen from the discussion above that the estimation of $\bar{\boldsymbol{Z}}_c$ is the

key to predicting the effective electroelastic moduli, while the estimation of

$\bar{\boldsymbol{Z}}_c$ requires the solution of ΔU_i. For a piezoelectric sheet containing a crack

of length $2a$ and subjected to a set of far fields $\boldsymbol{\Pi}_\infty$, the solution has been

given in Eq.(3.7.115). When there is no applied temperature load, Eq.(3.7.115)

becomes

$$\Delta U(x_1,\ 0) = (a^2 - x_1^2)^{1/2}\boldsymbol{C}\boldsymbol{\Pi}_\infty, \qquad |x_1| < a \tag{8.2.79}$$

where $\boldsymbol{\Pi}_\infty = \begin{bmatrix} \sigma_{31\infty} & \sigma_{33\infty} & D_{3\infty} \end{bmatrix}^T$ are the applied far fields, and the matrix $\boldsymbol{C}$ is

defined by Eq.(3.7.100), which depends on the material constants, i.e.

$C_{ij} = C_{ij}(\boldsymbol{E})$. Substituting Eq.(8.2.79) into Eq.(8.2.76) yields the expression of

$\bar{\boldsymbol{Z}}_c$ as

$$\bar{Z}_{ijc} = \frac{\pi\varepsilon}{4}\left\{[1 + H(i-3)]\boldsymbol{C}_i n_j + \boldsymbol{C}_j n_i\right\}\boldsymbol{\Pi}_\infty \tag{8.2.80}$$

where

$$\boldsymbol{C}_i = \begin{bmatrix} C_{i1} & C_{i2} & C_{i3} \end{bmatrix} \tag{8.2.81}$$

When all cracks are in the horizontal direction, noting that $n_1 = 0,\ n_2 = 1$

and

$$\bar{\boldsymbol{Z}}_c = \boldsymbol{P}\boldsymbol{F}_M\boldsymbol{\Pi}_\infty = \begin{bmatrix} \bar{Z}_{22c} & 2\bar{Z}_{12c} & \bar{Z}_{32c} \end{bmatrix}^T \tag{8.2.82}$$

we have

$$\boldsymbol{P} = \frac{\pi\varepsilon}{2}\boldsymbol{R}_c(\boldsymbol{E})\boldsymbol{E}_M \tag{8.2.83}$$

$$\boldsymbol{R}_c(\boldsymbol{E}) = \begin{bmatrix} C_{21}(\boldsymbol{E}) & C_{22}(\boldsymbol{E}) & C_{23}(\boldsymbol{E}) \\ C_{11}(\boldsymbol{E}) & C_{12}(\boldsymbol{E}) & C_{13}(\boldsymbol{E}) \\ C_{31}(\boldsymbol{E}) & C_{32}(\boldsymbol{E}) & C_{33}(\boldsymbol{E}) \end{bmatrix}, \quad \boldsymbol{E}_M = \begin{bmatrix} c_{33M} & 0 & e_{33M} \\ 0 & c_{44M} & 0 \\ e_{33M} & 0 & -\kappa_{33M} \end{bmatrix} \tag{8.2.84}$$

where $C_{ij}(E)$ are functions of as yet unknown material constants. In the following, the results of Eqs.(8.2.76)~(8.2.84) are used to establish five micromechanics approximation theories for estimating the effective electroelastic moduli.

(1) Dilute method.

For the dilute method, we have $C_{ij}=C_{ijM}=C_{ij}(E_M)$. The concentration factor $\boldsymbol{P}^{\mathrm{DIL}}$ is thus given by

$$\boldsymbol{P}^{\mathrm{DIL}}=\frac{\pi\varepsilon}{2}\boldsymbol{R}_c(\boldsymbol{E}_M)\boldsymbol{E}_M \qquad (8.2.85)$$

(2) Self-consistent method.

Self-consistent theory gives results with the same form as Eq.(8.2.85) except that $\boldsymbol{R}_c(\boldsymbol{E}_M)$ in Eq.(8.2.85) is replaced by $\boldsymbol{R}_c(\boldsymbol{E}^*)$

$$\boldsymbol{P}^{\mathrm{SC}}=\frac{\pi\varepsilon}{2}\boldsymbol{R}_c(\boldsymbol{E}^*)\boldsymbol{E}_M \qquad (8.2.86)$$

(3) Mori-Tanaka method.

For the Mori-Tanaka theory, we have the same form of $\boldsymbol{P}^{\mathrm{MT}}$ as in Eq.(8.2.48), i.e.

$$\boldsymbol{P}^{\mathrm{MT}}=\boldsymbol{P}^{\mathrm{DIL}}(\boldsymbol{I}+\boldsymbol{P}^{\mathrm{DIL}})^{-1}=\frac{\pi\varepsilon}{2}\boldsymbol{R}_c(\boldsymbol{E}_M)\boldsymbol{E}_M[\boldsymbol{I}+\frac{\pi\varepsilon}{2}\boldsymbol{R}_c(\boldsymbol{E}_M)\boldsymbol{E}_M]^{-1} \qquad (8.2.87)$$

(4) Differential method.

Similar to the formulation of differential theory in Subsection 8.2.2, we have

$$\varepsilon\frac{\mathrm{d}\boldsymbol{E}^{\mathrm{DS}}}{\mathrm{d}\varepsilon}=-\boldsymbol{E}^{\mathrm{DS}}\boldsymbol{P}^{\mathrm{DS}} \qquad (8.2.88)$$

with the initial condition

$$\boldsymbol{E}^{\mathrm{DS}}\big|_{\varepsilon=0}=\boldsymbol{E}_M \qquad (8.2.89)$$

Generally, Eq.(8.2.88) represents a set of 3×3 coupled nonlinear ordinary differential equations, which can also be solved with the well-known fourth order Runge-Kutta integration scheme.

(5) Generalized self-consistent method.

Similar to the treatment in Subsection 8.2.2, the generalized self-consistent method here is also based on the effective cracked medium model in Fig.8.2. The energy functional corresponding to the electroelastic problem can be defined as

$$J(\boldsymbol{U})=\frac{1}{2}\int_{S^M}\breve{\boldsymbol{Z}}^{\mathrm{T}}\boldsymbol{E}_M\breve{\boldsymbol{Z}}\mathrm{d}s+\frac{1}{2}\int_{S^{EM}}\breve{\boldsymbol{Z}}^{\mathrm{T}}\boldsymbol{E}^*\breve{\boldsymbol{Z}}\mathrm{d}s-\int_{-a}^{a}\Delta\boldsymbol{U}^{\mathrm{T}}\boldsymbol{\varPi}_\infty\mathrm{d}c \qquad (8.2.90)$$

where

$$\breve{Z} = \begin{bmatrix} Z_{22} \\ 2Z_{12} \\ Z_{32} \end{bmatrix} = \begin{bmatrix} U_{2,2} \\ U_{2,1} + U_{1,2} \\ U_{3,2} \end{bmatrix} = \breve{Z}_R \boldsymbol{\Pi}_\infty \tag{8.2.91}$$

and $\breve{Z}_R$ can be evaluated by substituting Eq.(3.7.113) or Eq.(3.7.114) into Eq.(8.2.91). For a piezoelectric sheet containing a crack of length $2a$ and subjected to a set of far fields $\boldsymbol{\Pi}_\infty$, the solution U has been given in Eqs.(3.7.113) and (3.7.114). When there is no applied temperature load, Eqs.(3.7.113) and (3.7.114) become

$$U_{(1)} = \mathrm{Re}\left[\overline{A} F(\overline{z}) \overline{B}^{-1} \right] \boldsymbol{\Pi}_\infty, \qquad x_2 > 0 \tag{8.2.92}$$

$$U_{(2)} = \mathrm{Re}\left[A F(z) B^{-1} \right] \boldsymbol{\Pi}_\infty, \qquad x_2 < 0 \tag{8.2.93}$$

If we denote $U_{(U)} = \mathrm{Re}\left[\overline{A} F(\overline{z}) \overline{B}^{-1} \right]$ and $U_{(L)} = \mathrm{Re}\left[A F(z) B^{-1} \right]$,

Eqs.(8.2.92) and (8.2.93) can be written in one equation as

$$U = U_R \boldsymbol{\Pi}_\infty = \begin{cases} U_{(U)} \boldsymbol{\Pi}_\infty, & x_2 > 0 \\ U_{(L)} \boldsymbol{\Pi}_\infty, & x_2 < 0 \end{cases} \tag{8.2.94}$$

Hence, with the generalized self-consistent method, U and $\breve{Z}$ can be assumed in the form

$$U = (\xi^M U_R^M + \xi^{EM} U_R^{EM}) \boldsymbol{\Pi}_\infty, \qquad \breve{Z} = (\xi^M \breve{Z}_R^M + \xi^{EM} \breve{Z}_R^{EM}) \boldsymbol{\Pi}_\infty \tag{8.2.95}$$

where ξ^M and ξ^{EM} are unknown constants which can be determined by taking the vanishing variation of the functional (8.2.90) with respect to ξ^M and ξ^{EM}. To find the solution ξ^M and ξ^{EM}, we consider first the cracked body subjected to a set of far fields $\boldsymbol{\Pi}_\infty = \begin{bmatrix} \sigma_{31\infty} & 0 & 0 \end{bmatrix}^{\mathrm{T}}$. Substituting Eq.(8.2.79) into Eqs.(8.2.97) and (8.2.95), we obtain the expression of U and $\breve{Z}$ as

$$\Delta U = \left(\begin{bmatrix} C_{11M} \\ C_{21M} \\ C_{31M} \end{bmatrix} \xi^M + \begin{bmatrix} C_{11}^* \\ C_{21}^* \\ C_{31}^* \end{bmatrix} \xi^{EM} \right) \sigma_{31\infty} = (c_{x(1)}^M \xi^M + c_{x(1)}^{EM} \xi^{EM}) \sigma_{31\infty} \tag{8.2.96}$$

$$U = \left(\begin{bmatrix} (U_R^M)_{11} \\ (U_R^M)_{21} \\ (U_R^M)_{31} \end{bmatrix} \xi^M + \begin{bmatrix} (U_R^{EM})_{11} \\ (U_R^{EM})_{21} \\ (U_R^{EM})_{31} \end{bmatrix} \xi^{EM} \right) \sigma_{31\infty} = (u_{x(1)}^M \xi^M + u_{x(1)}^{EM} \xi^{EM}) \sigma_{31\infty} \tag{8.2.97}$$

$$\breve{Z} = \left(\begin{bmatrix} (\breve{Z}_R^M)_{11} \\ (\breve{Z}_R^M)_{21} \\ (\breve{Z}_R^M)_{31} \end{bmatrix} \xi^M + \begin{bmatrix} (\breve{Z}_R^{EM})_{11} \\ (\breve{Z}_R^{EM})_{21} \\ (\breve{Z}_R^{EM})_{31} \end{bmatrix} \xi^{EM} \right) \sigma_{31\infty} = (z_{x(1)}^M \xi^M + z_{x(1)}^{EM} \xi^{EM}) \sigma_{31\infty}$$

$$(8.2.98)$$

Then, substituting Eqs.(8.2.96)~(8.2.98) into Eq.(8.2.90) and taking the vanishing variation of Eq.(8.2.90) with respect to ξ^M and ξ^{EM} yields

$$\begin{bmatrix} \Phi_{11}^{(1)} & \Phi_{12}^{(1)} \\ \Phi_{12}^{(1)} & \Phi_{22}^{(1)} \end{bmatrix} \begin{bmatrix} \xi^M \\ \xi^{EM} \end{bmatrix} = \frac{\pi a^2}{2} \begin{bmatrix} C_{11M} \\ C_{11}^* \end{bmatrix} \tag{8.2.99}$$

where

$$\Phi_{11}^{(1)} = \int_{S^M} (u_{x(1)}^M)^{\mathrm{T}} E_M u_{x(1)}^M \mathrm{d}s + \int_{S^{EM}} (u_{x(1)}^M)^{\mathrm{T}} E^* u_{x(1)}^M \mathrm{d}s \tag{8.2.100}$$

$$\Phi_{12}^{(1)} = \int_{S^M} (u_{x(1)}^M)^{\mathrm{T}} E_M u_{x(1)}^{EM} \mathrm{d}s + \int_{S^{EM}} (u_{x(1)}^M)^{\mathrm{T}} E^* u_{x(1)}^{EM} \mathrm{d}s \tag{8.2.101}$$

$$\Phi_{22}^{(1)} = \int_{S^M} (u_{x(1)}^{EM})^{\mathrm{T}} E_M u_{x(1)}^{EM} \mathrm{d}s + \int_{S^{EM}} (u_{x(1)}^{EM})^{\mathrm{T}} E^* u_{x(1)}^{EM} \mathrm{d}s \tag{8.2.102}$$

Solve Eq.(8.2.99) for ξ^M and ξ^{EM} and denote the solution as $\xi_{(1)}^M$ and $\xi_{(1)}^{EM}$. Finally, substituting the solution $\xi_{(1)}^M$ and $\xi_{(1)}^{EM}$ into Eq.(8.2.96) and subsequently into Eq.(8.2.82) yields three equations for the three components of $\boldsymbol{P}^{\mathrm{GSC}}$

$$P_{11}^{\mathrm{GSC}} f_{33M} + P_{13}^{\mathrm{GSC}} p_{33M} = \frac{\pi \varepsilon}{2} (C_{21M} \xi_{(1)}^M + C_{21}^* \xi_{(1)}^{EM})$$

$$P_{21}^{\mathrm{GSC}} f_{33M} + P_{23}^{\mathrm{GSC}} p_{33M} = \frac{\pi \varepsilon}{2} (C_{11M} \xi_{(1)}^M + C_{11}^* \xi_{(1)}^{EM}) \tag{8.2.103}$$

$$P_{31}^{\mathrm{GSC}} f_{33M} + P_{33}^{\mathrm{GSC}} p_{33M} = \frac{\pi \varepsilon}{2} (C_{31M} \xi_{(1)}^M + C_{31}^* \xi_{(1)}^{EM})$$

Similarly, assume $\boldsymbol{\Pi}_\infty = \begin{bmatrix} 0 & \sigma_{33\infty} & 0 \end{bmatrix}^{\mathrm{T}}$ and $\boldsymbol{\Pi}_\infty = \begin{bmatrix} 0 & 0 & D_{3\infty} \end{bmatrix}^{\mathrm{T}}$ and, using the procedure described above, we can finally obtain the following equations for the remaining six components of $\boldsymbol{P}^{\mathrm{GSC}}$

$$P_{12}^{\mathrm{GSC}} f_{55M} = \frac{\pi \varepsilon}{2} (C_{22M} \xi_{(2)}^M + C_{22}^* \xi_{(2)}^{EM})$$

$$P_{22}^{\mathrm{GSC}} f_{55M} = \frac{\pi \varepsilon}{2} (C_{12M} \xi_{(2)}^M + C_{12}^* \xi_{(2)}^{EM}) \tag{8.2.104}$$

$$P_{32}^{\mathrm{GSC}} f_{55M} = \frac{\pi \varepsilon}{2} (C_{32M} \xi_{(2)}^M + C_{32}^* \xi_{(2)}^{EM})$$

$$P_{11}^{GSC} p_{33M} - P_{13}^{GSC} \beta_{33M} = \frac{\pi\varepsilon}{2}(C_{23M}\xi_{(3)}^{M} + C_{23}^{*}\xi_{(3)}^{EM})$$

$$P_{21}^{GSC} p_{33M} - P_{23}^{GSC} \beta_{33M} = \frac{\pi\varepsilon}{2}(C_{13M}\xi_{(3)}^{M} + C_{13}^{*}\xi_{(3)}^{EM}) \qquad (8.2.105)$$

$$P_{31}^{GSC} p_{33M} - P_{33}^{GSC} \beta_{33M} = \frac{\pi\varepsilon}{2}(C_{33M}\xi_{(3)}^{M} + C_{33}^{*}\xi_{(3)}^{EM})$$

Thus, the concentration factor $\boldsymbol{P}^{GSC}$ can be determined by solving the nine equations above.

8.2.4 Effective thermal expansion and pyroelectric constants

To ascertain the relations between the thermal and electroelastic moduli of a cracked medium, similar to the treatment in Subsection 8.2.3, an auxiliary remote uniform temperature problem is considered in which the following boundary conditions are prescribed:

$$T(s) = T_{\infty} \qquad (8.2.106)$$

and

$$t(s) = 0 \quad \text{or} \quad U(s) = 0 \qquad (8.2.107)$$

When the boundary conditions (8.2.106) and (8.2.107) exist, it follows from the energy theorem [43]

$$\bar{\boldsymbol{\Pi}} = 0, \ \bar{T} = T_{\infty}, \ \bar{\boldsymbol{Z}} = \boldsymbol{\alpha}^{*}T_{\infty}, \ \bar{\boldsymbol{Z}}_{M} = \boldsymbol{F}_{M}\boldsymbol{\Pi} + \boldsymbol{\alpha}_{M}T \qquad (8.2.108)$$

For the boundary conditions (8.2.106) and (8.2.107), the corresponding fields are defined as

$$\bar{\boldsymbol{Z}} = 0, \ \bar{T} = T_{\infty}, \ \bar{\boldsymbol{\Pi}} = -\boldsymbol{\Gamma}^{*}T_{\infty}, \ \bar{\boldsymbol{\Pi}}_{M} = \boldsymbol{E}_{M}\boldsymbol{Z} - \boldsymbol{\Gamma}_{M}T \qquad (8.2.109)$$

where

$$\boldsymbol{\alpha} = \begin{bmatrix} \alpha_{11} & \alpha_{33} & \gamma_{3} \end{bmatrix}^{T}, \ \boldsymbol{\Gamma} = \begin{bmatrix} \lambda_{11} & \lambda_{33} & \rho_{3} \end{bmatrix}^{T} \qquad (8.2.110)$$

Making use of Eqs.(8.2.12) and (8.2.75), we have

$$\boldsymbol{\alpha}^{*}T_{\infty} = \boldsymbol{\alpha}_{M}T_{\infty} + \bar{\boldsymbol{Z}}_{c}, \ \boldsymbol{\Gamma}^{*}T_{\infty} = \boldsymbol{\Gamma}_{M}T_{\infty} - \boldsymbol{E}_{M}\bar{\boldsymbol{Z}}_{c} \qquad (8.2.111)$$

where $\bar{\boldsymbol{Z}}_{c}$ is defined by Eq.(8.2.76) in which ΔU is given by [see Eq.(3.7.115)]

$$\Delta U(x_{1}, 0) = -\boldsymbol{b}(a^{2} - x_{1}^{2})^{1/2} T_{\infty} \qquad (8.2.112)$$

with $\boldsymbol{b}$ being evaluated from Eq.(3.7.101). Substituting Eq.(8.2.112) into Eq.(8.2.76) yields the explicit expression of $\bar{\boldsymbol{Z}}_{c}$ as

$$\bar{\boldsymbol{Z}}_c = \begin{bmatrix} Z_{11} \\ Z_{22} \\ Z_{32} \end{bmatrix} = -\frac{\pi\varepsilon}{2}\begin{bmatrix} 0 \\ b_2 \\ b_3 \end{bmatrix} T_\infty \tag{8.2.113}$$

With the dilute and self-consistent methods, the substitution of Eq.(8.2.113) into Eq.(8.2.111) yields

$$\boldsymbol{\alpha}^{*\text{DIL}} = \begin{bmatrix} \alpha_{11}^{*\text{DIL}} \\ \alpha_{22}^{*\text{DIL}} \\ \gamma_{3}^{*\text{DIL}} \end{bmatrix} = \begin{bmatrix} \alpha_{11M} \\ \alpha_{22M} \\ \gamma_{3M} \end{bmatrix} - \frac{\pi\varepsilon}{2}\begin{bmatrix} 0 \\ b_{2M} \\ b_{3M} \end{bmatrix} \tag{8.2.114}$$

$$\boldsymbol{\alpha}^{*\text{SC}} = \begin{bmatrix} \alpha_{11}^{*\text{SC}} \\ \alpha_{22}^{*\text{SC}} \\ \gamma_{3}^{*\text{SC}} \end{bmatrix} = \begin{bmatrix} \alpha_{11M} \\ \alpha_{22M} \\ \gamma_{3M} \end{bmatrix} - \frac{\pi\varepsilon}{2}\begin{bmatrix} 0 \\ b_{2M}^{*} \\ b_{3M}^{*} \end{bmatrix} \tag{8.2.115}$$

For the concentration factors $\boldsymbol{P}$ and $\boldsymbol{Q}$, it can be shown that [16]

$$\boldsymbol{\Gamma}^{*} = (\boldsymbol{I} - \boldsymbol{P})\boldsymbol{\Gamma}_M, \quad \boldsymbol{\alpha}^{*} = (\boldsymbol{I} + \boldsymbol{Q})\boldsymbol{\alpha}_M \tag{8.2.116}$$

Comparing Eq.(8.2.114) with Eq.(8.2.116) we see that

$$\boldsymbol{Q}^{\text{DIL}} = -\frac{\pi\varepsilon}{2}\text{diag}\begin{bmatrix} 0 & \dfrac{b_{2M}}{\alpha_{33M}} & \dfrac{b_{3M}}{\gamma_{3M}} \end{bmatrix} \tag{8.2.117}$$

which yields $\boldsymbol{Q}^{\text{MT}}$ of Mori-Tanaka theory

$$\boldsymbol{Q}^{\text{MT}} = \boldsymbol{Q}^{\text{DIL}}(\boldsymbol{I} + \boldsymbol{Q}^{\text{DIL}})^{-1} \tag{8.2.118}$$

8.3　Micromechanics model of thermo-piezoelectricity with microvoids

In this section, the effective electroelastic behaviour of void-weakened 2D material is studied via the dilute, self-consistent, Mori-Tanaka, and differential micromechanics theories. For simplicity, all holes are assumed to have the same size and orientation. First the results of perturbed heat intensity, strain, and electric field due to the presence of voids are presented for two-dimensional piezoelectric plates with voids of various shapes, and then the above four micromechanics models can be established based on the perturbation results. These models are applicable to a wide range of holes such as ellipse, circle, crack, triangle, square and pentagon.

In the case of voids, Eqs.(8.2.21)~(8.2.24) are still applicable. It can be seen from Eq.(8.2.24) that the estimation of temperature, elastic displacement, electric potential, and their integration along the hole boundary is the key to predicting the effective material properties of void-weakened piezoelectric plate. To this end, consider an infinite sheet containing a hole of any one of various shapes, whose contour is described by [49]

$$x_1 = a(\cos\psi + \eta\cos m\psi), \quad x_2 = a(c\sin\psi - \eta\sin m\psi) \tag{8.3.1}$$

where $0 < c \leqslant 1$, and m is an integer. By appropriate selection of the parameters c, m, and η, we can obtain various shapes of voids, such as ellipse, square, and so on.

8.3.1 Effective conductivity

When a set of far-field $\boldsymbol{h}_\infty = \begin{bmatrix} h_{1\infty} & h_{2\infty} \end{bmatrix}^{\mathrm{T}}$ is applied on the voided infinite sheet above, the temperature change T at a point on the void boundary has been given in [49] as

$$T = -\frac{a}{k}\left[ch_{1\infty}\cos\psi + h_{2\infty}\sin\psi - \eta(h_{1\infty}\cos m\psi - h_{2\infty}\sin m\psi) \right] \tag{8.3.2}$$

$$\boldsymbol{k}^* = \boldsymbol{k}_M(\boldsymbol{I} - \boldsymbol{A}_{T0}v_2), \quad \boldsymbol{\rho}^* = \boldsymbol{\rho}_M(\boldsymbol{I} + \boldsymbol{B}_{T0}v_2) \tag{8.3.3}$$

where $\boldsymbol{A}_{T0}$ and $\boldsymbol{B}_{T0}$ are defined by

$$\overline{\boldsymbol{W}}_2 = \boldsymbol{\rho}_M\boldsymbol{B}_{T0}\boldsymbol{h}_\infty = \boldsymbol{A}_{T0}\boldsymbol{\rho}_M\boldsymbol{h}_\infty = \frac{1}{\Omega}\int_{\partial\Omega} T\boldsymbol{n}\,ds \tag{8.3.4}$$

Substituting Eq.(8.3.3) into Eq.(8.3.4) and integrating it along the contour of the void yields

$$\overline{\boldsymbol{W}}_2 = \boldsymbol{R}_T\boldsymbol{h}_\infty \tag{8.3.5}$$

where $\boldsymbol{R}_T$ is a 2×2 diagonal matrix whose components are

$$R_{T12} = R_{T21} = 0, \quad R_{T11} = \frac{c^2 + m\eta^2}{k(c - m\eta^2)}, \quad R_{T22} = \frac{1 + m\eta^2}{k(c - m\eta^2)} \tag{8.3.6}$$

Thus, from Eqs.(8.3.3)~(8.3.5), we find

$$\boldsymbol{A}_{T0} = \boldsymbol{A}_{T0}(\boldsymbol{k}_M, \boldsymbol{k}^*) = \boldsymbol{R}_T\boldsymbol{k}_M, \quad \boldsymbol{B}_{T0} = \boldsymbol{B}_{T0}(\boldsymbol{k}_M, \boldsymbol{k}^*) = \boldsymbol{k}_M\boldsymbol{R}_T \tag{8.3.7}$$

8.3.2 Effective electroelastic constants

Consider a piezoelectric plate containing a hole whose contour is defined by Eq.(8.3.1) and subjected a set of far fields $\boldsymbol{\Pi}_\infty$. The elastic displacement and electric potential at a point of the hole boundary has been given in [15]

$$U = x_1 \mathbf{Z}_{1\infty} + x_3 \mathbf{Z}_{3\infty} + \left[ac\mathbf{L}^{-1}\cos\psi - a\eta\mathbf{L}^{-1}\cos m\psi - ac\mathbf{S}\mathbf{L}^{-1}\sin\psi + \right.$$
$$\left. a\eta\mathbf{S}\mathbf{L}^{-1}\sin m\psi \right]t_{1\infty} - \left[a\mathbf{L}^{-1}\mathbf{S}^{\mathrm{T}}\cos\psi + a\eta\mathbf{L}^{-1}\mathbf{S}^{\mathrm{T}}\cos m\psi - \right.$$
$$\left. a(\mathbf{H} + \mathbf{S}\mathbf{L}^{-1}\mathbf{S}^{\mathrm{T}})(\sin\psi + \eta\sin m\psi) \right]t_{3\infty} \tag{8.3.8}$$

where

$$\mathbf{Z}_{1\infty} = \left[\varepsilon_{11\infty}\ \varepsilon_{13\infty}\ -E_{1\infty} \right]^{\mathrm{T}}, \quad \mathbf{Z}_{3\infty} = \left[\varepsilon_{31\infty}\ \varepsilon_{33\infty}\ -E_{3\infty} \right]^{\mathrm{T}}$$
$$\mathbf{t}_{1\infty} = \left[\sigma_{11\infty}\ \sigma_{13\infty}\ D_{1\infty} \right]^{\mathrm{T}}, \quad \mathbf{t}_{3\infty} = \left[\sigma_{31\infty}\ \sigma_{33\infty}\ D_{3\infty} \right]^{\mathrm{T}} \tag{8.3.9}$$

and $\mathbf{Z}_\infty = \mathbf{E}_M \boldsymbol{\Pi}_\infty$, $\mathbf{L}$, $\mathbf{S}$ and $\mathbf{H}$ are the well-known real matrices in the Stroh formalism, which is defined by Eq.(3.3.49), while $\mathbf{Z}_\infty$ and $\boldsymbol{\Pi}_\infty$ are

$$\mathbf{Z}_\infty = \left[\varepsilon_{11\infty}\ \varepsilon_{13\infty}\ \varepsilon_{33\infty}\ -E_{1\infty}\ -E_{3\infty} \right]^{\mathrm{T}}, \quad \boldsymbol{\Pi}_\infty = \left[\sigma_{11\infty}\ \sigma_{13\infty}\ \sigma_{33\infty}\ D_{1\infty}\ D_{3\infty} \right]^{\mathrm{T}}$$

$$\tag{8.3.10}$$

By substituting Eq.(8.3.8) into Eq.(8.2.24) and integrating it along the whole contour of the hole, we obtain

$$\bar{\mathbf{Z}}_2 = \mathbf{R}\boldsymbol{\Pi}_\infty \tag{8.3.11}$$

where $\mathbf{R}$ is a 5×5 symmetric matrix whose components are

$$R_{11} = f_{11}(c - m\eta^2) + (\mathbf{L}^{-1})_{11}(c^2 + m\eta^2)$$
$$R_{12} = (c - m\eta^2)\left[f_{13} - (\mathbf{L}^{-1}\mathbf{S}^{\mathrm{T}})_{12} \right]$$
$$R_{13} = (c^2 + m\eta^2)(\mathbf{L}^{-1})_{12} - (c - m\eta^2)(\mathbf{L}^{-1}\mathbf{S}^{\mathrm{T}})_{11}$$
$$R_{14} = (c^2 + m\eta^2)(\mathbf{L}^{-1})_{13}$$
$$R_{15} = (c - m\eta^2)\left[p_{31} - (\mathbf{L}^{-1}\mathbf{S}^{\mathrm{T}})_{13} \right]$$
$$R_{22} = (c - m\eta^2)f_{33} + (1 + m\eta^2)\left[(\mathbf{H})_{22} + (\mathbf{S}\mathbf{L}^{-1}\mathbf{S}^{\mathrm{T}})_{22} \right]$$
$$R_{23} = (-c + m\eta^2)(\mathbf{S}\mathbf{L}^{-1})_{22} + (1 + m\eta^2)\left[(\mathbf{H})_{21} + (\mathbf{S}\mathbf{L}^{-1}\mathbf{S}^{\mathrm{T}})_{21} \right]$$
$$R_{24} = (-c + m\eta^2)(\mathbf{S}\mathbf{L}^{-1})_{23}$$
$$R_{25} = (c - m\eta^2)p_{33} + (1 + m\eta^2)\left[(\mathbf{H})_{23} + (\mathbf{S}\mathbf{L}^{-1}\mathbf{S}^{\mathrm{T}})_{23} \right]$$
$$R_{33} = (c - m\eta^2)\left[f_{44} - 2(\mathbf{S}\mathbf{L}^{-1})_{12} \right] + (c^2 + m\eta^2)(\mathbf{L}^{-1})_{22} + (1 + m\eta^2)\left[(\mathbf{H})_{11} + (\mathbf{S}\mathbf{L}^{-1}\mathbf{S}^{\mathrm{T}})_{11} \right]$$
$$R_{34} = (c - m\eta^2)\left[p_{15} - (\mathbf{S}\mathbf{L}^{-1})_{13} \right] + (c^2 + m\eta^2)(\mathbf{L}^{-1})_{23}$$
$$R_{35} = (1 + m\eta^2)\left[(\mathbf{S}\mathbf{L}^{-1}\mathbf{S}^{\mathrm{T}})_{13} + (\mathbf{H})_{13} \right] - (c - m\eta^2)(\mathbf{L}^{-1}\mathbf{S}^{\mathrm{T}})_{23}$$
$$R_{44} = (c^2 + m\eta^2)(\mathbf{L}^{-1})_{33} + (c - m\eta^2)\beta_{11}$$
$$R_{45} = (m\eta^2 - c)(\mathbf{L}^{-1}\mathbf{S}^{\mathrm{T}})_{33}$$
$$R_{55} = (c - m\eta^2)\beta_{11} + (1 + m\eta^2)\left[(\mathbf{S}\mathbf{L}^{-1}\mathbf{S}^{\mathrm{T}})_{33} + (\mathbf{H})_{33} \right]$$

Thus, from Eqs.(8.2.20), (8.2.23) and (8.3.11), we have

$$A_0 = A_0(E_M, E^*) = RE_M, \quad B_0 = B_0(E_M, E^*) = E_M R \qquad (8.3.12)$$

In the following, the Eqs.(8.3.11) and (8.3.12) are used to establish various micromechanics models for the effective thermo-electro-elastic moduli.

8.3.3   Effective concentration factors based on various micromechanics models

1. Effective temperature field

Eqs.(8.3.5) and (8.3.7) can be used to establish micromechanics models for effective conductivity. First, we consider the dilute method. Since the interaction among voids is ignored in the dilute method, noting Eq.(8.3.7), A_{T0} and B_{T0} can be written as

$$A_{T0}^{\mathrm{DIL}} = R_T(k_M)k_M, \quad B_{T0}^{\mathrm{DIL}} = k_M R_T(k_M) \qquad (8.3.13)$$

Substituting Eq.(8.3.13) into Eq.(8.3.3) yields

$$k^{\mathrm{DIL}} = k_M\left[I - v_2 R_T(k_M)k_M\right], \quad \rho^{\mathrm{DIL}} = \rho_M\left[I + v_2 k_M R_T(k_M)\right] \qquad (8.3.14)$$

$$A_{T0}^{\mathrm{SC}} = R_T(k^*)k_M, \quad B_{T0}^{\mathrm{SC}} = k_M R_T(k^*) \qquad (8.3.15)$$

$$k^{\mathrm{SC}} = k_M\left[I - v_2 R_T(k^*)k_M\right], \quad \rho^{\mathrm{SC}} = \rho_M\left[I + v_2 k_M R_T(k^*)\right] \qquad (8.3.16)$$

In the Mori-Tanaka method, we assume that the average perturbed heat intensity $\bar{W}_2$ is related to the average heat intensity of the matrix $\bar{W}_1$ [15]

$$\bar{W}_2 = R_T(k_M)k_M\bar{W}_1 = A_{T0}^{\mathrm{DIL}}\bar{W}_1 \qquad (8.3.17)$$

Multiplying the both sides of Eq.(8.3.17) by v_1 and then substituting it into Eq.(8.2.46) yields

$$\bar{W}_2 = (v_1 I + v_2 A_{T0}^{\mathrm{DIL}})^{-1} A_{T0}^{\mathrm{DIL}} W_\infty = A_{T0}^{\mathrm{MT}} W_\infty \qquad (8.3.18)$$

Noting that A_{T0} is symmetric, A_{T0}^{MT} can be written as

$$A_{T0}^{\mathrm{MT}} = (v_1 I + v_2 A_{T0}^{\mathrm{DIL}})^{-1} A_{T0}^{\mathrm{DIL}} = A_{T0}^{\mathrm{DIL}}(v_1 I + v_2 A_{T0}^{\mathrm{DIL}})^{-1} \qquad (8.3.19)$$

Similarly, the concentration factor B_{T0}^{MT} can be obtained as

$$B_{T0}^{\mathrm{MT}} = B_{T0}^{\mathrm{DIL}}(v_1 I + v_2 B_{T0}^{\mathrm{DIL}})^{-1} \qquad (8.3.20)$$

With regard to the differential method, similar to the treatment in Subsection 8.2.2, we have

$$\frac{dk^{\mathrm{DS}}}{dv_2} = -\frac{k^{\mathrm{DS}} A_{T0}^{\mathrm{DS}}}{1-v_2}, \quad A_{T0}^{\mathrm{DS}} = R_T(k^{\mathrm{DS}})k_M \qquad (8.3.21)$$

Subjected to the initial condition

$$k^{\mathrm{DS}}\big|_{v_2=0} = k_M \qquad (8.3.22)$$

Eq.(8.3.21) represents a 2×2 coupled nonlinear differential equations which has a similar structure to that of Eq.(8.2.56).

2. Effective electroelastic moduli

Making use of Eqs.(8.3.11) and (8.3.12), the concentration factors and effective c electroelastic moduli corresponding to the following four micromechanics theories can be obtained and listed:

(1) Dilute method.

$$A_0^{\mathrm{DIL}} = R(E_M)E_M, \quad B_0^{\mathrm{DIL}} = E_M R(E_M) \tag{8.3.23}$$

$$E^{\mathrm{DIL}} = E_M \left[I - v_2 R(E_M)E_M \right], \quad F^{\mathrm{DIL}} = F_M + v_2 R(E_M) \tag{8.3.24}$$

(2) Self-consistent method.

$$A_0^{\mathrm{SC}} = R(E^*)E_M, \quad B_0^{\mathrm{SC}} = E_M R(E^*) \tag{8.3.25}$$

$$E^{\mathrm{SC}} = E_M \left[I - v_2 R(E^*)E_M \right], \quad F^{\mathrm{SC}} = F_M + v_2 R(E^*) \tag{8.3.26}$$

(3) Mori-Tanaka method.

$$A_0^{\mathrm{MT}} = A_0^{\mathrm{DIL}} (v_1 I + v_2 A_0^{\mathrm{DIL}})^{-1}, \quad B_0^{\mathrm{MT}} = B_0^{\mathrm{DIL}} (v_1 I + v_2 B_0^{\mathrm{DIL}})^{-1} \tag{8.3.27}$$

$$E^{\mathrm{MT}} = E_M \left\{ I - v_2 R(E_M)E_M \left[v_1 I + v_2 R(E_M)E_M \right]^{-1} \right\}$$

$$F^{\mathrm{MT}} = F_M \left\{ I + v_2 E_M R(E_M) \left[v_1 I + v_2 E_M R(E_M) \right]^{-1} \right\} \tag{8.3.28}$$

(4) Differential scheme.

$$\frac{\mathrm{d}E^{\mathrm{DS}}}{\mathrm{d}v_2} = -\frac{E^{\mathrm{DS}} A_0^{\mathrm{DS}}}{1 - v_2}, \quad A_0^{\mathrm{DS}} = R(E^{\mathrm{DS}})E_M \tag{8.3.29}$$

Subject to the initial condition

$$E^{\mathrm{DS}} \Big|_{v_2=0} = E_M \tag{8.3.30}$$

8.4 Micromechanics model of piezoelectricity with inclusions

8.4.1 Eshelby's tensors for a composite with an ellipsoidal inclusion

For problems of piezoelectricity with inclusions, Eqs.(8.2.15)~(8.2.20) can still be used to predict effective electroelastic properties. Evaluation of $\bar{Z}_2$ and the related concentration factor is the key to predicting the effective electroelastic properties. In this section, approaches presented in [4, 21, 50] are described to show how the micromechanics models can be derived. To this end, consider a piezoelectric composite consisting of an infinite domain D containing an ellipsoidal inclusion Ω defined by

$$\frac{x_1^2}{a_1^2} + \frac{x_2^2}{a_2^2} + \frac{x_3^2}{a_3^2} \leqslant 1, \quad \text{in } \Omega \tag{8.4.1}$$

where a_1, a_2, and a_3 are the semi-axes of the ellipsoid with the a_3 principle axis coincident with the x_3 axis. The assumption that the shape of the inclusion is ellipsoidal allows treatment of composite reinforcement geometries ranging from thin flake to continuous fibre reinforcement. Suppose that the inclusion Ω has electroelastic moduli E_I, while the matrix, $D\text{-}\Omega$, has electroelastic moduli E_M. The composite is subjected to a set of far fields Z_∞. Using the equivalent inclusion method for piezoelectric composites [4], the generalized stress in the representative inclusion can be written as

$$\boldsymbol{\Pi}_I = \boldsymbol{E}_I \boldsymbol{Z}_I = \boldsymbol{E}_I (\boldsymbol{Z}_\infty + \boldsymbol{Z}) = \boldsymbol{E}_M (\boldsymbol{Z}_\infty + \boldsymbol{Z} - \boldsymbol{Z}_*) \tag{8.4.2}$$

where $\boldsymbol{Z}$ represents the perturbation of the generalized strain in the inclusion with respect to the generalized strain in the matrix and $\boldsymbol{Z}_*$ is the fictitious eigenfield required to ensure that the equivalency of Eq.(8.4.2) holds. In Eq.(8.4.2), $\boldsymbol{Z}$ and $\boldsymbol{Z}_*$ are related through [4]

$$\boldsymbol{Z} = \boldsymbol{S}\boldsymbol{Z}_* \tag{8.4.3}$$

where $\boldsymbol{S}$ is the coupled electroelastic analog of Eshelby's tensor whose components can be expressed in terms of surface integrals over the unit sphere as [4]

$$S_{MnAb} = \begin{cases} \dfrac{a_1 a_2 a_3}{8\pi} E_{iJAb} \displaystyle\int_{|z|=1} \frac{1}{\zeta^3} \big[G_{mJin}(z) + G_{nJim}(z) \big] \mathrm{d}s(z), & M \leqslant 3 \\[4mm] \dfrac{a_1 a_2 a_3}{4\pi} E_{iJAb} \displaystyle\int_{|z|=1} \frac{1}{\zeta^3} G_{4Jin}(z) \mathrm{d}s(z), & M = 4 \end{cases} \tag{8.4.4}$$

where $|z| = 1$ is the surface of the unit sphere $\zeta = (a_1^2 z_1^2 + a_2^2 z_2^2 + a_3^2 z_3^2)^{1/2}$, and

$$G_{MJin}(z) = z_i z_n K_{MJ}^{-1}(z), \quad K_{MJ}(z) = E_{iJMn} z_i z_n \tag{8.4.5}$$

To perform the integration in Eq.(8.4.4), the unit sphere is parameterized as

$$z_1 = \frac{\xi_1}{a_1} = \frac{(1-\xi_3^2)^{1/2}\cos\theta}{a_1}, \quad z_2 = \frac{\xi_2}{a_2} = \frac{(1-\xi_3^2)^{1/2}\sin\theta}{a_2}, \quad z_3 = \frac{\xi_3}{a_3} \tag{8.4.6}$$

In general, for an anisotropic medium the integrals in Eq.(8.4.4) cannot be evaluated analytically. In this case, the integration is easily performed by Gaussian quadrature.

It should be mentioned that Huang [50] presented an equivalent formulation to Eq.(8.4.4) as follows:

$$S_{MnAb} = \begin{cases} \dfrac{1}{8\pi} E_{iJAb}(\bar{G}_{mJin} + \bar{G}_{nJim}), & M \leqslant 3 \\[2mm] \dfrac{1}{4\pi} E_{iJAb}\bar{G}_{4Jin}, & M = 4 \end{cases} \tag{8.4.7}$$

where

$$\bar{G}_{MJin} = a_1 a_2 a_3 \int_{|z|=1} \frac{1}{\zeta^3} N_{MJ}(\xi)\xi_i\xi_n D^{-1}(\xi)\,\mathrm{d}s(z) \tag{8.4.8}$$

with ξ_i being defined in Eq.(8.4.6), and $N_{Mj}(\xi)$ and $D(\xi)$ being the cofactor and the determinant of the 4×4 matrix $E_{iMJn}\xi_i\xi_n$, respectively [50]. The evaluation of $N_{Mj}(\xi)$ and $D(\xi)$ has been substantively discussed in [50] and we will not repeat it here as it is tedious and algebraic.

Note that S is a fourth order tensor and it is useful to use the generalized Voight two-index notation. With the two-index notion, the electroelastic Eshelby's tensor S_{MnAb} for an ellipsoidal inclusion in transversely isotropic piezoelectric materials can be expressed in the following form [21]:

$$[S_{mnab}] = \begin{bmatrix} S_{11} & S_{12} & S_{13} & 0 & 0 & 0 \\ S_{21} & S_{22} & S_{23} & 0 & 0 & 0 \\ S_{31} & S_{32} & S_{33} & 0 & 0 & 0 \\ 0 & 0 & 0 & S_{44} & 0 & 0 \\ 0 & 0 & 0 & 0 & S_{55} & 0 \\ 0 & 0 & 0 & 0 & 0 & S_{66} \end{bmatrix}, \quad [S_{mn4b}] = \begin{bmatrix} 0 & 0 & S_{19} \\ 0 & 0 & S_{29} \\ 0 & 0 & S_{39} \\ 0 & S_{48} & 0 \\ S_{57} & 0 & 0 \\ 0 & 0 & 0 \end{bmatrix} \tag{8.4.9}$$

$$[S_{4nab}] = \begin{bmatrix} 0 & 0 & 0 & 0 & S_{75} & 0 \\ 0 & 0 & 0 & S_{84} & 0 & 0 \\ S_{91} & S_{92} & S_{93} & 0 & 0 & 0 \end{bmatrix}, \quad [S_{4n4b}] = \begin{bmatrix} S_{77} & 0 & 0 \\ 0 & S_{88} & 0 \\ 0 & 0 & S_{99} \end{bmatrix} \tag{8.4.10}$$

where

$$S_{11} = S_{1111},\ S_{12} = S_{1122},\ S_{13} = S_{1133,}\ S_{19} = S_{1143},\ S_{21} = S_{2211},\ S_{22} = S_{2222}$$

$$S_{23} = S_{2233},\ S_{29} = S_{2243},\ S_{31} = S_{3311},\ S_{32} = S_{3322},\ S_{33} = S_{3333},\ S_{39} = S_{3343}$$

$$S_{44} = S_{2323} = S_{2332} = S_{3223} = S_{3232},\ S_{48} = S_{2342},\ S_{57} = S_{1341},\ S_{77} = S_{4141}$$

$$S_{55} = S_{1313} = S_{1331} = S_{3113} = S_{3131},\ S_{66} = S_{1212} = S_{1221} = S_{2112} = S_{2121}$$

$$S_{75} = S_{4113} = S_{4131},\ S_{84} = S_{4223} = S_{4232},\ S_{88} = S_{4242},\ S_{91} = S_{4311}$$

$$S_{92} = S_{4322},\ S_{93} = S_{4333},\ S_{99} = S_{4343}$$

$$(8.4.11)$$

In particular, the above electroelastic Eshelby tensors for an elliptic cylinder, a circular cylinder, and a penny-shaped inclusion in transversely isotropic piezoelectric solids have been obtained by Huang as [19]

(1) Elliptic cylinder $(a_1/a_2 = a, a_3 \to \infty)$.

$$S_{11} = \frac{a}{2(1+a)^2}\left(\frac{2c_{11}+c_{12}}{c_{11}} + \frac{2+a}{a}\right), \quad S_{12} = \frac{a}{2(1+a)^2}\left[\frac{(2+a)c_{12}}{ac_{11}} - 1\right]$$

$$S_{13} = \frac{c_{13}}{(1+a)c_{11}}, \quad S_{19} = \frac{e_{31}}{(1+a)c_{11}}, \quad S_{21} = \frac{a}{2(1+a)^2}\left[\frac{(1+2a)c_{12}}{ac_{11}} - 1\right]$$

$$S_{22} = \frac{a}{2(1+a)^2}\left(2a + \frac{3c_{11}+c_{12}}{c_{11}}\right), \quad S_{23} = aS_{13}, \quad S_{29} = aS_{19} \tag{8.4.12}$$

$$S_{44} = \frac{a}{2(1+a)}, \quad S_{55} = \frac{1}{2(1+a)}, \quad S_{66} = \frac{a}{2(1+a)^2}\left(\frac{a^2+a+1}{a} - \frac{c_{12}}{c_{11}}\right)$$

$$S_{77} = \frac{1}{1+a}, \quad S_{88} = \frac{a}{1+a}$$

(2) Circular cylinder $(a_1 = a_2, a_3 \to \infty)$.

$$S_{11} = S_{22} = \frac{5c_{11}+c_{12}}{8c_{11}}, \quad S_{12} = S_{21} = \frac{3c_{12}-c_{11}}{8c_{11}}, \quad S_{23} = S_{13} = \frac{c_{13}}{2c_{11}}$$

$$S_{29} = S_{19} = \frac{e_{31}}{2c_{11}}, \quad S_{44} = S_{55} = \frac{1}{4}, \quad S_{66} = \frac{3c_{11}-c_{12}}{8c_{11}}, \quad S_{77} = S_{88} = \frac{1}{2}$$

$$\tag{8.4.13}$$

(3) Penny-shaped inclusion $(a_1 = a_2 >> a_3, a_3 \to 0)$.

$$S_{44} = S_{55} = \frac{1}{2}, \quad S_{57} = S_{75} = S_{48} = S_{84} = \frac{c_{13}}{2c_{11}}, \quad S_{33} = S_{99} = 1$$

$$\tag{8.4.14}$$

$$S_{31} = S_{32} = \frac{c_{13}\kappa_{33}+e_{31}e_{33}}{c_{33}\kappa_{33}+e_{33}^2}, \quad S_{91} = S_{92} = \frac{c_{13}e_{33}-c_{33}e_{31}}{c_{33}\kappa_{33}+e_{33}^2}$$

8.4.2   Effective elastoelectric moduli

Substituting Eq.(8.4.3) into Eq.(8.4.2) yields the generalized strain in the inclusion Z_2 as

$$Z_2 = Z_\infty + SZ_* \tag{8.4.15}$$

Making use of Eqs.(8.4.2) and (8.4.15), Z_2 can be further written as

$$Z_2 = \left[I + SE_M^{-1}(E_I - E_M)\right]^{-1} Z_\infty \tag{8.4.16}$$

By comparing Eq.(8.4.16) with Eq.(8.2.20), we observe

$$A_2 = \left[I + SE_M^{-1}(E_I - E_M)\right]^{-1} \tag{8.4.17}$$

Similarly, the concentration factor B_2 can also be obtained as

$$B_2 = \left[I + F_M^{-1}(I - S)(F_I - F_M) \right]^{-1} \qquad (8.4.18)$$

Eqs.(8.4.17) and (8.4.18) provide the results of the concentration factors A_2 and B_2 by ignoring interaction among inclusions. Therefore, they represent concentration factors A_2^{DIL} and B_2^{DIL}.

For the self-consistent method, noting that each inclusion is assumed to be embedded in an infinite piezoelectric medium, Eqs.(8.4.17) and (8.4.18) become

$$A_2^{\mathrm{SC}} = \left[I + S^* E^{*-1}(E_I - E^*) \right]^{-1} \qquad (8.4.19)$$

$$B_2^{\mathrm{SC}} = \left[I + F^{*-1}(I - S^*)(F_I - F^*) \right]^{-1} \qquad (8.4.20)$$

With regard to Mori-Tanaka method, it can be shown that [4]

$$A_2^{\mathrm{MT}} = A_2^{\mathrm{DIL}}(v_1 I + v_2 A_2^{\mathrm{DIL}})^{-1}, \quad B_2^{\mathrm{MT}} = B_2^{\mathrm{DIL}}(v_1 I + v_2 B_2^{\mathrm{DIL}})^{-1} \qquad (8.4.21)$$

Finally, we discussion the differential scheme. Following Mclaughlin [47], the removal of a volume increment ΔV of the instantaneous configuration (thus a removal of $v_2 \Delta V$ of the reinforcing phase) leads

$$\mathrm{d}v_2 = \frac{\mathrm{d}V}{V}(1 - v_2) \qquad (8.4.22)$$

where V is the volume of the composite. Denoting $E^*(v_2 + \mathrm{d}v_2)$ as the effective electroelastic moduli at a reinforcement volume fraction of $(v_2 + \mathrm{d}v_2)$, use of Eqs.(8.2.18) and (8.4.22) leads [4]

$$\frac{\mathrm{d}E}{\mathrm{d}v_2} = \frac{1}{1 - v_2}(E_I - E^*)A_2^{\mathrm{DIF}} \qquad (8.4.23)$$

where

$$A_2^{\mathrm{DIF}} = \left[I + S^{\mathrm{DIF}} E^{*-1}(E_I - E^*) \right]^{-1} \qquad (8.4.24)$$

As in the self-consistent scheme, S^{DIF} is a function of E^* of the composite material at a reinforcement volume fraction of $(v_2 + \mathrm{d}v_2)$. Formally, Eq.(8.4.23) represents a set of 9×9=81 coupled nonlinear ordinary differential equations, in which

$$E^*(v_2 = 0) = E_M \qquad (8.4.25)$$

8.4.3   Effective thermal expansion and pyroelectric coefficients

As mentioned in Subsection 8.2.1, evaluation of the effective thermal expansion and pyroelectric coefficients requires information about the effective electroelastic moduli. To obtain the relationships between thermal and coupled electroelastic effects, Dunn [5] considered the following two auxiliary problems:

(1) Applied uniform electroelastic far-fields.

Consider a two-phase composite subjected to the boundary conditions (8.2.71) and (8.2.72). For the boundary conditions (8.2.71) and (8.2.72), the volume average fields and the phase and overall equations follow from the energy theorem [43], that is

$$\overline{{}^{\varPi}\varPi} = \varPi_\infty, \quad {}^{\varPi}\overline{T} = 0, \quad {}^{\varPi}\overline{Z}_I = F_I \,{}^{\varPi}\varPi_I, \quad {}^{\varPi}\overline{Z}_M = F_M \,{}^{\varPi}\varPi_M, \quad {}^{\varPi}\overline{Z} = F^* \varPi_\infty \qquad (8.4.26)$$

To distinguish the fields induced by different loading conditions, the left-superscript "$\varPi$" is used to represent fields associated with the applied far-field $\varPi_\infty$. With the boundary conditions (8.2.71) and (8.2.72), the average electroelastic concentration factors for each phase are defined in a way similar to that in Eq.(8.2.20) as

$$\overline{{}^{\varPi}\varPi}_1 = B_1 \varPi_\infty, \quad \overline{{}^{\varPi}\varPi}_2 = B_2 \varPi_\infty \qquad (8.4.27)$$

Likewise, for the boundary conditions (8.2.71) and (8.2.72), the corresponding fields are

$$\overline{{}^{Z}Z} = Z_\infty, \quad {}^{Z}\overline{T} = 0, \quad {}^{Z}\overline{\varPi}_I = E_I \,{}^{Z}Z_I, \quad {}^{Z}\overline{\varPi}_M = E_M \,{}^{Z}Z_M, \quad {}^{Z}\overline{\varPi} = E^* Z_\infty \qquad (8.4.28)$$

where the left-superscript "Z" denotes fields induced by loading conditions (8.2.71) and (8.2.72). The average electroelastic concentration factors for each phase are defined as

$$\overline{{}^{Z}Z}_1 = A_1 Z_\infty, \quad \overline{{}^{Z}Z}_2 = A_2 Z_\infty \qquad (8.4.29)$$

(2) Applied uniform temperature change.

Consider again the two-phase composite, but subjected to the boundary conditions (8.2.106) and (8.2.107). When the boundary conditions (8.2.106) and (8.2.107) are applied, the volume average fields and the phase and overall equations are as follows:

$$^{T}\overline{\varPi} = 0, \quad {}^{T}\overline{T} = T_\infty, \quad {}^{T}\overline{Z} = \alpha^* T_\infty$$
$$^{T}Z_I = F_I \,{}^{T}\varPi_I + \alpha_I T_I, \quad {}^{T}Z_M = F_M \,{}^{T}\varPi_M + \alpha_M T \qquad (8.4.30)$$

where the left-superscript "T" denotes fields induced by loading conditions (8.2.106) and (8.2.107). Those for the boundary conditions (8.2.106) and (8.2.107) are

$$^{T}\overline{Z} = 0, \quad {}^{T}\overline{T} = T_\infty, \quad {}^{T}\overline{\varPi} = -\varGamma^* T_\infty$$
$$^{T}\varPi_I = E_I \,{}^{T}Z_I - \varGamma_I T_I, \quad {}^{T}\varPi_M = E_M \,{}^{T}Z_M - \varGamma_M T_M \qquad (8.4.31)$$

With the boundary conditions (8.2.106) and (8.2.107), the average thermal concentration factors for each phase are defined as

$$^{T}\overline{\boldsymbol{\Pi}}_{1} = \boldsymbol{B}_{T1}T_{\infty}, \quad ^{T}\overline{\boldsymbol{\Pi}}_{2} = \boldsymbol{B}_{T2}T_{\infty}, \quad ^{T}\overline{\boldsymbol{Z}}_{1} = \boldsymbol{V}_{T1}T_{\infty}, \quad ^{T}\overline{\boldsymbol{Z}}_{2} = \boldsymbol{V}_{T2}T_{\infty} \qquad (8.4.32)$$

It is then necessary to establish relationships between effective thermal property and electroelastic property based on the above results. Based on the theorem of average strain energy [43]

$$\int_{\Omega} \boldsymbol{\Pi} \boldsymbol{Z} \mathrm{d}\Omega = \overline{\boldsymbol{\Pi}}\,\overline{\boldsymbol{Z}}\,\Omega \qquad (8.4.33)$$

and considering the electroelastic fields due to the boundary conditions (8.2.71) and (8.2.72), substituting Eq.(8.4.26) into Eq.(8.4.33) and then using Eqs.(8.4.26) and (8.4.30), we obtain

$$\int_{\Omega} {}^{T}\boldsymbol{\Pi}^{\Pi}\boldsymbol{Z}\mathrm{d}\Omega = \int_{\Omega_{I}} {}^{T}\boldsymbol{\Pi}_{I}(\boldsymbol{F}_{I}{}^{\Pi}\boldsymbol{\Pi}_{I})\mathrm{d}\Omega + \int_{\Omega_{M}} {}^{T}\boldsymbol{\Pi}_{M}(\boldsymbol{F}_{M}{}^{\Pi}\boldsymbol{\Pi}_{M})\mathrm{d}\Omega = 0 \qquad (8.4.34)$$

Similarly, substituting Eq.(8.4.30) into Eq.(8.4.33) and then using Eqs.(8.4.26) and (8.4.34) leads

$$\int_{\Omega} {}^{\Pi}\boldsymbol{\Pi}^{T}\boldsymbol{Z}\mathrm{d}\Omega = \int_{\Omega_{I}} {}^{\Pi}\boldsymbol{\Pi}_{I}(\boldsymbol{F}_{I}{}^{T}\boldsymbol{\Pi}_{I} + \boldsymbol{\alpha}_{I}T_{I})\mathrm{d}\Omega +$$

$$\int_{\Omega_{M}} {}^{\Pi}\boldsymbol{\Pi}_{M}(\boldsymbol{F}_{M}{}^{T}\boldsymbol{\Pi}_{M} + \boldsymbol{\alpha}_{M}T_{M})\mathrm{d}\Omega = \boldsymbol{\Pi}_{\infty}\boldsymbol{\alpha}^{*}T_{\infty}\Omega \qquad (8.4.35)$$

Similar manipulation with the quantities associated with the loading conditions (8.2.71) and (8.2.72) yields

$$\int_{\Omega} {}^{Z}\boldsymbol{\Pi}^{T}\boldsymbol{Z}\mathrm{d}\Omega = \int_{\Omega} {}^{T}\boldsymbol{Z}(\boldsymbol{E}^{*}{}^{Z}\boldsymbol{Z})\mathrm{d}\Omega = 0 \qquad (8.4.36)$$

$$\int_{\Omega} {}^{Z}\boldsymbol{Z}^{T}\boldsymbol{\Pi}\mathrm{d}\Omega = -\int_{\Omega_{I}} {}^{Z}\boldsymbol{Z}\boldsymbol{\Gamma}^{*}T_{\infty}\mathrm{d}\Omega = -\boldsymbol{Z}_{\infty}\boldsymbol{\Gamma}^{*}T_{\infty}\Omega \qquad (8.4.37)$$

Substituting Eqs.(8.4.26), (8.4.28), (8.4.27), and (8.4.29) into Eq.(8.2.17) leads to the relations

$$v_{1}\boldsymbol{A}_{1} + v_{2}\boldsymbol{A}_{2} = \boldsymbol{I}, \quad v_{1}\boldsymbol{B}_{1} + v_{2}\boldsymbol{B}_{2} = \boldsymbol{I} \qquad (8.4.38)$$

Enforcing the quantity on the right-hand side of Eq.(8.4.35) to the results of the middle integrals of Eq.(8.4.35), we obtain

$$\boldsymbol{\alpha}^{*} = v_{1}\boldsymbol{B}_{1}\boldsymbol{\alpha}_{1} + v_{2}\boldsymbol{B}_{2}\boldsymbol{\alpha}_{2} \qquad (8.4.39)$$

Similar manipulation for Eq.(8.4.37) yields

$$\boldsymbol{\Gamma}^{*} = v_{1}\boldsymbol{A}_{1}\boldsymbol{\Gamma}_{1} + v_{2}\boldsymbol{A}_{2}\boldsymbol{\Gamma}_{2} \qquad (8.4.40)$$

To express effective thermal property in terms of effective electroelastic property, we need to find relationships between the concentration factors $\boldsymbol{A}_{i}$ and $\boldsymbol{B}_{i}$ appearing in Eqs.(8.4.39) and (8.4.40) and the effective electroelastic moduli of the composite. To this end, substituting Eq.(8.4.26) into Eq.(8.2.17) and making use of Eq.(8.4.27) yields

$$\boldsymbol{F}^{*} = v_{1}\boldsymbol{F}_{1}\boldsymbol{B}_{1} + v_{2}\boldsymbol{F}_{2}\boldsymbol{B}_{2} \qquad (8.4.41)$$

which is similar to the expression in Eq.(8.4.39). Substituting Eq.(8.4.28) into Eq.(8.2.17) and making use of Eq.(8.4.29) yields

$$E^* = v_1 E_1 A_1 + v_2 E_2 A_2 \qquad (8.4.42)$$

Inserting Eq.(8.4.38) into Eq.(8.4.41) to eliminate B_2 in favor of B_1 then yields

$$v_1 B_1 = (F^* - F_I)(F_M - F_I)^{-1} \qquad (8.4.43)$$

Similarly, we have

$$v_1 A_1 = (E^* - E_I)(E_M - E_I)^{-1} \qquad (8.4.44)$$

Finally, substituting Eqs.(8.4.43) and (8.4.44) into Eqs.(8.4.39) and (8.4.40) we obtain

$$\alpha^* = \alpha_I + (F^* - F_I)(F_M - F_I)^{-1}(\alpha_M - \alpha_I) \qquad (8.4.45)$$

$$\Gamma^* = \Gamma_I + (E^* - E_I)(E_M - E_I)^{-1}(\Gamma_M - \Gamma_I) \qquad (8.4.46)$$

It can be seen from Eqs.(8.4.45) and (8.4.46) that α^* and Γ^* can be easily evaluated when the effective electroelastic moduli F^* and E^* are obtained in a manner such as the results presented in Subsection 8.4.2. In the following, Eqs.(8.4.39), (8.4.40), (8.4.45), and (8.4.46) are combined with the results of micromechanics theories obtained in Subsection 8.4.2 to obtain α^* and Γ^* of the composite.

From Eqs.(8.4.38)~(8.4.40), it is easy to prove that

$$\alpha^* = \alpha_M + v_2 B_2 (\alpha_I - \alpha_M), \qquad \Gamma^* = \Gamma_M + v_2 A_2 (\Gamma_I - \Gamma_M) \qquad (8.4.47)$$

Substituting Eqs.(8.4.17) and (8.4.18) into Eq.(8.4.47), the dilute method yields the effective thermal expansion and pyroelectric coefficients as

$$\alpha^* = \alpha_M + v_2 \left[I + F_M^{-1}(I - S)(F_I - F_M) \right]^{-1} (\alpha_I - \alpha_M) \qquad (8.4.48)$$

$$\Gamma^* = \Gamma_M + v_2 \left[I + S E_M^{-1}(E_I - E_M) \right]^{-1} (\Gamma_I - \Gamma_M) \qquad (8.4.49)$$

The substitution of Eqs.(8.4.19) and (8.4.20) into Eq.(8.4.47) yields the expressions of α^* and Γ^* for the self-consistent method as

$$\alpha^* = \alpha_M + v_2 \left[I + F^{*-1}(I - S^*)(F_I - F_M) \right]^{-1} (\alpha_I - \alpha_M) \qquad (8.4.50)$$

$$\Gamma^* = \Gamma_M + v_2 \left[I + S^* E^{*-1}(E_I - E_M) \right]^{-1} (\Gamma_I - \Gamma_M) \qquad (8.4.51)$$

With the Mori-Tanaka method, the insertion of Eq.(8.4.21) into Eq.(8.4.47) yields

$$\alpha^* = \alpha_M + v_2 B_2^{\mathrm{DIL}} (v_1 I + v_2 B_2^{\mathrm{DIL}})^{-1} (\alpha_I - \alpha_M) \qquad (8.4.52)$$

$$\Gamma^* = \Gamma_M + v_2 A_2^{\mathrm{DIL}} (v_1 I + v_2 A_2^{\mathrm{DIL}})^{-1} (\Gamma_I - \Gamma_M) \qquad (8.4.53)$$

For the differential method, E^* can be evaluated from Eqs.(8.4.23)~(8.4.25), while F^* is determined from the following equations [5]:

$$\frac{dF}{dv_2} = \frac{1}{1-v_2}(F_I - F^*)B_2^{\mathrm{DIF}} \qquad (8.4.54)$$

subjected to the initial conditions

$$F^*(v_2 = 0) = F_M \qquad (8.4.55)$$

where

$$B_2^{\mathrm{DIF}} = \left[I + F^{*-1}(I - S^{\mathrm{DIF}})(E_I - E^*) \right]^{-1} \qquad (8.4.56)$$

Substituting Eqs.(8.4.24) and (8.4.56) into Eq.(8.4.47) yields

$$\alpha^* = \alpha_M + v_2 \left[I + F^{*-1}(I - S^{\mathrm{DIF}})(E_I - E^*) \right]^{-1} (\alpha_I - \alpha_M) \qquad (8.4.57)$$

$$\Gamma^* = \Gamma_M + v_2 \left[I + S^{\mathrm{DIF}} E^{*-1}(E_I - E^*) \right]^{-1} (\Gamma_I - \Gamma_M) \qquad (8.4.58)$$

8.5   Micromechanics-boundary element mixed approach

It is noted that common to each of the micromechanics theories described in this chapter is the use of the well-known stress and strain concentration factors obtained through an analytical solution of a single crack, void, or inclusion embedded in an infinite medium. However, for a problem with complexity in the aspects of geometry and mechanical deformation, a combination of these micromechanics approaches and numerical methods such as finite element method and boundary element method (BEM) presents a powerful computational tool for estimating effective material properties. It is also noted from Section 8.2 that estimation of the integral (8.2.24), which contains unknown variables on the boundary only, is the key to predicting the concentration factor A_2 (or B_2). Therefore, BEM is very suitable for performing this type of calculation. In this section, a micromechanics-BE mixed algorithm is presented for analyzing the effective behaviour of piezoelectric composites. The algorithm is based on two typical micromechanics models (self-consistent and Mori-Tanaka methods) and a two-phase BE formulation. An iteration scheme is designated for the self-consistent-BE mixed method.

8.5.1   Two-phase BE formulation

In this subsection, a two-phase BE model is introduced for generalized displacements and generalized stresses on the boundary of the subdomain of each phase [34]. The two subdomains are separated by the interfaces between inclusion and matrix (see Fig.8.3). Each subdomain can be separately modelled by direct BEM. Global assembly of the BE subdomains is then performed by enforcing continuity of the generalized displacements and generalized stresses at the subdomain interface.

In a two-dimensional piezoelectric composite, the BE formulation takes the form [51]

$$c^{(\alpha)}(\xi)U_i^{(\alpha)}(\xi) = \int_{S^{(\alpha)}} [U_{ij}^{*(\alpha)}(x,\xi)T_j^{(\alpha)}(x) - T_{ji}^{*(\alpha)}(x,\xi)U_j^{(\alpha)}(x)]\mathrm{d}S(x) \quad (8.5.1)$$

where the superscript "(α)" stands for the quantity associated with the αth phase (α=1 being matrix and α=2 being inclusion), $T_i = \sigma_{ij}n_j$ (i=1,2), $T_3 = D_i n_i$ and

$$S^{(\alpha)} = \begin{cases} S+\Gamma, & \alpha=1 \\ S, & \alpha=2 \end{cases}, \quad c^{(\alpha)}(\xi) = \begin{cases} 1, & \text{if } \xi \in \Omega^{(\alpha)} \\ 0.5, & \text{if } \xi \in S^{(\alpha)} \ (S^{(\alpha)} \text{ smooth}) \\ 0, & \text{if } \xi \notin \Omega^{(\alpha)} \bigcup S^{(\alpha)} \end{cases} \quad (8.5.2)$$

$$[U_{ij}^*] = \begin{bmatrix} u_{11}^* & u_{12}^* & -\phi_1^* \\ u_{21}^* & u_{22}^* & -\phi_2^* \\ u_{31}^* & u_{32}^* & -\phi_3^* \end{bmatrix}, \quad [T_{ij}^*] = \begin{bmatrix} t_{11}^* & t_{12}^* & -\omega_1^* \\ t_{21}^* & t_{22}^* & -\omega_2^* \\ t_{31}^* & t_{32}^* & -\omega_3^* \end{bmatrix} \quad (8.5.3)$$

in which Γ and S are the boundaries of the representative area element (RAE) and inclusions, respectively (see Fig.8.3); u_{ij}^* and t_{ij}^* (i, j=1,2) denote, respectively, the displacement and traction component in the jth direction at a field point x due to a unit point force acting in the ith direction at source point ξ; u_{3i}^* and t_{3i}^* (i=1,2) represent the ith displacement and traction at x due to a unit electric charge at ξ; ϕ_i^* and ω_i^* (i=1,2) stand for the electric potential and surface charge at x due to a unit point force acting in the ith direction at ξ; and ϕ_3^* and ω_3^* denote the electric potential and surface charge at x due to a unit electric charge at ξ; These fundamental solutions are well documented in the literature and can be found in [51].

To obtain a weak solution of Eq.(8.5.1) as in conventional BEM, the boundary $S^{(\alpha)}$ is divided into a series of boundary elements. After performing discretization using various kinds of boundary element (e.g., constant element, linear element, higher-order element) and collecting the unknown terms to the left-hand side and the known terms to the right-hand side, as well as using continuity conditions at the interface S (Fig.8.3b), the boundary integral equation (8.5.1) becomes a set of linear algebraic equations

$$AY = P \qquad\qquad (8.5.4)$$

where Y and P are the total unknown and known vectors, respectively, and A is a known coefficient matrix.

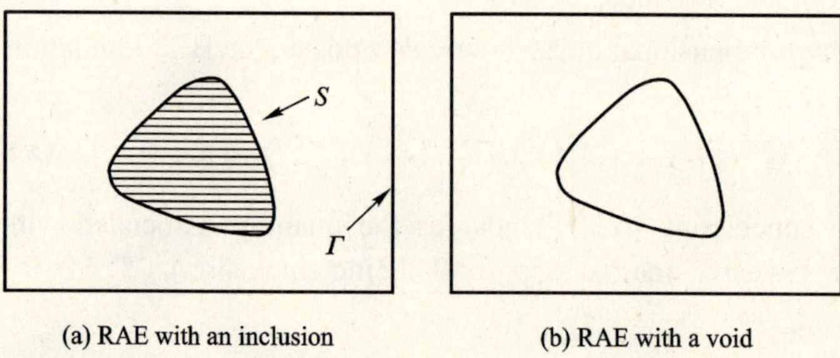

(a) RAE with an inclusion   (b) RAE with a void

Fig.8.3   RAE used in BE analysis

When the inclusion in Fig.8.3a becomes a hole, the boundary integral equation (8.5.1) still holds true if we take $\alpha=1$ only. In this case the interfacial continuity condition is replaced by the hole boundary condition: $T_j = 0$ along the boundary S (Fig.8.3b).

8.5.2   Algorithms for self-consistent and Mori-Tanaka approaches

(1) Self-consistent-BEM approach.

As stated in Subsection 8.2.2, in the self-consistent method, for each inclusion (or hole), the effect of inclusion (or hole) interaction is taken into account approximately by embedding each inclusion (or hole) in the effective medium whose properties are unknown. In this case, the material constants appearing in the boundary element formulation (8.5.1) are unknown. Consequently a set of initial trial values of the effective properties is needed and an iteration algorithm is required. In detail, the algorithm is:

(a) Assume initial values of material constants $E^*_{(0)}$.

(b) Solve Eq.(8.5.1) for $U_{(i)}$ using the values of $E^*_{(i-1)}$, where the subscript "(i)" stands for the variable associated with the ith iterative cycle.

(c) Calculate $A_{2(i)}$ in Eq.(8.2.20) by way of Eq.(8.2.24) and using the current values of $U_{(i)}$, and then determine $E^*_{(i)}$ by way of Eq.(8.2.18).

(d) If $\varepsilon_{(i)} = \left\| E^*_{(i)} - E^*_{(i-1)} \right\| / \left\| E^*_{(0)} \right\| \leqslant \varepsilon$, where ε is a convergent tolerance, terminate the iteration; otherwise take $E^*_{(i)}$ as the initial value and go to step (b).

(2) Mori-Tanaka-BEM approach.

With the Mori-Tanaka method, the concentration matrix A^{MT}_2 is given by the solution for a single inclusion (or void) embedded in an intact solid subjected to an applied strain field equal to the as yet unknown average field in the composite, which means that the introduction of inclusions in the composite results in a value of $\overline{Z}_2$ given by

$$\overline{Z}_2 = A^{DIL}_2 \overline{Z}_1 \tag{8.5.5}$$

where A^{DIL}_2 is the concentration matrix associated with the dilute model, which can be calculated by way of Eqs.(8.2.20), (8.2.24) and (8.5.1). Then, Eq.(8.4.21) is used to calculate A^{MT}_2. It can be seen from Eq.(8.4.21) that the Mori-Tanaka approach provides explicit expressions for effective constants of piezoelectric composites. Therefore, no iteration is required with the Mori-Tanaka-BE method.

References

[1] Newnham R E, Skinner D P, Cross L E. Connectivity and piezoelectric pyroelectric composites. Mater. Res. Bull., 1978, 13:526-536.

[2] Banno H. Recent developments of piezoelectric ceramic products and composites of synthetic rubber and piezoelectric ceramic particles. Ferroelectrics, 1983, 50:3-12.

[3] Grekov A A, Kramarov S O, Kuprienko A A. Effective properties of a transversely isotropic piezoelectric with cylindrical inclusion. Ferroelectrics, 1989, 99:115-126.

[4] Dunn M L, Taya M. Micromechanics predictions of the effective electroelastic moduli of piezoelectric composite. Int. J. Solids Struct., 1993, 30:161-175.

[5] Dunn M L. Micromechanics of coupled electroelastic composite: effective thermal expansion and pyroelectric coefficients. J. Appl. Phys., 1993, 73:5131-5140.

[6] Benveniste Y. Universal relations in piezoelectric composites and polarization fields, I: binary media: local fields and effective behaviour. J. Appl. Mech., 1993, 60:265-269.

[7] Benveniste Y. Universal relations in piezoelectric composites and polarization fields, II:

multiphase media: effective behaviour. J. Appl. Mech., 1993, 60:270-275.

[8] Chen T. Micromechanical estimates of the overall thermoelectroelastic moduli of multiphase fibrous composites. Int. J. Solids Struct., 1994, 31:3099-3111.

[9] Qin Qinghua, Yu Shouwen. Using Mori-Tanaka method for effective moduli of cracked thermopiezoelec-tric materials//Proc. of 9th International Conf. on Fracture, April 1-5, Sydney, 1997: 2211-2218.

[10] Mai Y W, Qin Qinghua, Yu Shouwen. Effective thermal expansion and pyroelectric constants for cracked piezoelectric solids//Chien W Z .Proceedings of the third international conference on nonlinear mechanics, August 17-20, Shanghai, P.R.China. Shanghai Univesity Press, 1998: 309-314.

[11] Benveniste Y, Dvorak G J. Uniform fields and universal relations in piezoelectric composites, J. Mech. Phys. Solids, 1992, 40:1295-1312.

[12] Benveniste Y. Exact results in the micromechanics of fibrous piezoelectric composites. Proc. R. Soc. London, 1993, A441:59-81.

[13] Benveniste Y. Micromechanics of fibrous piezoelectric composites. Mech. Mater., 1994, 18:183-193.

[14] Qin Qinghua. Using GSC theory for effective thermal expansion and pyroelectric coefficient of cracked piezoelectric solids. Int. J. Frac., 1996, 82:R41-R46.

[15] Qin Qinghua, Yu Shouwen. Effective moduli of thermopiezoelectric material with microcavities. Int. J. Solids Struc., 1998, 35:5085-5095.

[16] Qin Qinghua, Mai Y W, Yu Shouwen. Effective moduli for thermopiezoelectric materials with microcracks. Int. J. Frac., 1998, 91:359-371.

[17] Levin V M. The effective properties of piezoactive matrix composite materials. J. Appl. Mathe. Mech., 1996, 60:309-317.

[18] Levin V M, Rakovskaja M I, Kreher W S. The effective thermoelectroelastic properties of microinhomogeneous materials. Int. J. Solids Struc., 1999, 36:2683-2705.

[19] Huang J H. Micromechanics determinations of thermoelectroelastic fields and effective thermoelectroelastic moduli of piezoelectric composites. Mat. Sci. Eng. B, 1996, 39:163-172.

[20] Huang J H, Kuo W S. Micromechanics determination of the effective properties of piezoelectric composites containing spatially oriented short fibers. Acta Mater., 1996, 44: 4889-4898.

[21] Kuo W S, Huang J H. On the effective electroelastic properties of piezoelectric composites containing spatially oriented inclusions. Int. J. Solids Struc., 1997, 34: 2445-2461.

[22] Wu T L. Micromechanics determination of electroelastic properties of piezoelectric

materials containing voids. Mat. Sci. Eng., 2000, A280:320-327.

[23] Chen J, Wang B, Du S Y. A statistical model prediction of effective electroelastic properties of polycrystalline ferroelectric ceramics with randomly oriented defects. Mech. Mater., 2002, 34:643-655.

[24] Wojnar R. Homogenization of piezoelectric solid and thermodynamics. Reports on Mathematical Physics, 1997, 40:585-598.

[25] Eduardo L L, Federico J S, Julián B C, et al. Overall electromechanical properties of a binary composite with 622 symmetry constituents: Anti-plane shear piezoelectric state. Int. J. Solids Struc., 2005, 42:5765-5777.

[26] Hori M, Nemat-Nasser S. Universal bounds for effective piezoelectric moduli. Mech. Mat., 1998, 30:1-19.

[27] Poizat C, Sester M, Effective properties of composites with embedded piezoelectric fibres. Comput. Mat. Sci., 1999, 16:89-97.

[28] Beckert W, Kreher W, Braue W, et al. Effective properties of composites utilising fibres with a piezoelectric coating. J. Euro. Ceramic Soc., 2001, 21:1455-1458.

[29] Li Z H, Wang C, Chen C Y. Effective electromechanical properties of transversely isotropic piezoelectric ceramics with microvoids. Comput. Mat. Sci., 2003, 27:381-392.

[30] Berger H, Kari S, Gabbert U, et al. An analytical and numerical approach for calculating effective material coefficients of piezoelectric fiber composites. Int. J. Solids. Struc., 2005, 42:5692-5714.

[31] Li J Y. The effective electroelastic moduli of textured piezoelectric polycrystalline aggregates. J Mech Phy Solids, 2000, 48:529-552.

[32] Jiang C P, Tong Z H, Cheung Y K. A generalized self-consistent method for piezoelectric fiber reinforced composites under antiplane shear. Mech. Mat., 2001, 33:295-308.

[33] Wang H, Tan G W, Cen S, et al. Numerical determination of effective properties of voided piezoelectric materials using BNM. Eng. Anal. Boundary Elem., 2005, 29:636-646.

[34] Qin Qinghua. Material properties of piezoelectric composites by BEM and homogenization method. Composite Structures, 2004, 66:295-299.

[35] Li J Y, Dunn M L. Micromechanics of magnetoelectroelastic composite materials: average fields and effective behaviour. J. Intell. Mat. Sys. Struc., 1998, 9:404-416.

[36] Li J Y. The effective pyroelectric and thermal expansion coefficients of ferroelectric ceramics. Mech. Mater., 2004, 36:949-958.

[37] Lee J, Boyd J G, Lagoudas D C. Effective properties of three-phase electro-magneto-

elastic composites. Int. J. Eng. Sci., 2005, 43:790-825.

[38] Zhang Z K, Soh A K. Micromechanics predictions of the effective moduli of magnetoelectroelastic composite materials. Euro. J. Mech. A/Solids, 2005, 24:1054-1067.

[39] Berger H, Kari S, Gabbert U, et al. Unit cell models piezoelectric fibre composites for numerical and analytical calculation of effective properties. Smart Mat. Struc., 2006, 15:451-458.

[40] Xu Y L, Lo S H, Jiang C P, et al. Electroelastic behavior of doubly periodic piezoelectric fiber composites under antiplane shear. Int. J. Solids Struc., 2007, 44: 976-995.

[41] Saha G C, Kalamkarov A L, Georgiades A V. Asymptotic homogenization modeling and analysis of effective properties of smart composite reinforced and sandwich shells. Int. J. Mech. Sci., 2007, 49(2):138-150.

[42] Rödel J. Effective intrinsic linear properties of laminar piezoelectric composites and simple ferroelectric domain structures. Mech. Mater., 2007, 39(4):302-325.

[43] Nemat-Nasser S, Hori M. Micromechanics: overall properties of heterogeneous materials. Amsterdam: Elsevier, 1999.

[44] Hashin Z. The differential scheme and its application to cracked materials. J Mech Phy Solids, 1988, 36:719-734.

[45] Atkinson C, Clements D L. On some crack problems in anisotropic thermoelasticity. Int J Solids Struc, 1977, 13:855-864.

[46] Mori T, Tanaka K. Average stress in matrix and average elastic energy of materials with misfitting inclusions. Acta Metall, 1973, 21:571-574.

[47] Mclaughlin R. A study of the differential scheme for composite materials. Int. J. Eng. Sci., 1977, 15:237-244.

[48] Huang Y, Hu K K, Chandra A. A generalized self-consistent mechanics method for microcracked solids. J Mech Phy Solids, 1994, 42:1273-1291.

[49] Qin Qinghua, Mai Y W, Yu Shouwen. Some problems in plane thermo-piezoelectric materials with holes. Int J Solids Struc, 1999, 36:427-439.

[50] Huang J H. An ellipsoidal inclusion or crack in orthotropic piezoelectric media. J. Appl. Phys., 1995, 78:6491-6503.

[51] Ding H J, Wang G Q, Chen W Q. A boundary integral formulation and 2D fundamental solutions for piezoelectric media. Compu. Meth. Appl. Mech. Eng., 1998, 158:65-80.

Index